Experimental Techniques in Condensed Matter Physics at Low Temperatures

Advanced Book Classics

David Pines, Series Editor

Anderson, P.W., *Basic Notions of Condensed Matter Physics*

Bethe H. and Jackiw, R., *Intermediate Quantum Mechanics, Third Edition*

Feynman, R., *Photon-Hadron Interactions*

Feynman, R., *Quantum Electrodynamics*

Feynman, R., *Statistical Mechanics*

Feynman, R., *The Theory of Fundamental Processes*

Nozières, P., *Theory of Interacting Fermi Systems*

Pines, D., *The Many-Body Problem*

Quigg, C., *Gauge Theories of the Strong, Weak, and Electromagnetic Interactions*

Richardson, R. and Smith, E., *Experimental Techniques in Condensed Matter Physics at Low Temperatures*

EXPERIMENTAL TECHNIQUES IN CONDENSED MATTER PHYSICS AT LOW TEMPERATURES

Edited by

ROBERT C. RICHARDSON
ERIC N. SMITH

Cornell University
Ithaca, New York

The Advanced Book Program

Addison-Wesley
Reading, Massachusetts

ISBN 0-201-36078-0

Addison-Wesley is an imprint of Addison Wesley Longman, Inc.

Cover design by Suzanne Heiser

1 2 3 4 5 6 7 8 9 10-MA-0201009998
First printing,March 1998

Find us on the World Wide Web at
http://www.aw.com/gb/abp/

Editor's Foreword

Addison-Wesley's *Frontiers in Physics* series has, since 1961, made it possible for leading physicists to communicate in coherent fashion their views of recent developments in the most exciting and active fields of physics—without having to devote the time and energy required to prepare a formal review or monograph. Indeed, throughout its nearly forty-year existence, the series has emphasized informality in both style and content, as well as pedagogical clarity. Over time, it was expected that these informal accounts would be replaced by more formal counterparts—textbooks or monographs—as the cutting-edge topics they treated gradually became integrated into the body of physics knowledge and reader interest dwindled. However, this has not proven to be the case for a number of the volumes in the series: Many works have remained in print on an on-demand basis, while others have such intrinsic value that the physics community has urged us to extend their life span.

The *Advanced Book Classics* series has been designed to meet this demand. It will keep in print those volumes in Frontiers in Physics or its sister series, *Lecture Notes and Supplements in Physics*, that continue to provide a unique account of a topic of lasting interest. And through a sizable printing, these classics will be made available at a comparatively modest cost to the reader.

The lecture note volume *Experimental Techniques in Condensed Matter Physics at Low Temperatures* grew out of a seminar on this topic for Cornell University graduate students and research associates organized by Robert C. Richardson, who received the 1996 Nobel Prize in Physics as a codiscoverer of the superfluidity of liquid He^3, a discovery made possible by the development and application of some of the techniques described in these notes. The lectures presented in this volume, which has been edited by Richardson and his colleague, Eric Smith, represent an informal hands-on approach to measurements at low temperatures—from cooling and cryogenic techniques to thermometry. The material in these lectures continues to be invaluable for all scientists who carry out research on condensed matter systems at low temperatures, be they beginning graduate students or experienced researchers.

David Pines
Urbana, Illinois
March 1998

Credits

Figure 2.13. Reprinted by permission. © 1962 by John Wiley & Sons, Inc. From Dushman, S., *Scientific Foundations of Vacuum Technology.*

Figure 2.18. Reprinted by permission. © 1962 by John Wiley & Sons, Inc. From Dushman, S., *Scientific Foundations of Vacuum Technology.*

Figure 2.24. Reprinted by permission. © 1972 by North-Holland Physics Publishers. From Hudson, R. P., *Principles and Applications of Magnetic Cooling.*

Figure 2.25. Reprinted by permission. © 1972 by North Holland Physics Publishers. From Hudson, R. P., *Principles and Applications of Magnetic Cooling.*

Figure 2.30. Reprinted by permission. © 1974 by Academic Press, Inc. From Lounasmaa, O. V., *Experimental Principles and Methods Below 1K.*

Figure 2.31. Reprinted by permission. © 1979 by Oxford University Press. From White, G. K., *Experimental Techniques in Low-Temperature Physics.*

Figure 2.32. Reprinted by permission. © 1976 by Sussex University Press. From Betts, D. S., *Refrigeration and Thermometry Below 1K.*

Figure 2.33. Reprinted by permission. © 1974 by Academic Press, Inc. From Lounasmaa O. V., *Experimental Principles and Methods Below 1K.*

Figure 2.34. Reprinted by permission. © 1981 by North-Holland Physics Publishers. From Meijer, H. C., G. J. C. Bots, and H. Postma, *Physica* **107b**, 607.

Figure 3.15. Parts a, b, c, and d provided courtesy of Newport Corporation, Fountain Valley, CA, USA.

Figure 3.20. Reprinted by permission. © 1978 by American Institute of Physics and author(s). From Kirk, W. P. and M. Twerdochlib, *Rev. Sci. Instr.* **49**, 765.

Figure 3.21. Reprinted by permission. © 1978 by American Institute of Physics and author(s). From Kirk, W. P. and M. Twerdochlib, *Rev. Sci. Instr.* **49**, 765.

Figure 3.36. Reprinted by permission. © 1985 by American Physical Society. From Osheroff, D. D. and R. C. Richardson, *Phys. Rev. Lett.* **54**, 1178.

Figure 4.35. Reprinted by permission. © 1938 by Bell, London. From Bergman, L., *Ultrasonics.*

Figure 4.38. Reprinted by permission. © 1962 by John Wiley & Sons, Inc. From

Figure 5.15. Reprinted by permission. Lakeshore Cryotronics. From *Lakeshore Cryotronics, catalog.* Figure 5.16. Reprinted by permission. Lakeshore Cryotronics. From *Lakeshore Cryotronics, catalog.*

Figure 5.17. Reprinted by permission of the publishers, Butterworth & Company (Publishers), Ltd. © 1977. From Sample, H. H. and L. G. Rubin, *Cryogenics* **17**, 597.

Figure 5.18. Reprinted by permission. © 1975 by American Institute of Physics and author(s). From Lawless, W. N., *Rev. Sci. Instr.* **46**, 625.

Figure 5.19. Reprinted by permission. © 1975 by American Institute of Physics and author(s). From Lawless, W. N., *Rev. Sci. Instr.* **46**, 625.

Table 2.4. Reprinted by permission. © 1979 by Oxford University Press. From White, G. K. *Experimental Techniques in Low Temperature Physics.*

Contents

Chapter 1

Introduction

by Robert C. Richardson

1.1   The Origin of "Techniques"

The work which follows is not intended to be a comprehensive monograph on how to or why to do experiments at low temperatures. Instead, it is meant to be a supplement to such excellent books as those by White, Lounasmaa, and Betts. It grew out of a Special Topics course offered in the Physics Department of Cornell University during the Spring terms of 1981 and 1985. In order to understand something of what we have attempted let me give you a bit of the history of how "Techniques" evolved.

In 1981 a group of graduate students and research associates working in the Laboratory of Atomic and Solid State Physics (LASSP) felt it would be useful to have a seminar course on the experimental techniques used in studying specimens of matter at low temperatures. The students themselves took turns preparing the lectures twice each week during the term. With each lecture a 'hand-out', containing between 5 and 15 pages of usually handwritten notes and xerox copies of tables, was distributed. At the end of the term we made several hundred copies of the collection of notes, stapled them together, and called the result "Low Temperature Techniques—Spring 1981." The publication was sold for the copying cost in the LASSP Stockroom. It was marvelously successful, as such things go, and had to be 'reprinted' in at least five more batches of 100 by 1985. We still receive mail requests for the 1981 "Techniques" but, alas, it is out of print. The original notes were thrown out in a lab clean-up and copies of the copies of the copies have become quite illegible.

The reason for the success of "Techniques" was probably its informality. It contained the sort of advice a senior graduate student or post-doc gives a beginning student by word of mouth or sometimes in lab books or appendices to a PhD thesis. It was a great deal more informal than the latter. There was a lot of "don't use this brand but use that instead because ..." in it.

By 1984 there was a new batch of students and visitors at Cornell who had an interest in revising "Techniques." With the more widespread familiarity with word processing it was felt that it might be an easy matter for the contributors to prepare sections of a book which might be somewhat more legible than the original. In addition we were in the process of constructing our new Microkelvin Laboratory and had learned some more techniques for such things as vibration

1

isolation and rf shielding. In the spring of 1985 the new set of seminars was presented and the present book followed from that effort. Any proceeds from this book will go into a fund to be used for graduate student travel and entertainment.

We decided that we wanted to keep something of the informal style of the 1981 "Techniques" with lots of very frank advice and a conversational tone in the discussion. Some contributors succeeded in this more than others. Some of the recommendations are probably too frank and we risk the eternal enmity of more than a few equipment manufacturers. The contributors had a wide variation in the amount of their laboratory experience as well as in the amount of their practice in writing English. They ranged from first year graduate students for whom English was not the mother tongue to seasoned veterans such as Eric Smith. Despite encouragement to be colloquial in style and use the first person as frequently as possible, many sections are written in the third person passive style of a thesis chapter on apparatus. The contributions have been only lightly edited and remain, for the most part, as they were presented in class.

Some of the contributions will be more than a little mysterious to readers who have never lived in the United States or, better yet, visited Cornell. This is not just because of the informality. Many of the glues, gadgets, and epoxy brands are available only in the United States. In some cases we have given the addresses of sources of the more useful products. For the most part, the units used in the discussions are those that we use, a mixture of metric and English units. Most of the lengths are given in feet and inches ("). Most of the masses are given in kilograms. Magnetic fields are given in both Gaussian and MKS units. The mysterious psi is a pressure unit (pounds per square inch).

There are quite a number of references to people who have worked at Cornell in the past. Any time you run into an unfamiliar name it is safe to assume that the person is a past graduate student or visitor. Clark Hall is the building in which most of us work. The H corridor is the domain of the low temperature group and the C corridor is that of the Pohl Group. Both corridors are in the basement of Clark Hall.

The authors frequently credited the names of those from whom they first learned a technique. Many of these have never been published elsewhere. Others' techniques might have been published somewhere long ago in places by now forgotten. We apologize to those who might have invented and published a method that resides at Cornell in the folk memory only of the graduate students.

How innocent we were when the second project started. The more formal preparation of book sections was a formidable barrier over which a few of the participants in the course had difficulty hopping. Vena Kostroun, a very dedicated Cornell undergraduate, was employed to help in editing the collection of notes using the TEXprocessing scheme which had been recently adapted to the Prime computer system in the Cornell Materials Center. Through his efforts and the very helpful assistance of Douglas Neuhauser, the manager of the Multi-User-Facility of MSC, most of the seminar contributions were finally processed. Some of the contributions passed through the barrier by the tunneling method. We are deeply in the debt of Vena for this final production of the notes.

1.2 About the Content of the Book

The book is organized into Chapters which are quite uneven in length. Many of the topics in the 1981 edition of notes have been omitted or replaced with quite new approaches to the subject.

Chapter 2 is about the hardware for cooling material to low temperatures and some of the cooling techniques. In 1981 we had an extensive discussion about how to build your own dilution refrigerator. Nowadays, if at all possible, such a practice should be avoided. This time we discuss, instead, the problems encountered in making a dilution refrigerator run. Very reliable dilution refrigerators have been manufactured by Oxford Instruments and the SHE Corporation. Unfortunately SHE, now named BMT, no longer manufactures dilution refrigerators. However, it has recently licensed a firm in Munich, Cryovac, to produce models similar to the old SHE cryostats. We have no information at this time on how successful the new models of the SHE equipment have been.

David Cahill has described a 'dipper' cryostat which Eric Swartz invented for use in a Helium storage dewar. After a liquid Nitrogen precool one just dips the thing into the liquid Helium. Quite a number of successful copies of this cryostat for temperatures greater than 4 K have been propagated at Cornell. If one has easy access to 50 liter storage dewars with a large 'throat' his device is especially useful for measurements over a wide range of temperatures.

Chapter 3 on some special cryogenic design methods contains an updating of Eric Smith's recipe chapter of 1981. Most of Chapter 3 is quite new with this edition. The ideas about isolation from vibration and electro-magnetic signals have been incorporated in our new facilities. A variation of the Texas A&M gimbal mounted bellows has been especially successful for isolating our new cryostats from the vibrations of the roots blowers we use for pumping helium.

Chapter 4 is the longest division of the book. It contains discussions of some of the specialized instrumentation techniques we have used for experiments here at Cornell. Many other topics could have been included, but life is short. During the 1985 term David McQueeney gave a lecture on the use of computers to control experiments and manage data. His notes are not included here. Much of the way that we use DEC and Prime computers at Cornell is too specific to the local electronic architecture. David's compendium of data analysis and presentation techniques is sold as the software package called PLOT by New Unit, Incorporated of Ithaca, NY. In Clark Hall the program is called LT PLOT and it has proven extremely useful to a wide variety of experimentalists (and even theorists).

We replaced McQueeney's computer discussion with a section on electro-magnetic compatibility by John Denker. Of the 1981 notes, only this section has been included with little change. John came to Cornell after having co-founded a software company, APh Technological Consulting, Inc. APh, among other things, produced most of the successful Mattel electronic games in the early 1980s. His story about the production of the film *Jaws* is based upon the work done by APh in making a computer controlled shark function.

Chapter 5, on thermometry is the shortest. A great deal more could be usefully written about thermometry methods. Since the time of the course we have used the temperature dependence of gamma ray anisotropy from Cobalt at the encouragement of Oxford Instruments. The technique is very useful and past biases I have held against the method were wrong. Oxford Instruments successfully started up and demonstrated two new dilution refrigerators in our laboratory with essentially no other instrumentation than bourdon gauges and the gamma ray anisotropy. The use of electronic noise and resonance are other methods we have employed or are planning to employ but they are also omitted. If there is a future edition of "Techniques" these topics are likely to be included, along with a discussion of the use of devices made from high temperature superconductors.

1.3   Some Recommended General References

My three favorite sources of general information about low temperature techniques are the monographs *Experimental Principles and Methods Below 1K* by Lounasmaa, *Experimental Techniques in Low-Temperature Physics* by White, and *Refrigeration and Thermometry Below 1K* by Betts. All of these are more comprehensive than that which we have attempted here. Lounasmaa and Betts have written very enjoyable discussions of very low temperatures techniques and White's text is especially valuable for its discussion of cryostat design. A valuable Appendix to our 1981 "Techniques" was a set of tables which Ben Crooker had gathered in answering his Admission to Candidacy Examination. We considered appending a similar set of data to this edition but found that almost everything we wanted to include was in White's text. Thus, any new tables which we compiled were included in the sections prepared by the contributors.

1.4   Acknowledgement

We are indebted to a number of agencies for the stipend support of the graduate students and visitors who made the various contributions. The major support has come from the National Science Foundation through grants to the Cornell Materials Science Center, the Microkelvin Laboratory, and the Low Temperature Research Program.

Chapter 2

Cooling and Cryogenic Equipment

Nicholas P. Bigelow, David G. Cahill, Geoffrey Nunes, jr.,
Keith A. Earle, and Henry E. Fischer

2.1 Dewars and Magnets

by Nicholas P. Bigelow

This section is grouped into two separate divisions. The first is a discussion about the containers for cryogenic liquids, now called dewars by those in the trade. I will also talk about the basic concepts of a superconducting magnet design. It is by no means a complete tutorial but is a starting point on how to produce magnetic fields in a low temperature apparatus.

When visualizing a low temperature apparatus, one of the first things that usually comes to mind is a dewar: the insulated container which houses the experiment and serves to isolate it from the relatively hot environment of the laboratory. Dewars are available in a wide range of sizes and design styles and in an equivalently large range of prices. When setting up a low temperature apparatus, the wrong choice of a dewar can drastically increase the day-to-day operating costs and can sometimes cause problems which limit the attainable base temperature of the cryostat. On the other hand, there is no need to buy a dewar which has design features that are unnecessary and may be costly. The purpose of this section is to discuss some considerations that go into the design and purchase of a dewar.

This section was written with a clear bias toward helium research dewars, although most of the ideas are relevant to either research or storage dewars. I do give some brief advice about storage dewars at the end of Subsection 2.1.1.

2.1.1 Dewars

Heat Leaks There are two basic mechanisms for heat transport into the experimental region that one must consider when designing a dewar: thermal conduction and thermal radiation. For the simple case of two thermal reservoirs which are at temperatures T_h and T_l ($T_h > T_l$) and are linked by a single piece of material of uniform cross sectional area A, the heat flow can be described by Fourier's Law, $\dot{Q} = KA\nabla T$ where K is the thermal conductivity of the material and ∇T is the vector temperature gradient between the two reservoirs. In the case where the two reservoirs are separated by a vacuum, the heat transport between them can be described in terms of black body radiation between the two materials. The heat flux radiated from the surface of a body at temperature T is described by the Stephan-Boltzmann relation, $\dot{Q} = \sigma \epsilon A T^4$, where σ is the Stephan-Boltzmann Constant and ϵ is the emissivity of the material, and A is the available surface area for heat radiation. If we assume that each material's adsorptivity is roughly the same as its emissivity and, further, that the radiating surface areas of the two reservoirs are approximately the same, then the net heat transport between them is given by $\dot{Q} = \sigma(\epsilon_1 - \epsilon_2)A(T_h^4 - T_l^4)$. In designing the "perfect" dewar we wish to minimize the effective thermal conductivities of any materials which connect the experiment to the outside world and to minimize the thermal radiation which reaches the dewar's contents.

Insulation: Concepts As pointed out by Sir James Dewar, when he invented the thermos bottle (Dewar(1898)), the lowest thermal conductivity material that can be used to separate two thermal reservoirs is no material at all, namely a vacuum. In fact, this is the defining property of a dewar vessel; it is a double walled container where the inner and outer walls are separated by a vacuum.

An important question then is how good must this vacuum be? From simple kinetic theory of gasses one can show that for a gas in a container whose characteristic size L is much greater than the collisional mean free path l (i.e., $L >> l$), the thermal conductivity of the gas is relatively independent of pressure. This is the case where inter-particle collisions are much more frequent than particle-wall collisions. However, as the pressure decreases, and hence the mean free path increases, collisions with the container walls become more frequent than inter-particle collisions (that is, when $l >> L$) and the thermal conductivity of the gas can be shown to vary linearly with particle number and hence with pressure. Since the transition from constant thermal conductivity depends on the container size one can decrease the effective thermal conductivity of the gas by decreasing the container size. This is one of the important principles behind most forms of insulation.

Consider the problem of minimizing thermal radiation. The most obvious way to minimize the heat transferred from a hot body to a cold one is to minimize the emissivities of the two materials. From Wein's displacement law one finds that the dominant wavelength radiated by a body at temperatures less than 300 K is in the infrared (10 μm for 300 K), therefore one must be careful to consider emissivities evaluated at these wavelengths. Many materials which appear very reflective at optical wavelengths (and are assumed to be of low emissivity) can often turn out to have relatively high infrared emissivities. While highly polished metals such as copper, aluminum, silver or gold have been shown to be highly reflective in the infrared, in practice they quickly become tarnished or covered with fingerprints which greatly increase their emissivities. In fact most metals which have been exposed to the atmosphere for extended periods tend to exhibit very similar IR emissivities. As a result, it is not usually considered to be worth the time or expense to use exotic or highly polished materials in order to suppress thermal radiation.

To reduce the thermal radiation to the experiment one can try to reduce the temperature of the 'hot' body which is closest to the 'cold' experiment. This can be achieved in a variety of ways. By inserting a series of thermally floating shields (often called baffles) between the hot and cold surfaces, it is easy to show that the final heat transfer to the experiment is reduced by a factor of $n+1$, where n is the number of baffles used. One or more of the baffles can be thermally connected to some cold reservoir, for instance a liquid nitrogen supply, in order to force the baffle's temperature below its floating value. This thermal anchoring can have the advantage of reducing the number of baffles that are needed, and thereby the space between the room temperature walls and the experiment. In experimental dewars the thermal anchoring is often achieved by placing a liquid nitrogen filled 'jacket' at the top of the dewar which extends far enough down into the dewar wall to allow it to hold a reasonable amount of liquid. A thermal shield anchored to the bottom of this jacket is then extended the rest of the way down the dewar wall in the form of a thin metal shield, as depicted in Figure 2.1.

In practice, most manufacturers allow one surface of the nitrogen jacket to act as the inner surface of the dewar neck. Such a design has the advantage that it provides the user with a region at the top of the dewar which is at 77 K and to which the apparatus inserted into the top of the dewar can be heat sunk, decreasing the heat leak into the liquid helium bath at the expense of an increased liquid nitrogen boil-off rate. There are disadvantages to the nitrogen shielded dewar: the nitrogen jacket requires filling, often on a daily basis; a more serious problem arises because the liquid in the jacket is constantly boiling under its own vapor pressure. The boiling can produce vibrations in the cryostat causing problems in some experiments of a mechanical nature, or producing eddy current heating in experiments performed in high magnetic fields. There are more elaborate and expensive dewar designs which incorporate a shield that is cooled by small closed cycle refrigerators to 15 K, although these are used primarily in whole body nmr imaging magnet systems or in dewars for remote installations where long liquid helium bath lifetimes are particularly desirable. We know of one group that tried such a dewar on an adiabatic demagnetization cryostat and found that the vibrations produced by the expansion engine caused unacceptable levels of eddy current heating.

Insulation: Commercial Products   There are two principal types of insulation, in addition to the vacuum, which are available for use in the vacuum space of a dewar: so-called superinsulation and powdered insulation. Superinsulation is a thin metallized insulator, usually aluminized mylar of about .25 to

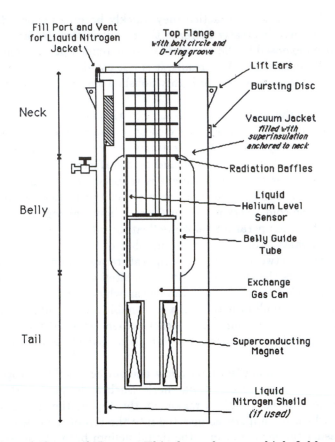

Figure 2.1. *A Research Dewar. This dewar houses a high field superconduct-*
ing magnet and a dilution refrigerator. The left side depicts a
nitrogen shielded design.

1.0 mils thick. The metallization is usually very reflective (high emissivity). Its
remarkable insulating properties can be discovered by wrapping one's hand in a
few layers. There is a sense of heating. To prevent adjacent layers from touching
each other, the material is often corrugated or embossed with small pimples.
Another common approach is to sandwich a very thin layer of fiberglass cloth
between adjacent layers.

Powder insulation consists of a fine grained insulating powder (typically
10 μm in diameter) which is poured into the vacuum space of the dewar. The
powder relies on the fact that the contact area between neighboring granules
is so small that the effective thermal paths are long and tortuous. The powders
used have very poor thermal conductivities; however, they have surprisingly large
emissivities. The successful attenuation of thermal radiation is achieved by the
large number of scatterers created by the powder. The scheme is analogous to

the attenuation of a light beam by a dense fog.

Both types of insulation act to suppress thermal radiation by the intermediate shield principle. The insulation also acts to reduce the effective cell size for any residual gas in the vacuum space, thereby suppressing the thermal conductivity of the gas. In a typical commercial superinsulated dewar there are about 50 layers of superinsulation, corresponding to a thickness of about one inch. The first few layers are the most effective in the attenuation of thermal radiation, however the subsequent layers are important for the suppression of thermal conductivity in any residual gas. One can define an effective thermal conductivity for these insulations, which in the case of superinsulation is about 10^{-6} W/cm-K between 300 and 4 K (White, 1959).

Although the use of most insulation can render the dewar almost serviceable even with 10^{-3} millitorr of residual pressure in the vacuum space, the jacket should be evacuated to at least 10^{-5} millitorr for practical use. To help matters along, it is standard practice to incorporate some sort of 'getter' into the vacuum jacket. Getters are effectively cryopumps that are attached to the inner wall of the jacket so that as the dewar is cooled, any residual gas is adsorbed. Getters often consist of activated charcoal which, in some economical nitrogen storage dewars, rattles about freely in the jacket.

Heat Leaks Down Dewar Neck: Conduction Having looked at the problem of heat input to the experiment through the dewar walls, consider the problem of heat leak into the bath by thermal conduction down the inner wall of the dewar. For a typical dewar this heat leak contributes about 5% of the total heat-leak into the liquid contents. The obvious way to suppress conduction is to use low thermal conductivity materials in construction. The trade-offs here are between tensile strength, mass, and permeability to helium gas (particularly at room temperature). Spun stainless steel or spun aluminum is frequently used as it can be made very thin without sacrificing strength. There is also a type of thin walled stainless tubing which is corrugated to increase the effective thermal path length, thereby decreasing heat leak and the boil-off. Unfortunately, I don't know of any commercial companies in the United States which offer this feature on their research dewars. Fiberglass can also be used, however, it is reasonably permeable to helium vapor at, or near, room temperature. To alleviate this problem, manufacturers will laminate a thin layer of metal onto the fiberglass, in a very similar fashion to the lamination of copper onto circuit board material. Although almost all of the dewars in our group seem to have their vacuum jackets eventually become soft, especially after they have been warm for extended periods, a working dewar will not indicate even the smallest leak when tested with a helium leak detector (less than 10^{-8}atm-cc/sec leak rate).

Most of the residual heat that is transported down the dewar wall does not reach the liquid bath but is absorbed by the enthalpy of the cold gas boiling off the bath and flowing past the walls. A variety of experiments have been performed to study the temperature profile of the gas exiting the dewar. For a cylindrical dewar with an open neck the vapor forms thermal strata with a stagnant boundary layer at the walls. Greater liquid helium boil-off rates are observed in dewars that are left untouched than in dewars which are periodically disturbed by jiggling. This is attributed to the fact that thermal exchange between the exiting gas and surrounding walls is much better in the presence of turbulent flow. The trick then is to induce turbulent flow in the gas as it flows past the walls.

Heat Leaks Down Dewar Neck: Radiation and Baffles To minimize the net heat flow down the neck of the dewar one has to suppress heat-leaks due to thermal radiation down the dewar neck. This problem can be solved by the use of shields or baffles in the dewar neck. The baffles are usually metal discs, the diameter of which is chosen to force the exiting gas to flow in a turbulent fashion past the walls. The dimensions must be chosen with particular care for dewars which contain large superconducting magnets, because if there is a magnet quench, large amounts of helium vapor must be able to escape very rapidly without interference from the baffles. Experiments have been performed to try to optimize the materials for construction of the baffles and for their placement in the dewar neck.

It is our experience that no more than four or five baffles are necessary and that almost any convenient metal works. Sheet aluminum is the most convenient material to use because it can be easily cut into the appropriate shape. It is also useful to make the baffles thick enough so that things can be mounted on them (say 1/16 of an inch). The baffles can be attached to the pumping lines of the experimental insert by small friction clamps or can be suspended from the top (room temperature) flange of the cryostat by metal rods or wires. Such baffles have been found to be most effective when the lowest is placed near the bottom of the LN_2 jacket or the 77 K region. If the baffles are placed too low in the dewar neck they have been observed to promote the onset of Taconis oscillations, and an associated increase in helium boil-off rate. Keep in mind that the baffles should be located with respect to the liquid helium level at its highest point.

One can also make effective baffles for the dewar neck from styrofoam sheets cut to form disc-shaped plugs. This design has the advantage that the plugs can easily be modified while in place as new wiring, etc. is installed. The plugs have the drawback that, over time in a given low temperature 'run,' their performance is degraded as they become filled with helium gas.

The blackbody radiation from 300 K to 4 K is about 50 mW/cm^2. Therefore if the experimental insert includes any straight open pipes that extend from room temperature into the helium bath such as pumping lines, it is important to remember to place radiation baffles in these also. The baffles can be made from thin sheets of metal cut into semicircles and suspended in the pipes so as to eclipse any direct path for thermal radiation from room temperature. The baffles can be held against the pipe's inner wall, dumping the radiation energy into the bath, or can be suspended in the line on a string or a thin wire. In order to completely block the path, at least three baffles are needed and will not significantly affect the gaseous flow impedance of the line. If the flow impedance is not important, a simple solution is to use balls of bronze wool which fit tightly into the pipe. If wires run through the pipe, it is handy to have a scheme for guiding wires safely around the baffles. It is also possible to slightly reduce the thermal radiation down such tubes by putting bends in them so that there is no direct path for room temperature 'light' into the cryostat; however, the use of light reflecting baffles is much more effective.

Planning Your Research Helium Dewar   Armed with a host of ideas and principles of dewar design, the experimentalist is then confronted with the question, "How should I design my dewar and what should I say on the phone to the sales person at the dewar company?"

For the sake of discussion assume that you are interested in a dewar which will house a dilution refrigerator, some sort of superconducting magnet and maybe a lambda refrigerator for the magnet. The lambda refrigerator is a device which permits the cooling of the liquid helium at the bottom of the dewar to temperatures well below the superfluid transition of ^{4}He. The poor thermal conductivity of liquid helium permits the operation of the surface of the liquid at 4 K so that, apart from a small additional consumption of liquid helium, the dewar behaves as if it were filled with ordinary liquid. Figure 2.1 depicts such a dewar. The diameter of the neck is chosen to be as narrow as possible in order to minimize the available area for thermal radiation. At the same time the neck is wide enough to allow the user ample room to insert the experiment. The dewar belly serves to act as a reservoir which increases the available liquid volume, and the tail section is usually narrowed to allow the use of a lambda refrigerator.

In choosing the length of the neck it is necessary to allow sufficient length for the exiting gas to make the transition from 4 K to 300 K so that none of the enthalpy of the gas is wasted on cooling the room or the helium recovery plumbing. Typically, this length is about 18 inches or so. At the bottom of this region lives the lowest radiation baffle, as discussed above. In order to maximize the available volume of liquid helium, the belly diameter is chosen to be as large as possible while still allowing sufficient clearance for the insulation in the surrounding vacuum jacket. Typically, at least one inch should be allowed. If you accidentally goof this part up, the manufacturer will let you know right away. If the dewar is to contain a superconducting magnet, it is standard practice to design a narrower tail section in the dewar in the region where the magnet will reside. This design is used because any excessive volume in the magnet region must be filled with liquid helium which essentially serves no purpose and is eventually wasted when the cryostat is warmed up. The tail section should begin somewhere above the top of the magnet; however, the exact location is only really important if the dewar is to include a lambda refrigerator for any magnet. In this case there must be a long enough region between the top of the refrigerator and the top of the tail to allow the helium trapped in that region to make the transition from 2.2 K to 4 K. An estimate of the needed length can be made by using the expected boil-off rate for the bath, which seems to hover at about 14 liters/day for most of the large cryostats in our lab, with the assumption that the equivalent heat leak is incident on the 'top' of the 4 K annular shaped region at the top of the tail. With this energy flux and the thermal conductivity of helium, one can estimate the required length needed to make the transition between 4 K and 2.2 K. Clearly, the length will depend on the clearance between the exchange gas can and the cryostat tail wall. It will be smaller for a closer fit. For a separation of 1 cm and a boil-off rate of about 15 liters/day this length must be about 10 cm. If you've already bought your dewar and have decided that it's time to install a lambda fridge, but realize that your dewar's inner diameter in the region of concern is a bit too stout, fear not. By forming a cylinder of styrofoam of the appropriate thickness it is possible to decrease the thermal conductivity in the transition region by enough to make almost any dewar serviceable.

The outer shell of the dewar can be made from a variety of materials. Whatever material is chosen, it must be able to meet the needs of strength, weight and manufacturing ease. The most common commercial materials are stainless steel and aluminum. The metal can be spun into the desired shape or it can be welded from stock. If the dewar is welded together it can be formed either using only plate stock or it can be made from plate and tube stock. If the dewar is to be constructed from rolled sheet metal stock it will have a seam that runs vertically along the dewar. Although such a construction is certainly not as aesthetically pleasing as the seamless designs made from piping or spun materials, it has the advantage that the dewar can be arbitrarily long with minimal problems associated with alignment of the inner wall of the jacket with respect to the outer wall. The welds are usually done by inert gas techniques and should be quite reliable.

The material used for the inner wall will often depend on the choice for the outer wall or vice-versa. In addition to meeting all of the requirements of the outer wall material, the inner wall must also be chosen so that the thermal conductivity does not allow an outrageous heat leak into the bath. Aluminum alone is not usually an acceptable choice for the inner wall. With aluminum, it is difficult to attain a small heat loss while still retaining enough strength to withstand the pressure of even one atmosphere. There are designs used frequently in storage dewars which use a fiberglass tube for the upper region of the dewar neck and aluminum for the main volume. If fiberglass is used, there will inevitably be joints which the manufacturer makes with some sort of epoxy. In this case it is particularly important to request that the dewar be tested at operating temperature before being shipped. We have had experience in our lab of receiving untested dewars which leaked at such a joint when cooled to 4 K for the first time. One important drawback to the hybrid aluminum and fiberglass designs is that aluminum corrodes when exposed to moisture for extended periods. If the dewar is casually warmed by exposure to room air, moisture condenses inside and corrodes the exposed surfaces, particularly the bottom metal cap. Because of its higher yield strength and low thermal conductivity a very nice material is stainless steel. However, when the experiments involve magnetic fields of high homogeneity, one must be careful to choose materials which will not distort the field. If such considerations are important, it's probably a good idea to include a field specification on the dewar when placing an order to the manufacturer.

Most dewars are supported from a bolt circle machined into a flange attached to the top of the dewar's neck. This flange probably has an O-ring groove in it to allow the bath to be connected to a sealed recovery helium system or to allow the user to pump on the bath space. To raise the dewar up to the support structure it is easiest to have a pair of support ears welded onto the dewar exterior, as shown in Figure 2.1. Alternately, it is possible to screw cable raising supports into the bolt circle on the top of the dewar. Such a method can prove to be inconvenient, so unless you can't afford the space, have the ears put on anyway.

Because 'no dewar is perfect,' there will come a time when the vacuum space will need to be pumped out. To make this easy, be sure to ask for a pump-out port at some handy place and to ask for a fixture and valve with connections that you understand (that is, connection compatible with fittings in your lab). If the pump-out port has a 'fixed-state' valve on it, then a bursting disc on the vacuum space is a must. Such discs are usually made from a thin metal foil with a nearby sharp edge, positioned in such a way that when the pressure in the vacuum space rises above an atmosphere, the foil will bend and be punctured by the sharp edge, and thereby venting the vacuum space into the room. Such a safety feature can be awfully important if some air manages to leak in the vacuum space and freeze while the dewar is cold, only to expand disastrously as the dewar is warmed.

Many manufacturers use a pressure relief design which acts both as a pump-out port and as a safety valve. A generic version of this is shown in Figure 2.2. The basic idea is that as long as the pressure in the vacuum space is less than an atmosphere, the differential pressure will hold the ball snugly in place. If, however, disaster should occur the ball will fall (or launch) out of its socket relieving the pressure. Such a design is clearly more elegant and compact than a flange and valve; however, such gadgets require a special fixture for connections to pumps for evacuating the jacket. Figure 2.2 depicts such a device.

One final word: rarely will a dewar meet the boil-off rate specified by the manufacturer, especially once an experiment with its associated pipes and wires is inserted. The manufacturer will have assumed that all of the enthalpy of the exiting gas would be available to suppress intrinsic heat leaks in the dewar.

There are many dewar manufacturers in the world today. For some reason, even during the time of an over-valued dollar, the least expensive dewars are made in the United States. Many U.S. manufacturers grew out of the demise of Cryogenic Associates, Inc. in Indianapolis, Indiana. At Cornell we have had a wide range of experience with most of these companies. Their performance has ranged from 'stellar' to 'abysmal.' At the end of this section there is a list of manufacturers with whom we have had satisfactory experiences.

Exotic Dewars   In addition to the design cases considered in the previous discussion, there are a variety of special purpose dewars that one might need. A fairly common example is that of the dewar tailored to fit between the pole faces of a room temperature magnet. The constraint in such a dewar is in the thickness

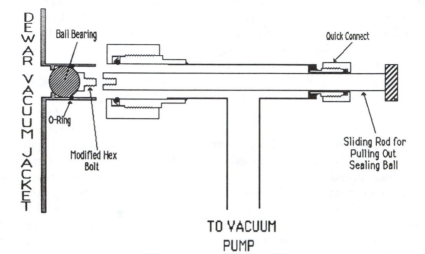

Figure 2.2.   *A Pressure Relief and Pump Out Port for Dewars. The pump out device is fabricated from a copper tee fitting and commercial quick connect fittings. A brass bolt is fitted to the dewar's sealing ball by machining a concave surface on the bolt heat with an end mill and epoxying the bolt to the ball.*

of the wall in the tail section. An identical problem arises in the case in which the dewar is designed for a room temperature experiment using a superconducting magnet. Such a dewar must have some sort of re-entrant bore, usually accessible from the dewar's bottom, that is often rather small. In order to use as thin a wall as possible, these dewars frequently incorporate a nitrogen cooled shield connected thermally to a liquid reservoir high up in the dewar's fat portion. Because of the close tolerances in the thin-walled portion it is often necessary to have some alignment structures to prevent the shield from touching the adjacent wall as the dewar is cooled or when it is tilted. Such spacing structures are especially important in larger dewars or in almost any re-entrant dewars. Even so, such dewars are quite sensitive to twisting and hanging forces that can distort their original shape.

Sometimes it may be necessary to have optical access to an experiment through the dewar wall. That is, one may want windows in the dewar jacket. Although optical access dewars of this sort are not uncommon, they are often rather complex and can be very unreliable. Frequently, the helium bath must be pumped below its superfluid point to reduce the effects of boiling bubbles on the liquid's index of refraction. The necessity of pumping aggravates the need for perfect seals on the windows. Therefore, the first design rule here is to be sure that windows into the helium bath are really necessary. One easy way to avoid the window problem is to use light pipes or optical fibers which enter the dewar through the room temperature flange at the top of the cryostat. If this won't work, how about changing the experiment? If it is a 4 K experiment and the sample doesn't need to be directly immersed in liquid helium, it can be mounted in the dewar's vacuum space on a flange which is in good thermal contact with the bath, avoiding the problem of making a seal to a window in contact with the helium bath. The sample can be accessible from a room temperature flange that uses conventional rubber O-rings and the window material can be easily changed. In any case, there is another problem that can arise. As the dewar is cooled, the various dewar and internal support materials contract. It is quite possible that the final result will be a sample which is no longer in the carefully aligned spot where it started. In fact, it may be that the sample will disappear from the window's field of view.

More likely than not, the architect of your lab didn't think about your dewar much, and you may decide that the space constraints of the room and the apparatus do not allow a single piece dewar to be used. The obvious solution is to use a dewar which is divided into a number of segments which can individually be put around the experiment and then bolted together. This design has the appealing feature that if the nature of the experiments changes a lot, the dewar can be modified too. Unfortunately, these dewars can suffer from a host of practical drawbacks. Most dewars of this sort are designed to have a vacuum space which is common to all segments. This means that not only will the dewar need to be evacuated before each use, but the use of any getters in the jacket will be futile, as they will quickly become saturated when exposed to air during the disassembled phase. Any super-insulation in the vacuum jacket adds massive amounts of surface area, which means that the dewar will demand a fairly long pump-down time. Low temperature seals are very difficult to make, particularly over such large dimensions as needed by segmented dewars. For this reason, a segmented dewar should be made in as few pieces as possible. If the dewar utilizes a nitrogen cooled shield, then more reliable designs will not segment the liquid nitrogen reservoir. In fact, the most desirable design will not segment the dewar at all.

Glass dewars can provide an economical low temperature environment, particularly for experiments above 1 K and for demonstrations. These dewars usually consist of two nested cylindrical dewars in which the outer dewar is filled with liquid nitrogen while the inner dewar contains liquid helium and the experiment. The inner surfaces of the glass walls are frequently aluminized or silvered to block thermal radiation although there is often a small stripe of glass left uncoated along the length of the dewar to allow the user to view the contents and the liquid levels. These dewars suffer from a number of weaknesses. They are obviously quite fragile. There are few mistakes more spectacular than breaking a glass dewar while it's full of liquid helium. Many such dewars are covered with plastic film or a plastic or metal mesh to catch the hurricane of glass that may erupt when the dewar is accidentally broken. If the user accidentally allows the neck of a glass dewar to become blocked either by closing the wrong valve or by allowing the opening to become sealed with ice, the dewar is dangerously suscep-tible to breakage. At room temperature glass is sufficiently permeable to helium that the dewar vacuum space of a liquid helium dewar may need to be evacuated regularly, especially if any residual helium gas from the last use was not flushed from the dewar. If a glass dewar is to be left unused for an extended period, and if it is at all possible, it is advisable to let the vacuum space up to atmosphere. This will minimize the hazard that can result from accidental implosion.

Storage Dewars For the most part, storage dewars are purchased from a manufacturer's standard product line far more cheaply than by a special design request. As a result, the consumer is more often confronted with various trade-offs between existing designs than with the problem of the design itself. There are some features that all storage dewars share in common. One of these is safety. The greatest hazard to the user from a dewar can be the explosion that results if the dewar is accidentally sealed from venting to the atmosphere. This is, in fact, one of the most common events to lead to the demise of a glass dewar. A typical scenario for such disaster is that the user accidentally leaves the dewar open to room air allowing the moist air to be condensed into the dewar and be frozen. To prevent such catastrophes, all commercially made dewars are required to incorporate some sort of pressure pop-off device. In the case of a helium storage dewar where not only moisture but a large fraction of the room air can be frozen by the dewar's contents, it is a common practice to incorporate a double walled re-entrant neck design. The basic idea is to provide two separate chambers in the dewar neck, only one of which is opened during use. If the main chamber is allowed to become blocked, a pressure relief vent in the other chamber will prevent disaster. This is usually achieved by using a coaxial geometry where the inner chamber is available for routine use. When disaster strikes and the primary neck is allowed to freeze shut one is often in the position of having to open it up, either because the only liquid helium available is trapped inside or, even worse, because whatever you had inserted into the dewar is now frozen inside. Often, the ice block can be melted or chipped away with a metal rod of the appropriate size and heat capacity. Sometimes it helps to heat the rod first, but watch out for the very cold plume of exiting gas you'll inevitably generate. A slightly more sophisticated design involves grinding some teeth-like shapes onto the end of the ice boring rod. A different tactic is to use a long piece of copper capillary, bent into a convenient shape, through which one blows hot helium gas while poking away. (This is a particularly useful technique for those days when you freeze the transfer syphon into your dewar.)

In choosing a liquid helium storage dewar, it may be useful to consider use of the dewar as more than just a storage vessel. The dewar can be used as a convenient low temperature environment for a trap of a small 'dipper' cryostat as described in Section 2.3, and it is important to consider the dimensions of the dewar neck and the operating depth of the liquid when the dewar is, say, half full. Other potential constraints might include mobility and ease of connection of gas recovery lines. A convenient size for a storage dewar might be sightly larger than the initial transfer volume needed for the greediest cryostat in your lab.

There are many manufacturers of helium storage dewars and many of the design features mentioned previously can help as a guide when calling a company. In our group, the helium storage dewars are about 60 liters in volume, seem to exhibit about 0.5 liters/day boil-off, cost about two kilobucks and offer 14 inches of liquid when full. These were manufactured by Cryo-Fab Inc. and are made entirely from stainless steel.

Liquid nitrogen storage dewars make much less stringent demands on the manufacturer and are available from an even larger array of sources. If the dewar is to house a trap in a refrigerator system the usable depth and holding times are of prime importance. Dewars used to transport small amounts of liquid should be convenient to move and feel well balanced. Many dewars tend to spit nitrogen out the neck when overfilled, making handles on the sides unusable. Others have asymmetrically located handles that do not allow one to pick the dewar up with its center of gravity in a reasonable place. Typical smaller dewars hold about 10 liters of liquid and cost on the order of $300 while a 160 liter dewar costs about $2000. For some applications, a beverage thermos bottle is more than adequate.

Transferring Techniques and Tools Liquid nitrogen can be transferred over short distances with relative ease. For transfer of liquid nitrogen over short distances, the tubes or pipes used need not always be insulated. Conventional funnels and rubber lab hose are often more than adequate.

Helium transfer, however, requires well insulated transfer plumbing. A typical transfer tube is made from two lengths of thin-walled stainless steel tubes about 3/16 and 1/2 inch diameter which are arranged coaxially so that the outer tube forms an annular vacuum jacket around the flow tube. The tube is usually bent in a U shape and equipped with a vacuum pump-out port for the jacket. Some sort of low thermal conductivity spacers, made from teflon or nylon, are usually needed between the two tubes. These can be easily made from small triangles or squares of thin sheets which are drilled in their centers to fit around the inner tube. The contact area at the corners of the squares is sufficiently small that the heat leak is acceptable.

Level Measurement Liquid nitrogen level detection is often much less critical than for helium. Many users are familiar enough with critical holding times of their dewars that level detection would be an unnecessary luxury. A clever level indicator can be fashioned from a piece of styrofoam and a wire. The foam floats on the liquid and the wire pokes out the neck indicating the level. Some larger storage dewars have built-in mechanical gauges. Level sensors can also be made from resistors and semiconductors that are selected to show usable change when immersed in liquid. A handy low level alarm can be fashioned in this manner.

The dielectric constants for both helium and nitrogen, though small for liquid helium, are sufficiently different from unity that a capacitive level sensor is practical. A level sensor can be fashioned from a pair of coaxial thin-walled stainless steel tubes held apart by either small teflon spacers or a spiral of thin monofilament nylon fishing line. A capacitance measuring circuit can then be used to measure the level as a linear function of capacitance. The dielectric constant measurement technique has the advantage that there is essentially no heat dissipated in the liquid by the measurement.

Until recently, the use of superconducting level detectors has been restricted to liquid helium. This design uses a piece of superconducting wire which has a transition point near 4.2 K. A current very near the critical current is driven through the wire so that only the portion of the wire immersed in liquid remains superconducting. The level is then given as a linear function of measurable resistance. This design will dissipate significant heat in the cryostat so measurements can only be made on an intermittent basis. The large difference in heat capacity between the helium gas and liquid can also be used for level detection. A carbon resistor driven with a constant current will equilibrate at high resistance when immersed in liquid. In the vapor just above the surface the heated resistor will rise to a higher temperature and the resistance will drop a great deal.

Cooling Down   Cooling a liquid nitrogen storage dewar can be a relatively straightforward event. Pour liquid in at the top and eventually the dewar cools and fills with liquid. On the other hand, cooling a liquid helium dewar, particularly an experimental dewar with a magnet in it, can be more than hard if done the wrong way. Each system will have different heat leaks, heat capacities, and paths for flow of entering liquid and exiting gas. The most effective and efficient method of cooling a cryostat is different for each system; nevertheless, the general approach is the same. As an example, consider a liquid nitrogen shielded experimental dewar that houses a dilution refrigerator and a large superconducting magnet. The dewar jacket has been evacuated and the refrigerator system is leak tight at room temperature. A Speer 220 Ω carbon resistor and some sort of heater coil sit at the bottom of the dewar and there is a sensibly located helium level detector in the belly.

The bulk of the stuff that you're trying to cool down will undoubtedly be metal. Therefore the dominant heat capacity of the system will have a temperature cubed fall-off at lower temperatures. Any pre-cooling will reduce the amount of helium needed to cool to 4 K. For pre-cool liquid nitrogen is very useful. The temperature of boiling nitrogen liquid, 77 K, is well into the Debye knee of the specific heat of most metals. Liquid nitrogen has 60 times the latent heat of evaporation of liquid helium. The gas carries more enthalpy on warming from 77 K to 300 K than helium and, most importantly, liquid nitrogen is about a factor of 25 cheaper than liquid helium. After pre-cooling to 77 K and removing the liquid nitrogen from the dewar the idea will be to cool the rest of the way to 4 K using helium, while getting the most from the enthalpy of the exiting gas.

In filling the dewar with either helium or nitrogen it is usually necessary to force liquid through a tube that reaches to the dewar's bottom. If the liquid is simply dumped in at the top, it will generally evaporate long before it reaches the magnet at the bottom. In most dewars the clearances are tight enough that the gas at the bottom may form a relatively static plug that further inhibits the cooling of the magnet. Once the dewar is filled with liquid and experiments are underway, the tube is not used for fill-ups. In fact, the vibrations resulting from forcing liquid to the dewar bottom can cause heating in most very low temperature cryostats. The most common solution to the problem is to have a tube that runs up from the bottom of the dewar, around the magnet and ends somewhere between the lowest two baffles. The top of the tube is outfitted with a fixture that allows for connection to the transfer tube. Threading is a workable solution although it is rather susceptible to freezing. A simple and practical connection can be made with a tapered conical teflon plug on the tip of the transfer syphon with a correspondingly tapered socket on the tube in the dewar. A long slow taper is desirable, allowing for more contact area on the mating surfaces. Depending on the lengths and clearances in the system, it may be useful to install a well-perforated guide tube along the intended path for the transfer syphon when it is inserted from the top. Such a tube can prevent the transfer 'stick' from becoming caught on the baffles during insertion and removal.

The direct connection to the bottom of the dewar is also important for removal of the liquid nitrogen after the pre-cool. The liquid nitrogen is removed from the dewar by pressurizing the bath space with helium gas at a pressure greater than one atmosphere to force the liquid out through transfer tube. For this purpose it is especially important that the tube reach all the way to the bottom so that as much liquid nitrogen as possible is removed.

A typical cool-down might proceed as follows. The dewar is filled with liquid nitrogen from a large storage dewar by forcing liquid through a thin walled stainless tube which is mated to the fitting in the dewar described above. The resistor at the bottom of the dewar is monitored to assure that the transfer tube is well-seated and that liquid is reaching the bottom. Some care is taken so as not to overfill the dewar, as this may cause undue stress on any epoxy joints and O-rings at the top of the dewar. If the dewar has a liquid nitrogen jacket it should also be filled.

If the apparatus contains a sealed 'exchange gas can' around the experiment and the refrigerators, it will be necessary to add some gas to the can to make thermal contact between the bath and the can's contents. Any acceptable gas that liquifies below 77 K is acceptable, and only a fraction of a torr is needed. In fact, the less gas the better as it must be removed for subsequent leak tests. The favorites in our lab are helium, hydrogen and neon. If a sensitive leak test will be made, helium is probably not a good choice as it will pollute the exposed surfaces and will be very difficult to completely remove.

If the apparatus includes a $PrNi_5$ nuclear demagnetization stage, hydrogen cannot be used as it reacts with the refrigerant in an irreversible manner. It is also useful to note that many gases of standard purity contain enough helium to make leak testing difficult. After the dewar is filled with liquid, the temperatures of the experiment are monitored. When the entire contents has equilibrated to 77 K, the exchange gas is pumped out with a diffusion pump and the can is linked to a helium leak detector. The refrigerator and sample systems are tested for leaks into the can as necessary, and all of the electronics in the cryostat are tested if possible.

If all systems appear OK, the bath space is pressurized with helium gas and the liquid nitrogen is forced out through the fill tube. While the gas is being forced out, the leak detector is monitored, so that any leaks uncovered by the dropping liquid level will be found. When the system incorporates a continuous-fill 1 K helium refrigerator, the refrigerator pumping line must be pressurized with pure helium gas throughout the entire cool-down in order to prevent nitrogen from entering the fill capillary. Otherwise, the liquid nitrogen may later freeze and render the fill line unusable. When this is the case, the pressure used to force the liquid nitrogen out of the dewar must be kept below that in the 1 K pot to avoid forcing liquid nitrogen into the fill capillary. If nitrogen does not come out when the bath space is pressurized, it is a good indication that the transfer 'stick' is not well seated in the coupling to the link to the dewar's bottom. After nitrogen has stopped coming out and the system is still leak tight, it is a good idea to seal and pump out the bath space. This serves two functions: it provides a control for the leak test, but more important, it evaporates most of the remaining small liquid nitrogen puddles and freezes the rest. This can save a lot of liquid helium, as the latent heat of solidification of nitrogen is quite high.

If the bath space doesn't pump down in a few minutes time, there is either a large leak or too much nitrogen was left in the dewar. If too much nitrogen is left in the dewar, it will eventually freeze and may block the bottom fill tube as well as freezing the magnet and the experiment into the dewar.

With everything still go, the next step is to back fill the bath space with DRY gas and prepare for helium transfer. A transfer stick suitable for liquid helium is connected to the cryostat, again with care to seat it well in the bottom fill tube. The other end of the syphon is then slowly inserted into the helium storage dewar. After sealing the syphon on both ends against pressure leaks to the room, an excess pressure is applied to the helium gas in the storage dewar. Liquid helium is thus forced into the experimental dewar. As the transfer begins, the bath resistor is monitored for flow to the bottom and exchange gas is put into the exchange gas can. To limit the size of gas charge put in the can, it can be useful to have a small volume trapped between two valves which can be filled with gas and which is then opened to the gas can.

The dos and don'ts on transfer rate should incorporate a few basic constraints. If the top plate and much of your helium recovery plumbing become heavily encrusted in ice, you are probably going too fast. If the transfer rate in a nitrogen jacketed dewar is too fast, the helium gas at the top of the dewar's neck may be sufficiently cold to cause the nitrogen to freeze and to condense room air and moisture into the jacket. If the pressure in the jacket falls below an atmosphere, the transfer rate is too fast. The first major hurdle in the cooling process remains the heat capacity of the magnet and refrigerators. As the system cools, the gas used for heat transfer in the vacuum space may be adsorbed on the surface of the exchange gas can when the exterior is too cold. In order to avoid this difficulty it is important to monitor the pressure in the exchange gas can as you transfer. To remedy the mistake of cooling the bath space too rapidly, it can be handy to have a heater attached to the can wall so that the prematurely frozen gas can be quickly revaporized. The first time this happens it is good to note the conditions as it occurs, in particular the exiting gas flow rate and the value of the bath resistor. The optimal transfer rate at this stage is usually with a mixed liquid and gas helium flow to maintain the exterior wall just below the conditions of exchange gas "freeze-out."

If the exchange gas can's contents continue to cool below the freezing point of whatever heat transfer gas is being used, there is trouble. There is either a leak into the can, or the exchange gas has been contaminated with helium. In very low temperature cryostats especially, residual helium can give rise to nasty heat leaks. If everything cools too rapidly you should be suspicious and test the apparatus for leaks between the helium bath and the exchange gas can.

Once you have squeezed all possible cooling from the exchange gas and the innards are at the gas' freezing point, it is possible to slow, or stop, the transfer and connect a leak detector to the gas can for final tests. It may also be a good idea to test any continuous fill refrigerator for flow by noting the over-pressure value and monitoring rate of decrease in pressure when the fridge is isolated from the pressure source. Any tests that can be done should be, as this is the last opportunity before investing the liquid helium into filling the bath.

If helium exchange gas was used; the preferred technique is to slow or stop the transfer when the bath resistor is just below the value corresponding to liquid accumulation. The exchange gas can is then connected to a diffusion pump and pumped on until a leak detector reads a value found to be acceptable for that system.

Subsequent liquid transfers require a different strategy. As mentioned earlier, the link for fluid to the bottom of the dewar is no longer used. The transfer has two stages: 1) getting the transfer tube cooled, into the cryostat and transferring liquid, and 2) choosing a liquid flow rate that will disrupt the cryostat as little as possible. The helium liquid transfer can be so violent that unless one is careful an unpleasantly large volume of liquid can be boiled off before the liquid level increases. This is largely due to the fact that the fluid that emerges from the transfer tube is essentially in two phases, having both liquid and gas components. To prevent this highly turbulent stream from impinging on the existing liquid's free surface the transfer tube should be inserted until the end is somewhere between the lowest two neck baffles. Some strategies include more aggressive phase separation of the transferred helium. One design utilizes a porous bronze sponge at the end of the transfer tube. This works by providing a low impedance path for the gas near the top of the sponge while allowing the liquid to flow the length of the sponge and pour off its end. Another approach is to pack the space between baffles with cotton or fiberglass to inhibit turbulent flow. One cryostat in our group has two transfer ports, one for initial transfers and one for refills.

The first stage of a transfer can be very critical, and as always, depends on the cryostat. The idea is to cool the transfer line before it is inserted down into the cryostat. This can be accomplished by starting the transfer slowly with the end of the tube high up in the baffles. Alternatively, the tube can be pre-cooled by transferring into the recovery lines, or the room, until a cold plume of gas begins to emerge from the business end of the tube. A third technique is to slowly insert the tube into the cryostat before lowering the other end into the storage dewar, then close the exit port for helium recovery and 'back transfer' OUT of the experimental dewar until the tube is cold. Finally, the transfer tube can be left in the dewar continuously. If this approach is taken it may be necessary to seal the storage dewar end carefully to prevent Taconis oscillations from developing.

How frequently transfers are made is ultimately determined by the dewar volume and boil-off rates. In any very low temperature system, a transfer is a disruptive event which causes vibrations and hence heating. Some of the cryostats in our lab behave better if liquid helium transfers are made more frequently than maximum holding time. Important parameters may include initial and final bath levels, the vibration rate after liquid helium refills, and the rates and techniques used in the liquid helium transfer.

2.1.2   Superconducting Magnets

We now shift to a somewhat different discussion about fundamental cryogenic hardware, one about superconducting magnets. In the following, I will discuss magnet accessories and home-built magnets. There are also discussions about magnet power supplies, their selection, and advice on how to make them work. As with the previous discussion, I have provided a list of magnet manufacturers. The list is essentially our in-house favorites list. In the following, I will assume that the reader has a rudimentary knowledge about superconductivity, but is otherwise a novice on the subject of superconducting magnets. It is the dream of all of us who use such magnets that the incredible developments on high T_c superconductors in 1986-7 will make major changes in the things which I say in the following.

Type II superconductors are able to support dissipationless currents at much higher magnetic fields than their Type I counterparts, and are therefore the choice in magnet design. The most common commercial materials as of 1987 are: NbTi, Nb_3Sn and NbGe, in order of increasing cost and maximum achievable field. Nb_3Sn and NbGe wires are much less ductile than NbTi and, hence, are more costly and require more care in assembly. In the highest field magnets, it is common to use a hybrid design in which a higher critical field wire is employed for the innermost windings, where the field is highest, and a less expensive lower critical field wire is used on the outside.

The wonderful thing about superconducting magnets is that they produce high fields over small volumes with very little inherent dissipation. An important question then is, what limits the fields and what causes losses in these magnets? A piece of wire made from a superconductor will exhibit a lower critical current at a given applied field strength when wound in a solenoid than when used as an isolated straight wire. This effect was quite unexpected in early magnet research and can be attributed primarily to two effects, mechanical forces on the wires and flux motion.

Many uses of superconducting magnets require a steady state at some fixed current and field. In such a case, the magnet is designed to operate in a persistent mode. In the persistent mode the magnet windings form a completely closed superconducting circuit. With a persisting current, the dissipation mechanisms in the magnet wire are entirely due to mechanical and flux flow effects.

Such magnets are energized through a persistent 'switch.' The switch is a small length of superconducting wire that is wrapped with a heater wire. When heated, the superconductor reverts to the highly resistive normal state. A power supply connected across the switch can then be used to increase or decrease the current flowing in the magnet. The persistent mode switch must be able to carry roughly twice the magnet's rated current when in the superconducting state.

Mechanical Stress and Thermal Stability of Magnets Consider an element of wire in a solenoid carrying a current just below the critical current. If the wire is moved within the applied field, there will be an emf induced in the wire. If the current in the wire has a component along the direction of the electric field then work will be done and energy will be released in the wire. Thermal fluctuations induced by the energy release may cause the wire to momentarily exceed the critical current locally and drive the wire to the normal state. As this occurs, additional energy is released when flux penetrates into the normal region. If the resulting heat cannot be removed more quickly than it is released, the normal region will propagate throughout the material in an avalanche fashion. On the other hand, if the heat capacity of the environment around the wire and the thermal contact to the surroundings are good, then the element can cool back below its critical temperature and return to the superconducting state with only a small dissipation of energy. The ability of the wire to recover from a local heat fluctuation plays an important role in the other major form of magnet disturbance: flux flow.

Flux Flow When a wire is wound in a magnet, the orientation of the induced magnetic field with respect to the current is different from the field around an isolated straight wire segment. As a result, the forces on the fluxoids trapped in the mixed state are quite different for the two. Furthermore, the screening currents that arise in the conductor due to a magnetic field perpendicular to the current flow can cause the net current density along one side of the wire to exceed the critical value, allowing flux to penetrate into a local region. The penetration of flux, in turn, gives rise to an emf and more energy dissipation. As in the case of a mechanical disturbance, the heat capacity of the wire and its thermal contact to its environment are critical in determining whether a disturbance related to flux flow can be stabilized.

To help suppress the degrading effect of these fluctuations on a magnet's performance, the thermal and electrical properties of the magnet wire are an important part of the design. A typical strand of magnet wire is usually a rather small gauge copper (or copper-nickel) wire in which many small filaments of the superconductor have been imbedded. The superconducting filaments thus imbedded in a normal metal matrix increase thermal stability and provide mechanical support. Copper is a nice matrix as it can readily be soldered in order to make an electrical connection to the superconductor. The smaller filaments in such a matrix are inherently more stable to various current fluctuations than large wires of pure superconductor because of their reduced heat capacities per unit length per unit surface area (thermal stabilization) and because image currents flow in the normal metal to oppose the magnetic field of fluxoids generated by large currents in the superconductor (electrical stabilization). The background normal metal in the matrix provides a low resistance shunt path for current around a section of filament that has momentarily been driven normal. During a flux jump the resistance of the surrounding normal metal is much lower than that of the hot and no longer superconducting filament. The local Joule heating is reduced. Unfortunately, when the normal part of the wire sees a changing magnetic field, screening currents are produced between filaments. These 'crossover currents' couple flux jumps and can persist for quite long times after changes in applied current. The filaments can be decoupled by twisting, in a manner similar to the reduction of induced currents in twisted pairs.

To enhance the thermal contact between the magnet's inner windings and the helium bath, it is common practice to form the wire with a rectangular cross section so that adjacent windings will stack more closely. Each wire may contain many hundred filaments, each of which is on the order of one hundred microns in diameter. As mentioned in the Recipes Section (3.1), it is possible to chemically etch away the matrix while leaving the bare filaments exposed.

The use of multi-filamentary wire contributes to the maximum field attainable as well as the long term stability of the field produced. It does, however, present a difficult problem at any joints that are made in the conductor. It is essentially impossible to preserve the identity of the filaments in a joint and they are typically all coupled over a small region. The region surrounding the joint will therefore exhibit a much higher flux flow rate and be less stable against thermal fluctuations. In fact, it is the joints in the magnet which are the limiting factor of the long term stability of most commercial magnet systems. For this reason most commercial high persistence magnets are wound using a single continuous wire, with the only joints at the persistent mode switch. A good superconducting joint will exhibit less than a $n\Omega$ of effective resistance.

Magnet Quenching Sometimes, when a large current has been imposed upon a magnet, a small region of the superconductor becomes normal and too much heat is generated. Once this happens, an avalanche of heating occurs. Since the normal section has a finite electrical resistance, it heats and drives the surrounding wire above the superconducting transition temperature. Once this avalanche process starts, there is nothing you can do to stop the process. The magnet has begun a field 'quench.' During the quench enough thermal energy is released to keep the magnet's windings normal until some of the energy stored in the magnet has been dissipated. Although such an event might look rather spectacular, it is quite rare that any permanent damage occurs. If the dewar is in a laboratory with a helium recovery system it is probably wise to have some sort of gas over-pressure protection system to either vent the helium to the atmosphere or trap the temporary excess in gas for later recovery.

Magnet Training If a newly wound magnet is ramped up in current until it quenches, it will freqeuntly quench at a greater current when the process is repeated. However, the rate of increase of the critical current will decrease with successive quenches. The magnet's performance will eventually stabilize with a maximum usable field and current. The process of teaching a magnet how to obtain larger critical currents is called training.

The mechanisms involved in training a magnet are not completely under-stood. They are probably special for each magnet. In high field magnets, the forces on the conductors can be very large. The radial outward pressure on the windings of a 10 tesla magnet can easily exceed 200 bar. These large stresses in the windings can force a motion of the wires, which then induces the quench. The local hot spot in the wire moves with a large velocity, approaching that of sound, until it gets pinned by the medium in which the magnet wires have been con-fined, usually some epoxy. During successive quenches new places around other smaller cavities in the medium confining the magnet windings suffer such motion until all flexible portions of the wire have been 'trained.' In order to decrease such mechanical instabilities and restrict motion of the wires, most magnets are wound under tension. The windings are almost always potted in epoxy resin or some other filler. Some designs include mechanical support for the windings in the form of fiberglass or metal bands that are wound tightly around the outside of the magnet.

How much quenching a given magnet may need in order to meet design parameters is an indication of the leeway in design. If a magnet requires much training, it may need to be retrained after each warm-up and cool-down. As the helium evaporated when the magnet is quenched can be rather costly, it is important that the magnet not require too much training. There is clearly a trade-off between stability and number of windings: the more turns, the lower the current for a given field and the less likely a quench. Most modern super-conducting magnets are made with little need for training. Manufacturers have learned how to pot the wire windings held under tension so there is little motion in response to even large currents. In our recent experience at Cornell, there has frequently been little increase in the maximum field achieved in successive current energizations after the first quenching of the field in a new magnet.

Quench Protection A high field superconducting solenoid can typically have tens of henries of inductance, and operate at currents on the order of one hundred amps. For example, most NbTi magnets produce 1 kgauss per ampere. A quick calculation shows that such a magnet might store on the order of 100,000 joules. Released into a bath of liquid helium, this energy will evaporate many liquid liters. Even more spectacular are the peak voltages that 20 henries can develop during the rapid change in currents that may occur during a quench. When a quench occurs, the largest voltage drops occur across the quenched region, as the back voltage across the magnet's inductance and the resistive voltage drop are exactly opposed. If the quench is occurring at one of the inner windings, it is possible to have arcing between adjacent layers which can permanently damage the magnet. To help minimize chances of arcing, successive winding layers are usually insulated with fiberglass or some polymeric insulator like mylar or kapton.

A clever approach to quench protection is have the energy released during a quench be distributed throughout the magnet or, if possible, removed from the magnet altogether. The simplest scheme is to connect a small normal resistance across the terminals of the magnet so that if it is driven normal all of the stored current will be dissipated in the normal shunt. We have frequently used diodes connected to oppose the usual current direction for such a purpose. As long as the potential energizing the magnet is less than a volt or so, the diodes appear to have an large impedance compared to the superconducting solenoid. However, when the solenoid begins to dump its current with an opposite voltage sense across the diodes the resistance is small compared to the magnet wire, and so all of the stored current is dumped in the liquid helium bath through the diodes, rather than through the magnet.

Magnet Design   The most common magnet topology is the solenoid. The solenoid can provide a uniform field over a substantial volume with a minimum in total stored energy. To account for finite size effects and fringing fields, magnets are designed using computer programs. The field profile for a particular winding distribution is numerically calculated using the Biot-Savart law. Important input parameters include wire dimension, winding distribution, as well as the magnet's overall physical size. Any design can be perfected by iterative calculations until the most practical solution is found. Before a design makes it off the drawing board, it should also be analyzed for quench stability. This is usually done with a computer modeling program that checks to see if a design can sustain the stress of an uncontrolled quench at design conditions.

Magnet Homogeneity   Many experiments place constraints on the homogeneity of the magnet's field. In most magnets, finite size effects are the dominant cause of inhomogeneities. The homogeneity can be tailored, to some degree, by varying the winding distribution; however, there is a practical limit to attainable results. A respectable homogeneity for a single winding magnet is about one part in a million field deviation over a one cubic centimeter volume in the center of the magnet. Although magnet homogeneities a factor of one hundred better than this can be achieved, the cost increases rapidly. When purchasing a high field magnet, price and overall size increase rapidly with bore diameter and field homogeneity. These limits are not too surprising when one realizes that a typical nine tesla magnet ten inches long having a two inch bore will have about 30,000 turns, yielding an average contribution of 3 gauss per turn. Moreover, as the magnet's intrinsic homogeneity increases, the perturbation of the field profile by the sample becomes increasingly important. High homogeneity magnets are also more delicate in that any disturbance of the windings may degrade the homogeneity. An effective and more economical approach to achieving the most homogeneous fields is to have additional trim or 'shim' coils that can be used to fine-tune the field profile.

The field pattern in the magnet can be expressed as a Taylor series expanded in three spatial coordinates. The first spatially dependent terms will be linear gradients in the 3 coordinates, while the next highest terms will involve quadratic terms in the coordinates as well as cross-terms. In designing shimming coils for a magnet, the best strategy is to design coils that cancel the gradients one term at a time. The linear field needed to cancel the linear gradient terms is easily produced with an 'anti-Helmholtz' pair. In commercial magnet systems, these are usually wrapped around the outside of the primary coil, are superconducting and are referred to as 'x, y, and z'. The next most significant gradient term, beyond linear, is usually the quadratic term in z, the field axis. Commercial magnets are often offered with four, six, or eight correction coils.

If field gradients in excess of a few tens of gauss are needed, gradient coils designed for low duty cycle, high current pulses can be used. If these are used on a very low temperature apparatus, eddy current heating from the pulsed gradient coils will freqeuntly produce too much eddy current heating.

When purchasing a high homogeneity magnet, it is important to discuss testing and calibration with the manufacturer. Many manufacturers rely on experience and accurate modeling programs to predict ultimate performance. They are seldom equipped to measure field profiles very accurately. The simplest schemes for measuring field profile utilize the Hall effect. These are limited in accuracy by the size of the detectors, which are usually at least a few millimeters in diameter. The most accurate calibration of both the absolute field and the homogeneity can be determined through nmr measurements using a small probe inserted in the magnet. Most magnet makers have no facilities for nmr calibration of their products.

In order to take full advantage of a magnet's homogeneity it is important that the sample be located at the magnet's center. As a result of differential thermal contraction in the cryostat, the careful alignment of a magnet may not be preserved as the system is cooled.

Nuclear Demagnetization Coils   Nuclear demagnetization magnets do not need to be nearly as homogeneous as magnets designed for spectroscopy. Therefore, the magnets can be cheaper, with fewer windings than those required for precise spectroscopies. Frequently, such magnets are designed to have fields that decrease quite rapidly in the vertical direction away from center. If the apparatus contains only a demagnetization coil, and the coil is to be situated close to the mixing chamber of a dilution refrigerator, it may be important to have the coil designed so that the field at the mixing chamber is a minimum. This will reduce eddy current heating, as well as effects on the refrigerator's thermometry.

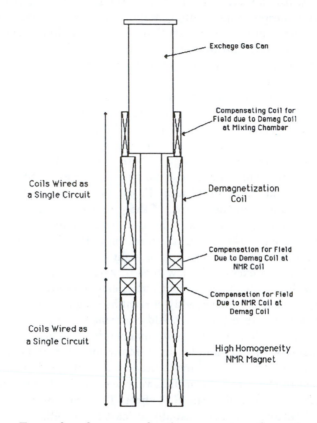

Figure 2.3. *Examples of superconducting magnet configurations. This system incorporates a high field NMR magnet as well as an adiabatic demagnetization coil. The configuration is shown with the demag coil above the NMR coil. The demag coil incorporates bucking coils to minimize the field at the mixing chamber as well as to minimize interactions between the two magnets.*

If the experiment includes a high field nmr magnet also, the coils can be designed to have as little interaction as possible. Figure 2.3 depicts two frequently used configurations.

Magnet Power Supplies In addition to supplying about one hundred amps into a low resistance highly inductive load, a power supply for energizing a superconducting magnet must satisfy a number of demands. Often a commercial magnet is specified to have a maximum current ramp rate, presumably based on thermal and mechanical stability of the coil. The voltage across the magnet will be determined by the ramping rate and the coil's inductance. For this reason, it is often necessary to have some sort of ramping circuit to control the power supply. If the supply is to be used in a steady state, the long term drift in the supply's regulation circuit will be especially critical.

For the case where the magnet being energized is to be ramped up, left persistent, and then ramped down at a later date, it is important that the power supplies output current be measurable to at least a part in 10,000. When the power supply is ramped back up and the magnet's persistent switch opened, there will be a substantial voltage developed across the switch if the currents are not the same. If the currents are sufficiently different and the voltage large, the power supply may be damaged. For this reason, the output of the supply is usually shunted with a LARGE diode which passes the current when the magnet voltage becomes high enough. If the magnet is ramped up at a rate which requires voltage higher than the diode's turn-on value, the ramping rate will be limited. In our group it is common strategy to install a pair of diodes back-to-back across the magnet, in the bath. In this case, a mismatch between magnet and power supply will turn on one of these diodes and the magnet's energy will be dissipated in the bath. The advantage here is that the diodes need not be rated for such high power as those needed for room temperature protection. In addition, the current path is closed inside of the dewar, protecting the user from hazardous voltages that could arise in the external wiring if the current path were to be broken. When there is a mismatch between magnet and supply, there is no quench. Instead, the magnet dumps its energy into the protection circuit and the current decays or increases smoothly to match that of the external supply.

If the supply is to be used in a demagnetization system the output stability and filtering are very important because of eddy current heating that field variations produce. At the end of a demagnetization interval, the current ramp rate may be below 25 mA per hour. This slow rate imposes restrictions on the power supply's inherent current stability as well as the stability of the ramping control circuit. Although precision linear circuits can be built with the desired stability, it is often more convenient to implement the ramp in digital circuitry and control the power supply through a digital to analog converter (D-to-A). For a one hundred amp dynamic range, most demagnetization cryostats would require a 16 bit D-to-A. If the ramper is designed to be controlled by a computer, it is important to design the circuit so that nothing too wild happens when the computer crashes. At present we know of no commercial power supplies which incorporate a digital interface with this accuracy. If the controlling circuit or computer has less than the needed resolution, it may be possible to switch power supplies during the demagnetization when the current is below a certain value. A more sensitive low current supply can then be used for the more delicate, low current, low ramp rate phase. Sources for supplies are given at the end of the section.

Magnets Cooled by Lambda Refrigerators The critical field and current for a magnet are a function of the magnet's temperature. A cooled magnet can be used to obtain larger magnetic fields. In practice, this can be achieved in either of two ways. The entire helium bath can be pumped to lower its temperature, or just a portion of the bath can be cooled. The trick involved in the latter scheme is achieved by a device known as a lambda fridge. The lamda fridge is essentially a pumped helium refrigerator that is immersed in the liquid helium bath near the magnet. The refrigerator cools the helium around the magnet while the free surface of the liquid above the magnet remains at 4 K. The constraints placed on a dewar by the use of a lambda fridge have been discussed in earlier parts of this section.

If the magnet is to be operated below 4 K it must be designed to be stable at the higher fields and under larger stresses. Most manufacturers will not guarantee their products under these conditions. It is probably reasonable to gamble on the successful operation of their magnets at higher fields and lower temperatures anyway.

Magnet Leads   The design of magnet current leads which span the temperature range between that of the magnet and room temperature while minimizing both thermal conductivity losses and the I^2R Joule heating in the wires is a delicate compromise.

If the magnet is energized with current and then left in a persistent mode for extended periods, removable leads can be used. The strategy in such a case is to have leads which are inserted into the top of the cryostat which mate with connectors in the bath space located at about 4 K. The fittings on the top of the cryostat can be made from quick-connectors and the magnet lead covered with an insulating tube which slides into the dewar from the top.

If the current supplying leads must remain in the cryostat continuously, the thermal conductivity of the electrtical leads can impose a terrible price on the running cost of the apparatus. The most efficient method for reducing costs of liquid helium is to cool the leads with the helium vapor as it boils off of the bath. Unfortunately, the low electrical resistivity of metals desired for current conduction goes hand-in-hand with a high thermal conductivity down the same wires.

One approach to vapor cooling of the leads is to simply increase the surface area of the lead by using sheet metal in the bath region which wraps almost half of the way around the circumference of the dewar neck. Because of the baffles in the dewar neck, the leads will be imposed in the flow path of the helium vapor emerging from the liquid bath. The potential between the leads is quite small under almost all conditions, and the leads can safely be insulated with a very thin layer of kapton or mylar.

A more active approach to vapor cooling is to design a lead which forms a flow path for the exiting helium vapor. When the lead is in use, the normal helium flow path is sealed. Exiting vapor is diverted from cooling the dewar neck and forced to flow through the lead. Ken Efferson at American Magnetics has one design solution which is commercially available and which is nicely described in his design article (Efferson, 1967). If the magnet leads are being used heavily, as is the case for a demagnetization cooling magnet, the efficiency of the leads

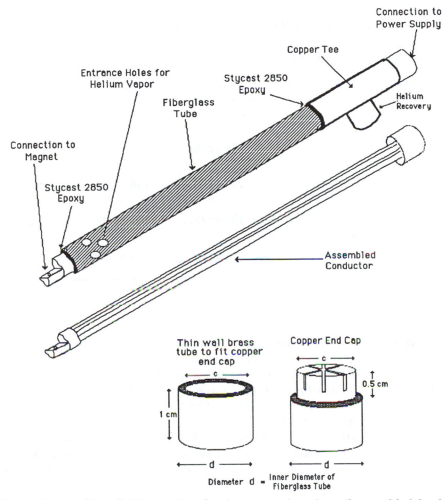

Figure 2.4.   *Cornell Magnet Leads. A perspective view of assembled lead as well as end pieces for brass conductor strips. The conductive element of the lead is also shown prior to assembly into fiberglass housing.*

In our group at Cornell we favor the design shown in Figure 2.4. The outer tube, which guides the vapor flow is made from thin wall fiberglass. The conductor is made from thin brass shim stock which is cut into 1/4 inch-wide strips with a razor blade or on a paper cutter. Rough data for boil-off rates of three different designs are given in Table 2.1.

The fabrication procedure for the leads is the following: First determine the amount and thickness of metal foil to be used. Assuming that the foil strips

Table 2.1. *Boil-Off Rates for Cornell Magnet Leads*

Current (amps)	foil thickness net width length	0.005" 6" 24"	0.003" 10" 24"	0.001" 10" 18"
100		367 cc/hr	253 cc/hr	
70		228 cc/hr	166 cc/hr	218 cc/hr
50		183 cc/hr	146 cc/hr	159 cc/hr
30		135 cc/hr	107 cc/hr	100 cc/hr
0		96 cc/hr	100 cc/hr	74 cc/hr

will be grouped in at most eight bundle stacks, determine the inner diameter of the fiberglass flow tube. After slicing the sheet up, divide the strips into stacks taking care to have the strips neatly stacked in parallel bundles. The ends of the bundles can be crimped together with some copper foil at each end. The end caps are made from two pieces as shown. The final outer diameter is chosen to fit inside the fiberglass flow tube.

The caps are attached to the bundles by inserting one end of each bundle in a slot in the copper piece and then slipping the sleeve around the bundle, forming a small well around the bundle base. This well is filled with flux, heated, and filled with silver solder. A nice fit of the bundles in the slots and of the sleeve over the copper end will help to ensure a nice final result. The finished product must be carefully cleaned of residual flux. A collection of small holes are machined into one end of the fiberglass tube to allow vapor to enter the lead. The room temperature end of the lead uses a modified copper tee fitting, drilled out to slip over the fiberglass tube and the conductor's cap. It is useful to have valves at each lead so that flow can be stopped and to prevent Taconis oscillations when the lead is not in use. Joints to the fiberglass tube are made using Stycast 2850 epoxy.

The choice of brass for conductor above 4 K was based on the trade-off between the decreased thermal and electrical conductivity which occurs in all metals, particularly in alloys. It turns out that pure metals, when used in highly optimized magnet leads are unstable against burn-out due to thermal runaway at their design currents, a problem that may not be apparent until the lead is ruined (Lock, 1969).

If vapor cooled leads are used it is important to have a relief path for gas to exit the dewar should the magnet quench. The ends of the leads should be safely above the highest bath level point at all times. Just above or just below the lowest baffle seems to be best.

Below the first baffle, and in the liquid bath, the simplest lead is a low resistivity copper bar. The bar can be insulated with fiberglass cloth sleeving that will not inhibit liquid from cooling the bar. A superconducting wire able to carry the full rated current should be soldered to the bar along its whole length.

In order to monitor the voltage across the magnet during energizing and de-energizing, it is handy to have a separate set of sense wires attached to the magnet at the connection to the current leads.

Home Built Magnets If the experiment requires a modest field, a small size, or an unusual field profile, it may be sensible to wind the magnet yourself. Magnet design programs are available from a variety of sources and are common in a scientific environment. They are all essentially Biot-Savart integrations.

In addition to the notions presented so far, there are a few practical hints that may be useful. The magnet form can be machined from aluminum, or for smaller magnets, from cast epoxy stock. The surface where the wire is wound must be very smooth to prevent damage to the wire. A layer of Kapton or Mylar can reduce problems. The magnet should be wound on a lathe or a coil winder which is set up to slowly rotate the former and help guide the wire into uniform well controlled layers. When winding the magnet, the wire should be held under some tension, however, there are two warnings: if the wire snaps during winding, the magnet will have to be rewound, and wire may be wasted. In addition, if the coil former's coefficient of thermal contraction is not well matched to the wire's, the windings may snap when the magnet is cooled. This can be particularly frustrating when the magnet has been potted in epoxy and the wire cannot be salvaged. For 30 gauge copper matrix wire a tension of 0.5 newtons seems close and for stiffer wires up to 10 N may be needed.

It is a good idea to leave lots of extra wire for leads. If single filament wire is used, nice joints can be made by spot welding. The Recipes section contains advice on superconducting joints. If the magnet incorporates some unusual winding configurations to shape the fields, it may be worth having separate wires for different regions of the coil. This will allow the field profile to be fine tuned at the cost of the additional heat loads resulting from more wires to room temperature. As with any magnet, when the magnet is placed near any magnetically active materials, its field profile will be affected.

If the magnet will not be used in a liquid helium bath, but rather will be cooled by connection to a refrigerator in a vacuum, the stability against quench damage may be degraded by the reduced thermal contact with the coolant.

As with any superconductor, sensitive experiments may make trapped flux an important consideration when the magnet is cooled through its transition.

Operating Tips The moisture that condenses on the magnet if it is casually warmed can be ruinous to the coil, causing distortions in the winding locations. Usually rapid warming of the magnet is quite harmless. Do not become paralyzed by worries over this issue unless there are special homogeneity requirements for the magnet.

When the magnet is energized and in a persistent state, care must be taken in shutting down the power supply that is connected across the coil. Any transients that may occur in the power supply might cause the persistent current switch to go into the normal state, dumping the magnet's current. This is important in initial choice of power supplies.

To prevent the disasters associated with current mismatch between magnet and power supply when the magnet is switched in and out of the persistent mode, it is important to know the current value at which the magnet was set, as well as its decay rate over time. If there is any doubt, it is often possible to test for mismatch by 'tickling' the persistent current switch with a short burst of current somewhat below the value required to drive the switch normal. If any voltage appears across the switch when this is done, there is probably a mismatch. Change supply currents and try again. Small differences are not usually a problem, as the magnet and supply will slew to match currents. If the voltage is above the turn on of protection diodes, or above the supply's capabilities, the magnet may begin to rapidly ramp down. When this occurs it may be possible to 'catch' the current by manually ramping the supply current down while monitoring the voltage across the magnet. If the currents can be matched, they might stabilize.

Some demagnetization magnets have been built which incorporate two separate persistent switches in series. One of the switches is shunted by a copper bar. When the switch across the shunt is opened, the magnet decays at a rate convenient to the demagnetization. By sensing voltage across the shunt, current mismatch can also be detected.

2.1.3   Sources and Suppliers of Useful Equipment

Sources of Research Dewars

- Kadel, 1627 E. Main St., Danville, IN 46122 (317) 745-2798. Kadel manufactures inexpensive superinsulated dewars made primarily from rolled and welded aluminum. Their dewars do not look quite as attractive as those made by Precision Cryogenic Systems, but they seem to work just as well.
- Precision Cryogenic Systems, Inc., 11717 W. Rockville Rd., Indianapolis, IN 46234 (317) 272-0888. Precision's superinsulated dewars of aluminum-and-fiberglass construction are both well made and attractive. The dewars are light and probably as efficient as any that do not use a liquid nitrogen jacket. Precision is an excellent company to turn to for the repair of dewars made by the now defunct Cryogenic Assoicates. They will also cheerfully repair defective dewars made by other manufacturers.
- Cryofab, Inc., 540 Michigan Ave., Kenilworth, NJ, 07033 (201) 686-3636. Cryofab builds excellent stainless steel research dewars at a fair price. In

our laboratory we have a number of superinsulated stainless steel dewars
they have made. The only complaint is the weight, an intrinsic problem in
large metal dewars.

- International Cryogenics, Inc., 2319 Distributors Drive, Indianapolis, IN
  46241 (317) 247-4777. ICI produces a superinsulated product which appears similar to that of Precision. We have had better luck with dewars
  made by Precision Cryogenics.
- Oxford Instruments, NA, 3A Alfred Circle, Bedford, MA 01730 (617) 275-
  4350. Oxford Instruments manufactures dewars to house the magnets and
  dilution refrigerators they sell. Their dewars for superconducting magnets
  used in nmr facilities are especially efficient in helium consumption. Oxford's prices tend to be much higher than those of many American dewar
  makers.
- Janis Research Company, Inc., 2 Jewel Drive, P. O. Box 696, Wilmington,
  MA 01887 (617) 657-8750. Janis manufactures stainless steel dewars of a
  very good quality. Janis has an especially good reputation for the production of dewars with reliable optical windows.
- Cryo Industries of America, Inc., 24 Keewaydin Drive, Salem, NH 03079
  (603) 893-2060. We have had no experience with this company but they do
  manufacture a wide variety of stainless research dewars.
- Pope Scientific, Inc., Menomee Falls, WI 53051 (414)251-9300. If you must
  use a glass dewar, this is a good source. The glass dewars they have made
  for us have all been satisfactory at a very reasonable price.
- RMC/Cryosystems, Inc., 1802 West Grand Rd., Suite 122, Tucson, AZ
  85749 (800) 882-2796. The company offers a variety of specialized dewars
  and cryostats including 'dippers' for storage dewars as described in section
  2.2.
- U. S. Cryogenics, Box 5733, Florence, SC (803) 664-2827. US Cryogenics
  makes one of the least expensive superinsulated research dewars on the
  market. Their dewars appear to be very efficient in helium consumption.

Sources of Superconducting Wire

- Supercon Inc., 830 Boston Turnpike Rd., Shrewsbury, MA 01545 (617)
  842-0174
- Oxford Superconductors, 600 Milik St., Carteret, NJ 07008 (201) 541-1300.
- Magnetic Corporation of America, 179 Bear Hill Rd., Waltham, MA 02154.
  They may no longer manufacture superconducting wire.
- Molecu Wire Corp., P. O. Box 495-1, Farmingdale, NJ 07727 (201) 938-
  9473. This company specializes in applying insulating coating to wire and
  may be willing to consider smaller special application jobs.

Sources of Superconducting Research Magnets

- American Magnetics, Inc., P.O. Box 2509, 105 Mitchell Rd., Oak Ridge,
  TN 37831 (615) 482-1056. A reliable company which usually offers one of

the lowest prices, especially for NbTi solenoid magnets used for demagnetization.

- Oxford Instruments, NA, 3A Alfred Circle, Bedford, MA 01730 (617) 275-
  4350. Oxford makes very elegant magnets and is probably the best source
  for magnets with very homogeneous fields. For simple solenoids, Oxford's
  prices are usually quite high.
- Cryomagnetics, Inc., 795 Oak Ridge Turnpike, P.O. Box 548, Oak Ridge,
  TN 37831 (615) 482-9551. A reliable company which usually offers one of
  the lowest prices, especially for NbTi solenoid magnets used for demagnetization.
- Cryogenics Consultants Limited, Box 416, Warwick, NY 10990 (914) 986-
  4090. Frequently a 'best buy', especially for very large field magnets. Beware of late deliveries!
- Intermagnetics General Co., P.O. Box 566, Guilderland, NY 12084 (518)
  456-5456. A manufacturer to approach for special magnets with unusual
  field configurations, large fields, or very homogeneous fields. For simpler
  solenoids IGC's prices tend to be very high.

Sources of Magnet Power Supplies

- In addition to the magnet manufacturers listed above we have had success
  with two HP series power supplies. The older 62XX series linear supplies
  seem to perform nicely, as well as the newer 6011A series switching supplies. The newer supplies are available with GPIB interfaces, although the
  resolution is not adequate for most demagnetization applications.

Sources of Magnet Lead Materials

- Almac, 1640 Emerson Rd., Rochester, NY 14606 (800) 462-6400. Fiberglass tube can be purchased for use in making the Cornell Magnet Lead.
  Occasionally small quantities of very thin walled tube are available as production overruns.

Useful References for Magnet Design

- Stekly, Z. J. J., 1965, *Rev. Sci. Instr.* **36** 1291.
- Maudsley, A. A., 1984, *J. Phys. E* **17** 216. This article discusses field pro-
  filing with nmr.
- Israelsson, U. E., and C. M. Gould, 1984, *Rev. Sci. Instr.* **55** 1143.
- Smith, T. I., 1973, *J. Appl. Phys.* **44** 852. This article has a discussion of
  the effects on a solenoid's field profile due to a cylindrical superconducting
  shield.
- Wilson, M. N., 1983, *Superconducting Magnets* (Oxford Univ. Press, Ox-
  ford). The is a very useful book on magnets.
- Sauzade, M. D., and S. K. Kam, 1973, *Advances in Electronics and
  Electron Physics* **34**. (Academic Press, New York). This article
  contains a useful discussion on field homogeneity and shimming.

2.2   Cryostats for Storage Dewars

by David G. Cahill

The purpose of this section is to describe, in some detail, the construction and operation of a cryostat designed to be inserted into a ^{4}He storage dewar. Although an insertable cryostat is not a new technique for low temperature experiments, I think that the reader will agree that this particular design is easily constructed, adaptable to many applications, and convenient to use. The cryostat described below was designed by Eric Swartz. The design is based on an original model due to Eric Smith.

The cryostat has been used to produce temperatures from 1.3 K to 300 K. Uses of this design have included measurements of thermal conductivity and heat capacity, testing of ultrasonics equipment, and the development of a cryogenic heat switch.

2.2.1   Design Overview

The principal reason for constructing an "insertable" cryostat is the savings of time and money that this type of cryostat produces, savings that any laboratory can appreciate. The cryostat allows quick turn-around times; typically, a sample can be mounted and ready to measure in a few hours. Eliminating transfers of liquid nitrogen and ^{4}He saves both time and liquid ^{4}He.

Since our applications in the Pohl research group center mostly on measurements of thermal properties, we designed a cryostat that could produce a wide range of temperatures with good temperature control and thermometry. Temperatures below 4.2 K are achieved by pumping on a ^{4}He pot. The pot can be filled from helium gas admitted through a valve or the pot can be filled by a capillary running from the storage dewar into the pot. Two constantan resistance heaters supply heat for controlling the temperature of the cryostat. Temperature measurements are provided by four resistance thermometers: two carbon resistors for temperature control in different temperature ranges and two secondary standards, germanium for temperatures below 30 K and platinum for above 30 K. Eleven twisted-pairs of constantan and brass wire provide electrical connections to the experiment.

The obvious constraint in this design is the dimensions of the ^{4}He storage dewar; the neck of our 60-liter storage dewars is 2 inches in diameter. The distance from the top of the neck to the bottom of the storage dewar determines the length of the pumping lines. To accommodate the small diameter of the neck and maximize the volume that is available for experiments, we use a grease seal on the vacuum can that takes up only a small diameter and allows easy access to the experiment. This seal is amazingly reliable, at least after the experimenter has gained some experience in using it. It will described in more detail later.

2.2.2 Construction Details

Three types of metals make up the bulk of the cryostat: #304 stainless steel tubing, plate brass, and OFHC copper. We used stainless tubing for pumping lines because of its high strength, small heat capacity, and poor thermal conductivity. The low thermal conductivity and thin walls of the pump-line minimize the heat leak into the storage dewar. The OFHC copper used in the pot, radiation shield, and experiment stage of the cryostat has a very high thermal conductivity; this property minimizes thermal gradients and shortens the time constants associated with reaching thermal equilibrium. Brass makes a logical choice for the vacuum fittings, flanges, and vacuum can because it is easy to machine.

Plumbing Figure 2.5 shows the exterior of the cryostat. For strength and reliability, all the joints in the plumbing were hard soldered using Eutectic 1801 solder and flux. We sandblasted the joints after soldering to remove the excess flux and make the job look more professional. A thorough check for leaks, including a check for cold-leaks, should be performed after the cryostat is assembled and before the wiring is installed.

Three hook-up tubes are connected to the vacuum pump-line: two tubes for the electrical connections and one for the vacuum pump. Two tubes lead from the ^{4}He pump-line, one for the ^{4}He pump and a second tube, oriented vertically, provides access to the pot for radiation shields, ^{4}He level meter, and any unforeseen needs.

The tri-clover flange seals (similar to KF flanges) are at the top of the ^{4}He storage dewar. The pump-lines pass through holes in this flange that are fitted with rubber O-rings. The O-rings make a tight enough fit to support the weight of the cryostat.

Radiation shielding in the vacuum line is not crucial; we used a short length of brass wool inserted in the vacuum can side of the vacuum pump-line. To extend the temperature range of the cryostat and the lifetime of ^{4}He in the pot when operating below 4.2 K, a more sophisticated radiation shield was inserted in the ^{4}He pump-line. We constructed this radiation shield from copper discs soldered to a length of Cu-Ni capillary; the Cu-Ni capillary has low thermal conductivity, it is easy to solder to, and it is fairly stiff. The diameter of the discs is about 7/16 inch. Six discs separated by a few inches and placed about in the middle of the ^{4}He pump-line seems to the do the job pretty well. The Cu-Ni capillary extends the entire length of the cryostat and can be removed easily.

Below the λ-point of ^{4}He in the ^{4}He pot, a major heat load on the cryostat is the film creep. To facilitate the operation of the cryostat in this temperature range, an impedance to the film creep should be fitted into the ^{4}He-pot pump-line. A practical impedance can be constructed by drilling a small hole in a thin disc of metal and attaching the disc to the ^{4}He-pot pump-line just above the pot. If the disc is thin, pumping speed will not be affected by this addition.

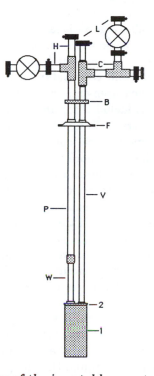

Figure 2.5.   *Exterior view of the insertable cryostat.*
B - pump-line brace, brass
C - copper couplings to fit
F - ⁴He dewar triclover flange, brass
H - hookups, stainless 1/2" diameter, .035" wall
L - KF flange, O-rings and clamps, brass
P - ⁴He pot pump-line, 1/2" diameter, .020" wall
V - vacuum can pump-line, stainless. 3/8" diameter, .010" wall
W - weak thermal link to pot, stainless. 1/2" diameter, .010" wall
1 - brass vacuum can
2 - brass vacuum fitting

Wiring This design uses two 19-pin feed-throughs purchased from Detoronics for electrical connections between the cryostat and the outside world. These connectors are inexpensive and can be soldered to the brass flanges. By using a hot-plate and Eutectic 157 solder, we were able to get very reliable, vacuum tight connections.

We wired the cryostat with 14 twisted-pairs of 3-mil constantan wire, 4 twisted-pairs of 5-mil brass wire, and two single lengths of brass wire. In addition, a few twisted-pairs of each type were included as extras. Inevitably some of the wires will short or break and the extras will be needed. An extra 6 inches of wire at both ends of the wire is also a good idea. The 3-mil constantan wire is used for all electrical connections that do not require a large current. The low thermal conductivity of this wire reduces the heat load on the cryostat. The relatively high electrical resistance of 3-mil constantan, approximately 30 ohms/foot, is not a problem. Four-lead measurements can eliminate the errors due to lead resistances. The brass wire is included for any high current needs. For example, the temperature control heaters are connected to brass leads. Cu wires of 1.5-mil diameter have about the same thermal conductivity and electrical resistance as the 5-mil brass. We found the brass more convenient since the 1.5-mil Cu is quite fragile. Another alternative to the brass wire is 10-mil constantan wire.

After leaving the Detoronics connectors, each set of 19 wires is protected by a woven nylon sheath until the wires enter the vacuum can. The sheath protects the wires from breakage and failure of the formvar insulation as the wires pass over any rough edges in the plumbing. A simple way to thread the wires through the vacuum pump-line is to first thread two weighted strings through the vacuum can side of the pump-line and out through the tubes running to the electrical connectors. Then each set of 19 wires is tied to a string and pulled down into the vacuum can.

Heat sinking of the leads occurs in two steps. First, wrapping the leads around a copper post soldered to the top part of the vacuum fitting absorbs most of the heat load associated with cooling the leads from room temperature to 4.2 K. Second, the leads are wrapped around the ^{4}He pot about 6 times, positioning them carefully to create as good a thermal contact as possible. This means eliminating configurations where leads are wrapped on top of other leads. If the leads were wrapped on bare metal, any breaks in the insulation would short the leads to ground and render them useless. To prevent this, we covered the metal surfaces with cigarette-paper prior to heat sinking the leads. We discovered, the hard way, that GE varnish does attack formvar insulation. Consequently, Stycast 1266 epoxy was used to hold the wires in place. This method worked very well.

As mentioned above, we took many precautions to prevent damage to the formvar insulation on the wires we used in this cryostat. Some consideration should be given to the use of wire insulated with polythermaleze, a somewhat more rugged compound than formvar. Unfortunately, the strength of polythermaleze also results in a major inconvenience; concentrated sulfuric acid is required to strip the insulation.

Inside the vacuum can Figure 2.5 shows a drawing of the apparatus inside the vacuum can. A short segment of 10-mil-wall stainless steel tubing connects the pot to the top half of the grease fitting and the ^{4}He pump-line. This thin-walled section helps to thermally isolate the pot from the ^{4}He bath. The stage, holding the four thermometers and one heater, is mounted on three posts on top

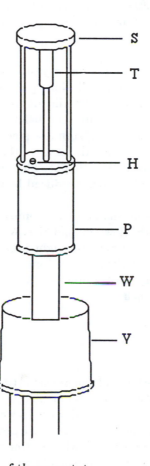

Figure 2.6.   *Interior view of the cryostat.*
H - constantan heaters for temperature control
P - ^{4}He pot
S - experiment stage
T - thermometers: two carbon, one germanium and one platinum
W - thermal weak-link, stainless steel 1/2" diameter, .010" wall
V - seal for vacuum can

of the pot. A radiation shield slips over the stage and fits tightly around the end of the pot.

All components inside the vacuum can are constructed of OFHC copper. The temperature control heaters consist of a several feet of 3-mil constantan wire wrapped around a copper form. The copper "button" is 3/8 inch in diameter and 1/8 inch thick. To improve the temperature stability of the cryostat, the stage heater is placed as close to the temperature control thermometers as is practical.

Specific experiments often require a special stage and mounting hardware. These customized platforms are attached to the stage by means of a 1/4-20 threaded copper rod.

Continuous-fill line When we run the cryostat at temperatures below 4.2 K, we usually fill the pot by admitting helium gas through a valve at the top of the cryostat. Some applications, an experiment with a very large heat load or an experiment that needs a very long run time below 4.2 K, will require a continuously filled pot. Two short lengths of stainless steel capillary, 1/16-inch-OD and 27-mil-ID, provide paths from the ^{4}He bath to the inside of the pot, and from inside the vacuum can to inside the ^{4}He pot. This capillary is normally sealed off on the storage dewar end. The seal is produced by soldering in a piece of copper wire.

To use the continuous-fill method, the copper wire plug is replaced by an appropriate impedance to the flow of liquid ^{4}He. A 60 cm length of 16-mil-OD, 4-mil-ID, Cu-Ni capillary should have the right impedance for most applications. Cutting the capillary is best done by scribing it with a razor blade and then gently bending the capillary until it breaks. The capillary coils up nicely around diameters as small as 1/8 inch.

2.2.3   Operation of the Cryostat

The vacuum can We have found the grease fitting quite reliable as long as we obey the following rules. We use a non-aqueous silicon grease to seal the vacuum can. Both sides of the seal are covered with just enough silicon grease so that your finger feels only grease and no brass. After pressing the two parts together, we pump out the vacuum can while slowly rotating the can by about 180 degrees.

The quality of vacuum required depends on the experiment. If the vacuum can is to stay in liquid ^{4}He for the entire run, one millitorre of pressure at room temperature is good enough. In this case, we place activated carbon in the bottom of the can which adsorbs any left over gasses as soon as the cryostat is inserted into the storage dewar. For experiments at temperatures above about 60 K, we usually pull the vacuum can up into the neck of the dewar to reduce the amount of heating required for temperature controlling and thereby minimize the ^{4}He consumption. In this situation, the activated-carbon cryopump is a bad idea because activated carbon can outgas ^{4}He when it reaches 30 K. When we omit the activated carbon cryopump, we pump out the vacuum can to about 2×10^{-5} torr.

2.2.4   Cool Down

To conserve ^{4}He, the cryostat is first cooled down to 77 K in a large liquid nitrogen dewar. This operation takes about two hours if ^{4}He exchange gas is

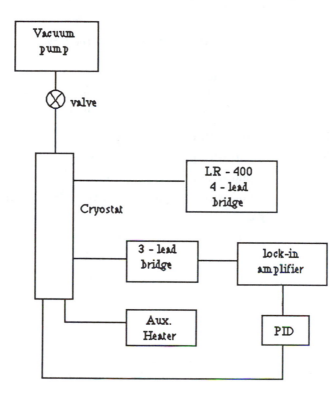

Figure 2.7.   *Block diagram of the temperature control and temperature measurement system.*

placed in the pot. After insertion into the ^{4}He storage dewar, the cryostat will reach 4.2 K in about another two hours.

Reaching temperature below 4.2 K requires that either a continuous-fill line be used or that the pot be filled with liquid ^{4}He by condensing ^{4}He gas. If the continuous-fill is used, care must be taken during the cool-down to prevent frozen air from plugging the impedance. A 5 psi overpressure of very high purity ^{4}He gas in the pot should prevent this blockage. About 20 minutes is required to fill the pot by condensing gas admitted through the ^{4}He pump-line.

Temperature Control and Measurement Figure 2.6 shows a block diagram of a typical temperature control and temperature measurement system. The PID temperature controller is an in-house design that is quite capable of temperature stabilities on the order of .01%. PID stands for for Proportional, Integral, Differential. The design of temperature controller is due to David McQueeny.

The PID controller drives the constantan heater attached to the stage. When high temperatures are required, above 60 K, we often use a D.C. current source to drive the heater mounted on the pot. By adjusting the D.C. current and the

Table 2.2. *Typical Cooling Power of the Cryostat.*

Temperature	gas, liquid, or vacuum in the pot	maximum heat load (one-shot)	maximum heat load (continuous)
1.3 K	liquid with pumping	0 mW	0 mW
1.6 K	liquid with pumping	.5 mW	0 mW
2 K	liquid with pumping	1 mW	2 mW
3 K	liquid with pumping	5 mW	10 mW
5 K	gas	1 mW	1 mW
5 K	liquid - no pumping	10 mW	10 mW
10 K	gas	10 mW	20 mW
25 K	gas	30 mW	40 mW
>50 K	gas	100 mW	—

height of the cryostat in the dewar, a coarse temperature control is achieved for temperatures too high for the power output or the temperature controller.

Heat Load on the Cryostat The cooling power of the cryostat, or the limit to the heat that can be produced by an experiment, is a function of the liquid ^{4}He depth, the temperature of the cryostat, the pumping speed, and the pressure of the exchange gas in the ^{4}He pot and pump-line. Table 2.2 comes from my experience with thermal conductivity measurements and is only meant as an order-of-magnitude guide to the heat loads that can be handled by this cryostat. Gas in the pot refers to an exchange gas pressure on the order of 1/10 atm. Higher pressures produce wild fluctuations in temperature.

2.3 Dilution Refrigerators

by Geoffrey Nunes, jr., and Keith A. Earle

The dilution refrigerator is without question the work-horse of low temperature physics. It can also be the bane of a low temperature physicist's existence. The purpose of this section is to enumerate the various options one has in purchasing a dilution refrigerator, and to give some pointers on how to run and trouble-shoot the one you buy. You will notice that we have said nothing about building one yourself. For that, you're on your own . . .

2.3.1 How a Dilution Refrigerator Works

For an understanding of how a dilution refrigerator really works, we refer the reader to the exellent text by O.V. Lounasmaa (1974). To get the basic idea, read on. Figure 2.8 shows the phase diagram of ^{3}He-^{4}He mixtures. If you lower the temperature of any solution of more than 6% ^{3}He sufficiently, the mixture will separate into two phases. One of these phases will (at very low temperatures) be almost pure ^{3}He. The other phase will be mostly ^{4}He, but at even at T=0, will have a 6% ^{3}He impurity. This last point is the key to the whole works.

Consider Figure 2.9, which illustrates the poor man's or U-tube model of a dilution fridge. If the liquid-vapor interface on the ^{3}He poor side of the tube is kept at about .7 K, most of the vapor will be ^{3}He. Pumping on this vapor will remove ^{3}He from the liquid on the right-hand side of the tube, and destroy the equilibrium between the two phases. In order to restore equilibrium, ^{3}He atoms will "evaporate" across the phase boundary from left to right. But in order for an atom to migrate in this direction, something must supply the associated latent heat. If you have been clever about things, that something could be your sample.

So much for simple models. Figure 2.10 gives an idea of what things really look like. Note that much of the terminology is drawn from the fine art of producing moonshine. In keeping with which, the ^{3}He-^{4}He solution that runs the fridge is commonly referred to as "mash."

2.3.2 Purchasing a Dilution Refrigerator

At the current writing, your options for purchasing a dilution refrigerator are mainly limited to what Oxford Instruments will sell you. The West German firm Cryovac will be manufacturing dilution refrigerators based on the old SHE system, but we have no information on the performance and quality of their product. Since our recent (and on the whole, positive) experience here at Cornell has been exclusively with Oxford, the following discussion pertains mainly to them.

Oxford has a wide product line, and will sell you a lot, including the proverbial bridge (resistance bridge). Their smallest model is the 75 (all Oxford model

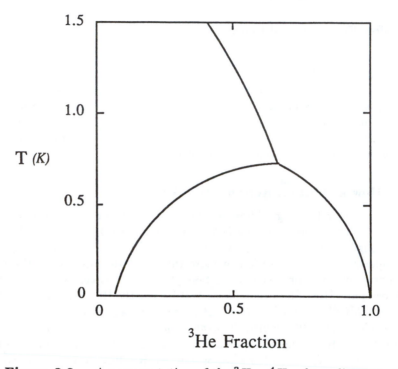

Figure 2.8. *A representation of the ^{3}He - ^{4}He phase diagram.*

numbers denote the fridge's nominal cooling power at 100 mK) with which no one at Cornell has had any direct experience. The next size up is the Model 200. While this is one of Oxford's smaller refrigerators, you should bear in mind that its cooling power is actually quite large on the scale of home built machines.

If you are primarily interested in working with small solid state samples at temperatures down to 5 mK, give serious thought to purchasing a top-loading Model 200. The top loading feature lets you mount your sample to a little copper slug with eight electrical contacts, fasten the slug to the end of a long stick, and screw it directly into the already cold mixing chamber. The fridge sort of burps a little, and keeps on running (if you've done it right . . .). This procedure allows you to cool a sample to millikelvin temperatures, make measurements on it, take the sample out and have the next one cooling in under 24 hours. There is one of these machines operating in Cornell's Microkelvin Lab, and so far, it has been used to measure the resistivity, dielectric constant and magnetic suceptibility of various samples. Prototype slugs for torsional oscillator and vibrating reed experiments have also been built, and plans are under way to allow the slug to be loaded into a region of high magnetic field.

Oxford's line of dilution refrigerators is rounded out by the Models 400, 600,

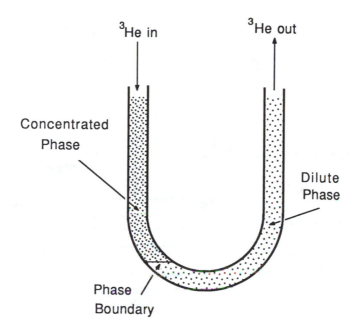

Figure 2.9.   *U-tube Model of Dilution Refrigerator.*

and 1000, all of which may be purchased with the top loading option, and all of which have a typical base temperature of 5 mK. They have also built two Model 2000s, one of which has been in operation at MIT for quite some time, and one of which is presently being installed here at Cornell. The Model 400 is ideal for pre-cooling a PrNi$_5$ or small copper bundle, though a Model 200 would typically pre-cool either in about a day. If you want to pre-cool a really large copper bundle in a really large magnetic field, buy a 600 or 1000. Oxford probably has no interest in producing more 2000s, though if you really need the muscle, you could always ask.

Once you have decided on the size of the refrigerator you want to buy, the next thing to consider is how much of it to buy from Oxford. The most straightforward option (it only costs money) is to buy a turn-key system from Oxford, and have them commission it. Oxford will then send over a team to install the fridge, a support for the fridge, a dewar for the fridge, the pumps, the plumbing, the wiring, the gas handling system—everything but the grad students. They will then send over a different team to leak check the fridge, get it running, and make sure it meets its guaranteed specifications.

If you opt for this route, be aware of some shortcomings in Oxford's designs. In the past, they have done a less than spectacular job on vibration isolation. This is only a problem if you are planning high magnetic field work, or want

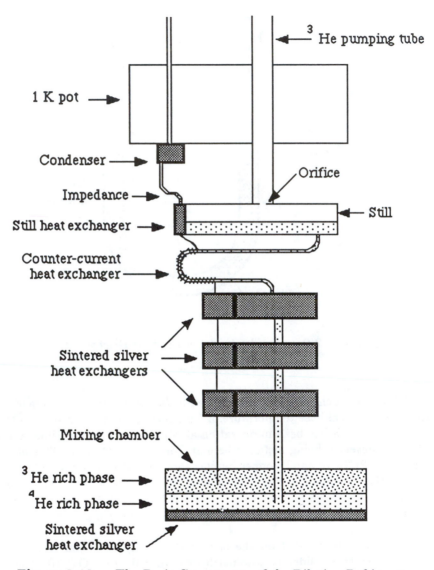

Figure 2.10. *The Basic Components of the Dilution Refrigerator.*

to use the fridge to pre-cool a demag stage. Their standard ^{3}He gas handling system has no provision for pre-circulating the mash (more on this later, but it is important). Their standard ^{4}He gas handling system has no provision for measuring the overpressure in the 1 K pot during an initial cool-down (see Nick Bigelow's section for an explanation of this procedure). They will also, unless you specifically instruct them otherwise—and you should most definitely do so—wire up the electrical connectors on your cryostat with straight ribbon cable, rather than twisted pairs. Don't get burned on this one. One of us spent a month replacing the un-twisted wiring in his cryostat. While the exterior of Oxford's cryostats is always shiny and bright, if you look inside any of it, you may find corrosion, old solder flux, and filth of indeterminate origin. When Oxford's chief engineer last visited Cornell, he promised that all of these flaws would be fixed in future editions.

Since each refrigerator is essentially custom made, Oxford will cheerfully modify yours in any way you like, but the additional charges add up fast. They will also design the entire system around architectural obstructions or incorporate such things as a shielded room. Their favorite way to package their gas handling system is as an enclosed column with all the pumps and interconnecting plumbing inside. This may look nice, but heaven help you if you have to get inside to find and fix a leak. If you request it, they will "open up" the system into a flat panel that allows you easy access to the back. At last report, this option was still quite expensive. Other additions you might like to consider are extra fill capillaries, high frequency co-ax, a lambda fridge, and extra tubes from the top plate into the vacuum can. For a while at least, Oxford designed their fridges to have a stainless steel mixing chamber bottom. Good thermal contact to the mash was then limited to the threaded copper hole in the center of the mixing chamber. This rather dramatically limits the number of things you can bolt directly to the mixer, so you might want to request a copper bottom (with a layer of sinter on it). The Cornell Microkelvin machines were "stretched" to allow more room between the 4 K flange and the 1 K pot, and between the 1 K pot and the still shield for cryo-electronics and such.

You do not, of course, have to buy the whole kit 'n' caboodle from Oxford. The Cornell Microkelvin Lab has vibration isolation tables designed and built by Wolfgang Sprenger, dewars purchased from Cryofab, and "home built" Kirk and Twerdochlieb (1978) style double-gimbals for the pumping lines. Oxford provided, installed, and commissioned all the rest. A different group at Cornell bought only the cryostat and ^{3}He gas handling system, and none of the commissioning and installation services. You can save a lot of money this way, but be prepared to spend a lot of time and effort getting the thing going. Yet another group bought only a Model 200 dilution unit (i.e., only what's below the 1 K pot in Figure 2.10) and built the rest of the cryostat themselves.

If you take this last route, the next section gives some useful information on how to design the still pumping line. Oxford will give candid recommendations on what pumps to use if you choose to buy them yourself. If you do not buy Oxford's gas handling system (if you're looking to cut costs, here is an excellent place to do so) Figure 2.11 illustrates the very nice manifold designed by Tom Gramila for the dilution refrigerator he built. You might also look at the one illustrated in Lounasmaa (1974) for ideas on the utility pumping system, which should have a reasonably fast diffusion pump, and should easily connect to every pumpable volume in, on, and around your cryostat.

One final note: every Oxford refrigerator we have ever dealt with or heard

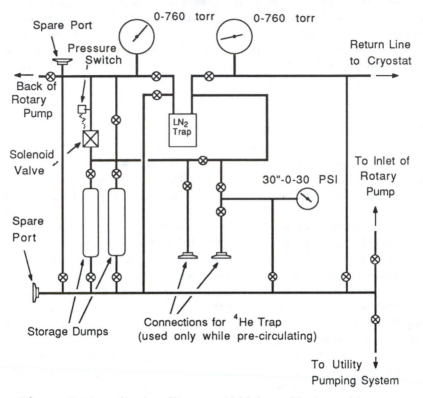

Figure 2.11.   *Gas handling manifold for a dilution refrigerator.*

about has leaked, most often on the first cool down. The usual culprit is an indium O-ring, which while tight during initial testing in England, loosened up in transit. You could probably save yourself some grief by tightening every one of the bloody things before even thinking about a cool-down. (Maybe they should read Eric Smith's section on how to make lead O-rings . . .) After about a month of operation, Dave McQueeney's fridge developed a leak in one of the heat exchangers. He managed to fix it himself, but to their credit, Oxford did offer to take the dilution unit back, replace the exchanger stack, and re-test the whole unit.

2.3.3   Starting a Dilution Refrigerator

In the following, we shall refer as much as possible to the discussion on "Cooling Down" in Section 2.1. We shall also assume that you have a fully tested and commissioned fridge, with reasonably calibrated diagnostic resistance thermometers on the 1 K pot, still and mixing chamber. It is also a good idea to have one on at least the lowest temperature heat exchanger, if not on all of them.

With the fridge still at room temperature and the vacuum can pumped out, check all the electrical systems. Hook up a leak detector to the vacuum can, evacuate the 1 K pot, and back fill it with about a 5 psi overpressure of pure helium gas. If the pot leaks, your detector will scream for mercy. Check to see that the pot fill line is not plugged by sticking it into a beaker of freon, acetone, or your favorite alcohol. Compare the rate at which it bubbles with the canonical value for your fridge. (Of course, if you are doing this for the first time, what you just measured is the canonical value.) Leave a helium supply attached to the pot to maintain the overpressure. Raise the dewar around the cryostat, pump out the helium bath space, and back fill it with helium gas. If you still have no leak signal, switch your leak detector over to ^{3}He (a handy option, if you have it) and admit a small amount of mash to both the return and still sides of the fridge. If you get a signal now, it is guaranteed that your fridge leaks. Finally, pump all the helium gas out of the bath space, and look for ^{3}He in there to see if either the return or still pumping line leak. (If your leak detector detects only ^{4}He, skip this step; the background signal will be enormous. You may want to skip it anyway, except on the very first cool-down of a new refrigerator.)

If you've found no leaks, suck the mash back out, and check the outgassing rate by watching the still pressure rise. Then put an atmosphere of mash, or pure helium gas if you prefer, into the return line, and again watch the pressure rise in the still. After subtracting off the outgassing rate from the previous test, check this number against its historical value. If it is too small, the return line may be partially plugged. You may, of course, do this test in the other direction: put the mash in the still and watch the pressure rise in the return line.

If everything is still A-OK, cool the cryostat to 77 K (see Section 2.1). Somewhere around this point, you should begin to pre-circulate (wash) the mash. That is, use your sealed rotary pump to draw mash out of the storage tanks (dumps), through a charcoal filled liquid nitrogen trap, through a bronze wool filled 4 K trap and back into the dumps. It is best to let this go on overnight, while your cryostat equilibrates at 77 K and you sleep. In the morning, put all the mash back in the dumps, seal off the traps, and warm them up to see what you've caught. If the pressure in the nitrogen trap is very high, then you've caught a lot, and you probably have a leak in your gas handling system. If only the helium trap shows much pressure, then you've just cleaned a lot of hydrogen out of your mash, which is what you were trying to do. (The hydrogen probably came from cracked diffusion pump oil.)

Before you blow out the liquid nitrogen, re-connect the leak detector to the vacuum can. If a leak developed in the 1 K pot, you'll know about it. Once you're convinced that all the liquid nitrogen is out of your cryostat, you should repeat all the room temperature leak tests, as well as the condenser capillary throughput test. You should get a much larger throughput this time around. If things are still proceeding without a hitch, you are probably lying, but go back to Section 2.1 anyway, cool until your exchange gas freezes or must be pumped out, leak check again, and fill the bath with liquid helium.

Once you are certain the liquid level has reached the fill line intake, remove the overpressure from the 1 K pot and start pumping on it. The pot should fill and cool to its usual base temperature, and should have a pressure of several torr in the pumping line. If either one of these conditions is not met, the fill line is probably plugged or plugging. Try putting a gradually increasing amount of heat into the pot. At some point you will exceed the cooling power of the pot, and it will run dry. If the fill line is, in fact, plugged, the amount of heat required to run the pot dry will be much smaller than usual. Of course, you could be fooled. If you have just modified your cryostat in some way, such as adding fill capillaries or high frequency co-axial cable, you may have dramatically increased the heat leak to your 1 K pot.

Once you are satisfied that the 1 K pot is up to snuff, you can condense the mash into the fridge. There are two approaches to this operation. The first is to open the mash tanks to the still pumping line, and run the 1 K pot as hard as possible. If you open the mash tanks slowly, and keep the pressure in the still line beween 50 and 100 mbar, all the mash will condense into the still. Keep an eye on the temperature of the 1 K pot. If you try to condense the mash too quickly, you will run it dry. This procedure has the additional advantage that any crap still in the mash will plate out on the inside of the still line.

Once the mash has condensed, make sure all possible inlets to the rotary pump are sealed off, open the back of the pump to the liquid nitrogen cold trap and the rest of the return line, and start the pump. Crack open the valve that allows gas in the still line to bypass the booster pump and go directly into the rotary pump. (You didn't design one into your system? It's too late now.) You are now circulating. Keep an eye on the pressure at the back of the rotary pump, making sure it does not exceed an atmosphere, as you gradually open the bypass valve. Otherwise, you may blow the shaft seals on your pump, lose a lot of mash, kick yourself, and generally feel miserable.

If all goes well, when the valve is fully open, the pressure in front of the rotary pump should be below 1 mbar. If this is the case, switch over to the booster pump, close the bypass valve, and apply the requisite amount of heat to the still. (If you don't heat the still, its temperature will drop, causing the pressure in it to drop, causing the circulation rate to drop, which will cause your cooling power to go away.) As Oxford instruments so glibly states in their manuals, "the refrigerator should now cool to its base temperature." If it actually does so without a hitch, you have led a very good life.

You can also condense the mash while you're circulating it. Open the back of the rotary pump to the return line, open the bypass valve, and crack the valve between the dumps and the still line. Throttle this valve, and the bypass valve if necessary, to keep the pressure in front of the rotary pump around 3 to 5 mbar. Again, once all the mash has been drawn out of the dumps and condensed into the fridge, the pressure in front of the rotary pump should fall below 1 mbar.

2.3.4   Tuning and Troubleshooting

One of the more useful diagnostic tools for checking the operation of your fridge is a helium gas flow meter. You can buy a complete system from Hastings or MKS, but it's much cheaper to just buy the sensor head, which has a linear, 0 to 5 volt output, and build your own circuit to read it.

The first task you face in starting up a new fridge, or re-starting one that has had a large leak, is figuring out how much mash you need, and what percentage of it should be ^{3}He. If you bought your fridge from Oxford, they will tell you this. Otherwise, you can use the volumes of the various parts of the fridge to get a rough starting estimate. If you don't know these volumes, try a 25% mixture for starters, and guess wildly on the total amount of mash. You would like the phase boundary in the mixing chamber, and the dilute phase liquid surface in the still.

If you have too much mash, the liquid level on the dilute side will be up in the still line, past the orifice designed to restrict superfluid film flow. If this is the case, recondensing helium gas will present a large heat load to the still. If the pressure in the still is high, but it won't cool below a degree, even with the still heater off, you should suspect this problem.

If you have too little mash, the liquid level will be down inside the counter-current heat exchanger. In that case, because of the reduced liquid surface area, you will not be able to circulate much ^{3}He. This symptom is identical to what you would see if the return line were plugging, but in this case the return line pressure will still be low. The still temperature will also be higher than you would expect, given the pressure in the pumping line. As a final check, heat the still. If the circulation rate does not change, but the temperature of the still runs away, then you have pegged the problem.

Once you have the amount of mash more or less under control, the next step is to find the phase boundary. If it is not in the mixing chamber, it is up in the heat exchangers (if it's not even that close, you're in trouble). The trick is to find which side of the mixing chamber it's on. Open the back of the rotary pump to the dumps and close off the return line. You are now "one-shotting" the fridge. Watch the resistor on the mixing chamber. If it does nothing for a long while, and then suddenly begins to cool, you have too much ^{3}He. If you pump all the ^{3}He out of your fridge and never see the mixing chamber cool, then you have too little ^{3}He in your mixture. If you've got it right, the mixing chamber should start to cool immediately, and should remain cold for several minutes before beginning to warm as the phase boundary moves out. You can easily see how an additional resistor on the last heat exchanger could really help.

If a working refrigerator ceases to do so, it is usually for one of three reasons: a touch, a plug, or a leak. A touch is essentially a thermal short circuit between two parts of the fridge that should be at very different temperatures. It is usually caused by contact between one of the heat shields and the vacuum can or another shield, and the symptoms are poor base temperature and poor cooling power. One of the more useful diagnostic tools for hunting down this sort of problem is a touch detector. Decide on the most likely region for two shields to come into contact if things get misaligned on assembly. Put a layer of kapton tape on one of the shields, and put some copper tape on top of that. Then run wires from the copper tape and the other shield out the top of the cryostat. Monitoring the resistance between the two wires will alert you if things get jostled enough on assembly and cool-down to cause a problem.

A plug is most often caused by dirty mash, or a leak somewhere in the room temperature part of the circulation system. The first sign of a plug is a drop in the circulation rate, and an increase in the return line pressure. If you clean your traps often enough, you may get an early warning in the form of more stuff in the traps than usual. The inevitable complete plug and forced warm-up can be forestalled by frequently cleaning the traps, and when you get desperate enough, by a "warm transfer." The idea is to let the bath level boil below the vacuum can flange (or the lowest point in the return line), and then start a transfer without properly cooling the stick. You can also just blow room temperature gas from a bottle down into the bath space. The idea is to melt the crud that is forming the plug, and get it to redistribute itself more uniformly in the return line. It's a good idea to pump on the return line and pressurize the still as you do this. That way, when the plug melts, the stuff will move in the right direction. Warm transfers are a measure of last resort, and don't always work.

A leak is one of the easier problems to detect, and one of the hardest to fix. The symptoms are much the same as for a touch, but there will be helium in the gas can. If there's ^{3}He in the gas can, you know it's the fridge (or a ^{3}He experimental cell if one is mounted on the apparatus). If there's only ^{4}He, then it could be either the vacuum can, or the 1 K pot. If, when you warm back up to room temperature, the leak is gone, the mark of the demon has been visited upon you. The most utterly painful, but most straightforward way to find the leak is to open up the fridge at some convenient point, and seal off the open ends. Cool down again, and find out which half of the system leaks. If you repeat this process long enough, you will eventually narrow down the location of the leak. Be sure the components you're testing are thermally anchored to the 1 K pot so they get cold enough for the leak to show up. If the thing leaks at liquid nitrogen temperatures, life is a little easier. You can just dunk the bare fridge in a wide neck nitrogen dewar. If you truly suspect that the leak occurs only when things are below the ^{4}He superfluid transition, you can try to manipulate a temperature gradient in such a way as to localize the leak somewhat. This trick requires lots of heaters and well calibrated thermometers, and has a near zero sucess rate.

If all of the above has not been enough to re-ignite your interest in, say, plasma physics, good luck, and may Murphy's shadow never darken your door.

2.4   Pumps and Plumbing

by Geoffrey Nunes, jr.

2.4.1   Introduction

The ultimate performance of any cryogenic refrigerator depends quite heavily on the capacity of its pumps and on the geometry of its plumbing. The aim of this chapter is to present information relevant to the design of such pumping systems. To that end, the first section contains some of the formulae used to describe the flow of gases in tubes, while the next section applies all that to the actual design of pumping lines. The third section contains a sampling of the types of pumps commonly used, as well as pricing information, and the fourth section discusses some of the important precautions one can take in putting together a complete system. The final section contains further design examples, as well as some references and a list of manufacturers.

2.4.2   Pumping Formulae

There are two regimes for which the flow of gas in a tube is well defined: viscous and molecular (the latter is often called the Knudsen regime). In the case that the mean free path of a molecule in the gas is much less than the diameter of the tube, the flow is viscous. In the case that the mean free path is much greater than the tube diameter, the flow is molecular. In the case of the average cryogenic pumping line, things are somewhere in between. More specifically, if L is the mean free path and a is the tube radius, the flow is considered viscous when $L/a < .01$ and molecular when $L/a > 1$. Note that the ill-defined region covers two orders of magnitude.

The mean free path of a molecule in a gas is given by

$$L = \frac{1}{\sqrt{2}\pi n d^2} \tag{2.1}$$

where n is the density of the gas and d is the equivalent hard-sphere diameter of the molecule. For helium, $d = 2.2$ Å (Kennard, 1938). By a simple application of the ideal gas law, Equation 2.1 can be written in terms of the pressure and temperature of the gas:

$$L = \frac{kT}{\sqrt{2}\pi d^2 P}. \tag{2.2}$$

For helium this expression reduces to

$$L = 4.8 \times 10^{-5} \frac{T}{P} \tag{2.3}$$

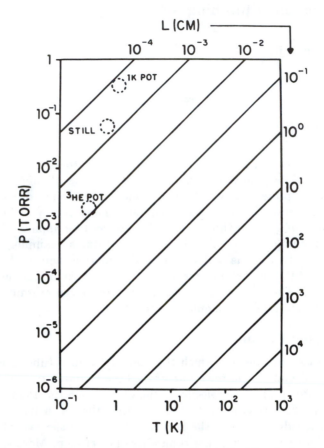

Figure 2.12.   *P vs. T curves for several values of L. The dashed circles indicate the approximate regions of operation for various refrigeration systems components.*

where L is in centimeters and P is in torr. Figure 2.12 gives P vs. T curves for several values of L, along with the approximate regions of operation for various parts of a refrigeration system.

If the flow is viscous, a second important parameter describing the flow is, not surprisingly, the viscosity. The theoretical (hard-sphere) viscosity of a gas is given by (Kennard, 1938):

$$\eta = \frac{2}{3}\sqrt{\frac{mkT}{\pi^3 d^4}}. \tag{2.4}$$

For ^{4}He this is

$$\eta = 7.5 \times 10^{-6}\sqrt{T}.$$

To get the theoretical viscosity of ^{3}He, just multiply by the square root of 3/4.

Real life is not so simple, and a more accurate, if less convenient expression for the viscosity of ^{4}He is (Kennard, 1938)

$$\eta = 5.18 \times 10^{-6} T^{.64} \tag{2.5}$$

where the units of viscosity are gm/cm/sec. A complete tabulation of the viscosity of ^{4}He from 2500 K to 1.25 K can be found in Thermophysical Properties of Matter, Volume 11 (Touloukian, 1975). Note that the viscosity depends only on the temperature, and not on the density of the gas.

Whatever the particular regime, the flow is usually described in terms of the volumetric flow rate, or throughput:

$$Q = P\frac{dV}{dt} \tag{2.6}$$

where dV/dt is the volume of gas crossing a plane perpendicular to the flow per unit time, and P is the pressure at which the flow rate is measured. Application of the ideal gas law shows that

$$Q = kT\dot{N} = Q_m\frac{kT}{m} \tag{2.7}$$

where N is the flow rate in molecules per second, and m is the mass of a gas molecule. Q is thus directly proportional to Q_m, the mass flow (in gm/sec). This latter quantity is actually the more useful for cryogenic applications, as will be discussed in the next section. The conductance F of a channel is defined by

$$F = \frac{Q}{P_1 - P_2} \tag{2.8}$$

where P_1 is the pressure at the upstream end of the channel and P_2 is the pressure at the downstream end.

Here are a couple of examples for tubes of circular cross section, with radius a and length l (Dushman, 1962). In the viscous case, the flow is described by the Pouseille equation:

$$Q = \frac{\pi a^4}{16\eta l}(P_1^2 - P_2^2) \tag{2.9}$$

$$= \frac{\pi a^4}{8\eta l}P_a(P_1 - P_2).$$

In this case the conductance increases linearly with P_a, the average pressure in the tube. For molecular flow, the conductance of a long tube is given by

$$F = \frac{4a^3}{3l}\sqrt{\frac{2\pi kT}{m}}. \tag{2.10}$$

Note that the conductance increases as the cube of the radius while decreasing only linearly as the length of the tube is increased. This means that you can make the line long enough to get your pump out of the room without having to put in a ridiculously large diameter pipe.

Both of these equations are somewhat approximate, so that most calculations have a "back of the envelope" quality to them. There is, however, one correction to the case of molecular flow that is not too laborious to make. Equation 2.10 is only valid in the case of long tubes, so that if the tube is "short," i.e., if the ratio of l/a becomes less than about 5, the actual conductance can be significantly reduced. The correction for this effect is straightforward. Let F_t be the conductance calculated according to Equation 2.10. The actual conductance is then given by

$$F = \frac{3l}{8a}KF_t \tag{2.11}$$

where K is Clausing's factor. Figure 2.13 shows a plot of K vs. l/a. As an example, consider a tube that is 10 cm long and 2 cm in diameter. Then l/a is 10, and K (from Figure 2.13) is about .2, so that the actual conductance is three fourths the value obtained if end effects are neglected.

Under most circumstances, the flow in a pumping line will be neither viscous nor molecular, but somewhere in the "transition region." It turns out, however, that in this particular case, you win. (Our famous Irish friend was caught napping.) Flow in the transition case is (almost) always greater than or equal to what you would calculate assuming molecular flow. Knudsen showed that the conductance of a tube throughout the transition region is accurately given by

$$F = F_v + \alpha F_m \tag{2.12}$$

where F_m is the conductance calculated according to Equation 2.10 and F_v is the conductance in the case of viscous flow (i.e., the coefficient of $(P_1 - P_2)$ in Equation 2.9). α is a function of T, P, η, and the radius of the tube. The interested reader should consult Dushman (1962) or Roth (1976) for more details on the exact form of α and the derivation of Equation 2.12. For our purpose, it is sufficient to note that α is never smaller than .81 and never larger than 1, so that it is not stupid to make the approximation that $\alpha = 1$ (Denker, private communcation).

If one defines P_x as the pressure at which $F_v = F_m$ (for a given tube radius) and uses Equation 2.4 for the viscosity, it is not difficult to show that

$$F = \frac{1}{Z\eta}(P_x + P_a) \tag{2.13}$$

where $\quad Z = \frac{8l}{\pi a^4} \quad$ and $\quad P_x = \frac{64\sqrt{2}kT}{9\pi^2 ad^2} \cong \frac{kT}{ad^2}.$

P_a and T are the average pressure and temperature in the tube, respectively. Figure 2.14 compares this result with that given by Knudsen, as well as with the molecular and viscous conductances separately.

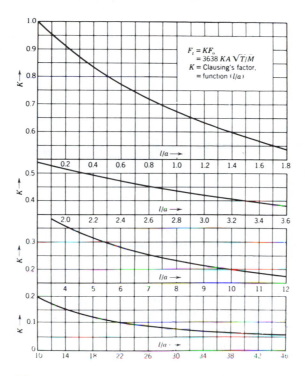

Figure 2.13. *Plot of Clausing's factor, K, as a function of 1/a. Scales for K and 1/a are indicated for each curve. [Dushman, S. Scientific Foundations of Vacuum Technology, (John Wiley & Sons, NY), 1962.]*

Most often, one is not interested in the conductance so much as the pressure drop along the pipe. If you know the pressure at one end (as determined, say, by the still temperature) and want to know the pressure at the other, it is trivial to combine Equations 8 and 13 to get:

$$P_2 = \sqrt{(P_1 + P_x)^2 - 2\eta \dot{N} k Z T} - P_x. \qquad (2.14)$$

Given a particular mass flow and the pressure at the upstream end of the tube, one can easily use Equation 2.14 to calculate the pressure at the downstream end. If the flow is close to being molecular, however, this equation is only a good approximation when the temperature along the tube is constant. In cryogenic pumping lines, of course, the temperature gradients are large, but the flow is usually far enough from being molecular for Equation 2.14 to be adequate.

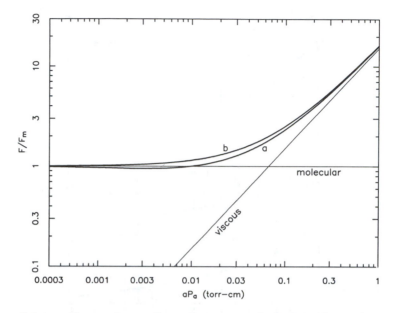

Figure 2.14. *Comparison of conductance calculations in various regimes. Curve a is the exact result and curve b is the more tractable approximation. P_a is the average pressure in the tube, and a is the tube radius.*

2.4.3   Design of Cryogenic Pumping Lines

In designing the pumping line for a refrigerator, it is simplest to first assume that the flow will be molecular. This assumption allows you to get a rough handle on what the line will look like before applying the transition flow corrections. It is especially useful in designing a ^{3}He cryostat, as the flow in that case will be relatively close to molecular. Using Equations 7 and 8, Equation 2.10 can be written as

$$Q_m = \frac{4a^3}{3l}\sqrt{\frac{2\pi m}{k}}\left(\frac{P_1}{\sqrt{T_1}} - \frac{P_2}{\sqrt{T_2}}\right) \tag{2.15}$$

where it is now assumed that the temperature as well as the pressure varies from one end of the tube to the other.

Following a suggestion originally made by Walton (1966), we now model the pumping line as a series of tubes, each with a different diameter, as in Figure 2.15. If the gas flow is in a steady state, then Q_m will be the same throughout the

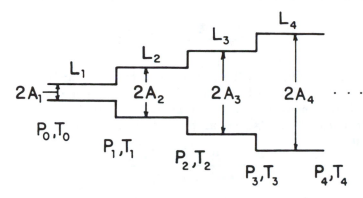

Figure 2.15.   *Simple model of a graduated pumping line.*

whole series and we can write Equation 2.15 explicitly for each tube:

$$\frac{P_0}{\sqrt{T_0}} - \frac{P_1}{\sqrt{T_1}} = A\frac{l_1}{a_1^3}$$

$$\frac{P_1}{\sqrt{T_1}} - \frac{P_2}{\sqrt{T_2}} = A\frac{l_2}{a_2^3}$$

$$\vdots$$

$$\frac{P_{n-1}}{\sqrt{T_{n-1}}} - \frac{P_n}{\sqrt{T_n}} = A\frac{l_n}{a_n^3}.$$

where $\qquad A = \frac{3}{4}Q_m\sqrt{\frac{k}{2\pi m}}.$

It is not difficult to see that if you add all these equations together, only the terms involving the pressure and temperature at the extreme ends of the entire pumping line remain:

$$\frac{P_0}{\sqrt{T_0}} - \frac{P_n}{\sqrt{T_n}} = A\sum_{i=1}^{n}\frac{l_i}{a_i^3}. \qquad (2.16)$$

Since the final one or two segments of pipe will be at room temperature, the term involving T_0 will be much greater than the term in T_n, so that the latter may simply be ignored:

$$\frac{P_0}{\sqrt{T_0}} \approx \frac{3}{4}Q_m\sqrt{\frac{k}{2\pi m}}\sum_{i=1}^{n}\frac{l_i}{a_i^3}. \qquad (2.17)$$

In convenient units (P in torr, Q_m in gm/sec and lengths in cm) this expression becomes

$$\frac{P_0}{\sqrt{T_0}} = 1.2 \, Q_m \sum_{i=1}^{n} \frac{l_i}{a_i^3} \tag{2.18}$$

if the refrigerant is ^{3}He. (For ^{4}He just multiply by the square root of $4/3\ldots$)

Equation 18 is essentially a constraint equation for use in designing a refrigerator. If you know the ultimate temperature and cooling power you wish to achieve, then P_0, T_0 and Q_m are all fixed. You need only design the pumping line so that

$$\frac{P_0}{1.2\sqrt{T_0}Q_m} > \sum_{i=1}^{n} \frac{l_i}{a_i^3}. \tag{2.19}$$

Of course, the smaller you make the sum, the easier time you will have getting the fridge to perform as planned. Note that this equation is not the only constraint around. The need to minimize heat conduction down the pumping line and the rather restricted geometry of a helium dewar will also affect the design.

As an example of how all this works, suppose that you want to design a ^{3}He fridge operating at .3 K with an effective cooling power of 50 microwatts (see Figure 2.16). The effective cooling power in this instance is defined as the rate at which heat is removed from the experimental cell. Therefore the total cooling power of the fridge must be equal to this effective cooling power plus any extraneous heat loads.

The latent heat of evaporation of ^{3}He is 21 J/mol = 7 J/gm at .3 K, so that the total cooling power of the fridge is related to the mass flow by

$$\dot{Q} = 7 \, Q_m. \tag{2.20}$$

Note that $\dot{Q}$ is a heat flow (with units of Joules/sec), not a mass flow. Barring unfortunate circumstances, the two largest heat loads in a circulating fridge are thermal conduction down the pumping line from the 1 degree pot and the enthalpy of the returning liquid ^{3}He. If this returning liquid is assumed to be at 1.1 K, then it will deliver about .85 J/gm to the ^{3}He pot, so that the heat load is

$$\dot{Q}_r = .85 \, Q_m. \tag{2.21}$$

The thermal conduction down the pumping line is given by

$$\dot{Q}_p = \frac{2\pi t a}{l} K \Delta T \tag{2.22}$$

where K is the thermal conductivity and t is the wall thickness of the pumping line. Taking $K = .8$ mW/cm/K for stainless steel at these temperatures, and $\Delta T = .8$ K gives

$$\dot{Q}_p = .004 \frac{at}{l}. \tag{2.23}$$

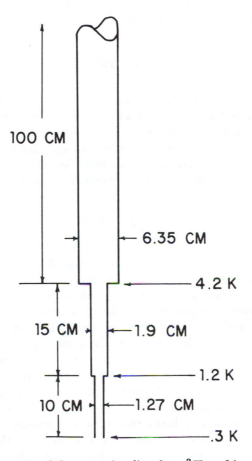

Figure 2.16.   *Schematic of the pumping line for a ^{3}He refrigerator. The upper end of the large pipe is assumed to be at room temperature.*

Combining Equations 20, 21, and 23 yields the following approximate expression for the mass flow in the fridge:

$$Q_m = .16\left(\dot{Q}_e + .004\,\frac{at}{l}\right). \tag{2.24}$$

$\dot{Q}_e$ is the effective cooling power we are tying to design the fridge for. (50 microwatts in this example. Note that the units are mixed: the heat flow is expressed in watts, while everything else is in centimeters and grams.)

Suppose now that geometry restricts the line between the .3 K and 1.1 K pots to be 10 cm long by 1.27 cm in diameter. A typical wall thickness for such a tube is .025 cm (.010 in). Putting these numbers into Equation 2.24 gives $Q_m = 1.8 \times 10^{-5}$ gm/sec. The vapor pressure of ^{3}He at .3 K is 1.9×10^{-3} torr, so $P_0/\sqrt{T_0} = 3.5 \times 10^{-3}$ torr$/\sqrt{K}$. Putting all of this into Equation 2.19 gives the requirement that

$$\sum_{i=1}^{n} \frac{l_i}{a_i^3} < 324 \text{ cm}^{-2}. \tag{2.25}$$

It turns out that this is a relatively easy criterion to meet. The first tube has already been specified, so suppose that there are only two other sections of tubing inside the dewar, a 1.9 cm diameter section that is 15 cm long, and a 6.35 cm diameter section that is 1 meter long. (Any tubing outside the dewar can safely be neglected, since space restrictions are much less severe and the radius of the tube can be made quite large.) For this particular pumping line

$$\sum_{i=1}^{n} \frac{l_i}{a_i^3} = 60 \text{ cm}^{-2} \tag{2.26}$$

which provides an ample margin for error over 324 cm^{-2}. It should be noted that well over half of this number is contributed by the narrowest tube, while the 1 meter long section accounts for only about 5%, which tells you where to look if you're trying to improve the performance of a refrigeration system.

To find the actual pressure at the room temperature end of the pumping line, it is necessary to return to Equation 2.14. By applying this equation successively to each section of tube (starting at the low temperature end where the pressure is known) one finds that the pressure at the upper end of the pumping line is about 10^{-3} torr, a factor of two lower than that in the ^{3}He pot. (This result was obtained by assuming the temperature gradient along each tube to be linear, and by using a computer to break each tube into smaller and smaller constant temperature segments. The pressure drop calculated in this fashion converges quite rapidly, and to a final value that is not very different from the result of a non-iterative calculation.) How to choose a set of pumps appropriate for a fridge with this pressure drop and mass flow will be discussed in the next section.

2.4.4   Pumps

The three types of pump most commonly used in low temperature experiments are rotary, Roots, and diffusion. The parameter used to describe the performance of all three types is S, the pumping speed. S has units of volume per unit time and is defined by

$$S = \frac{Q}{P} = \frac{dV}{dt} = \frac{kT}{mP}Q_m \tag{2.27}$$

so that the speed of a pump depends on the pressure at which it is operating, though the exact nature of that dependence varies from one type to another. In all cases, S goes to zero at some limiting value of the pressure, which may be

Table 2.3.   *Gas Flow Conversion Table*

To convert from	to	multiply by
Liters/second	cubic feet/min.	2.12
	cubic meters/hr.	3.60
Liters/min.	cubic feet/min.	0.0353
	cubic meters/hr.	0.060
Cubic meters/hr.	liters/second	0.2778
	liters/min.	16.67
	cubic feet/min.	0.589
Cubic feet/min.	liters/second	0.4719
	liters/min.	28.32

between 10^{-2} torr and 10^{-10} torr, depending on the type of pump. The usual convention in specifying the speed of a pump is to take its speed at an atmosphere of pressure if it is a rotary pump, and to just take its maximum speed if it is a Roots or diffusion pump. The units that manufacturers use for S follow some extremely obscure convention. Table 2.3 contains some of the factors necessary to convert, for example, a cubic foot per minute to a liter per second.

The parameter that is actually relevant in refrigerator design is Q, the throughput of the system. Obviously, the throughput of the entire refrigerator can be no greater than that of the pump, the latter being given by S times the pressure at the pump's inlet.

Rotary pumps are a basic ingredient of any refrigeration system. Figure 2.18 shows a cross sectional view of a standard, single stage, rotary vane pump. Gas enters at C, is compressed between the central cylinder and the casing, and is ejected at D. The whole thing is sealed and lubricated by a liberal amount of pump oil. Figure 2.19 shows some typical speed vs. pressure curves for rotary pumps. Pumps can be bought either single or double stage, and with either belt or direct drives. A two stage pump is essentially a pair of single stages connected in series, and can achieve pressures as low as 10^{-4} torr. Single stage pumps are just as fast at higher pressures, but can only pump down to about 10^{-2} torr. Most manufacturers seem to be emphasizing direct drive over belt driven pumps these days. Aside from the obvious advantage of never needing their belts replaced, direct drive pumps also seem to make the task of vibration isolation a little easier. Figure 2.19 shows some typical prices for two stage, direct drive pumps. Note that you can buy pumps that have been especially sealed for

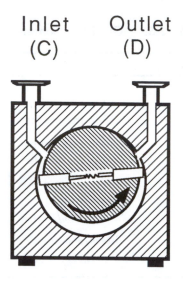

Inlet   Outlet
(C)   (D)

Figure 2.17.   *Cross sectional view of a rotary pump. As the cylinder rotates, air entering at C is compressed against the casing, ejected through a valve, and exits the pump at J. The vanes are held tightly against the casing and sealed with a film of pump oil.*

helium service, but it adds quite a bit to the price. What you get for that money depends on the manufacturer. Some have redesigned the casing to accommodate real O-rings (e.g., Alcatel), while others have simply taken their standard design and coated all the gaskets with some kind of goo (e.g., Edwards), which seems to work about as well. Note that Alcatel and Varian pumps have the same prices and speeds. Edwards' pumps have a very large back volume, which makes them less suitable for refrigerators which use ^{3}He as a coolant. The rumor mill has it that Varian's rotary pumps are actually Alcatels in sheep's clothing (so to speak . . .)

If your budget is tight, it is possible to buy an unsealed pump and seal it yourself. There are no real rules to this game. It is enough that the pump be leak-tight, as determined by a leak detector in modest working order. (A leak rate below 10^{-5} atmosphere-cc/sec can easily be handled by an average sized cold trap.) Two of the most important places to check are the shaft seals and the oil level sight windows. In both cases, the seal may have been made with a fairly porous material, and should be replaced with mylar or kapton. The housings for the shaft seals may have to be replaced as well, as the thin aluminum used in some pumps won't allow you to tighten the seal hard enough. Another perennial trouble spot is the ballast gas valve (used when pumping noxious chemicals). Even manufacturer sealed pumps have this little feature, and if it's not tight it can let quite a lot of air into your system. Be sure to check around the bolts

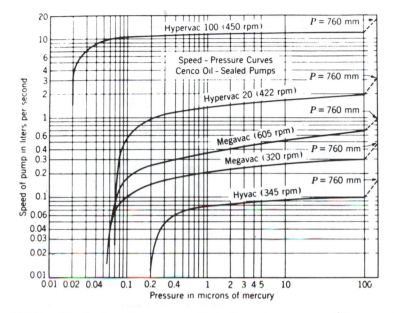

Figure 2.18. *Typical performance curves for rotary pumps (these particular ones are manufactured by Cenco). Note the sharp drop in pumping speed at 10^{-4} torr. [Dushman, S. Scientific Foundations of Vacuum Technology, (John Wiley & Sons, NY), 1962.]*

that hold the halves of the pump together (as well as those that hold the innards of the pump to the casing—on some pumps these go through to the outside!). If they leak, putting copper washers at either end may work, but if not, spread on a little of your favorite sealant (e.g., Leak Lock from Highside Chemicals). The cast pieces of the pump pose somewhat more of a problem, as they may be porous enough to admit helium. If that is the case, try soaking them in epoxy or gluing metal foil to the larger areas.

Since rotary pumps alone cannot approach the speed required for a ^{3}He or dilution fridge, the usual practice is to use a Roots or diffusion pump in combination with a rotary backing pump. The method of operation of a Roots pump (sometimes called a mechanical booster or blower) is depicted in Figure 2.20. The two rotors are machined to extremely close tolerances, and use no lubrication. Thus, if the pump is allowed to run for very long at too high an inlet pressure, the rotors will heat, expand, and seize. Since that would be an expensive mistake, there are a couple of approaches that manufacturers take to prevent it. Leybold-Heraeus pumps come with a pressure switch that prevents the pump from turning on if the pressure is too high. Edwards pumps have something they call a "hydrokinetic drive" that automatically reduces the pump's speed if the in-

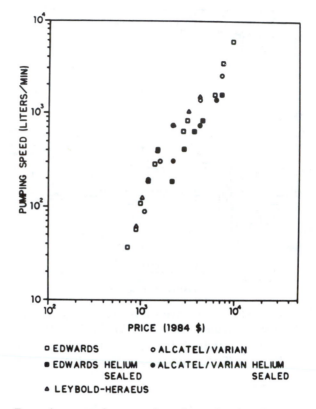

Figure 2.19.   *Pumping speeds as a function of price for two stage rotary pumps. All the pumps included here are direct drive.*

let pressure is too high. This feature means that the pump can be started at any pressure, which reduces the initial pump-down time by quite a bit. The other side of this coin, however is that the motor must remain outside the vacuum space, which requires a rotating seal. Some of the Leybold-Heraeus and Alcatel pumps are available with a "canned" motor, which eliminates the rotating seal and allows for a (theoretically anyway) smaller leak rate. This option adds about $500 to the price of a Leybold-Heraeus pump, and about $2000 to that of an Alcatel.

Diffusion pumps work by trapping the pumped gas molecules in a high velocity stream of oil. Figure 2.21 depicts a relatively standard design. (Vapor booster pumps work in the same fashion, and are essentially diffusion pumps designed for high throughput applications.) Diffusion pumps offer a high pumping rate, an extremely low ultimate pressure, and a relatively small cost. Figure 2.22 compares the prices and pumping speeds of some diffusion and Roots pumps.

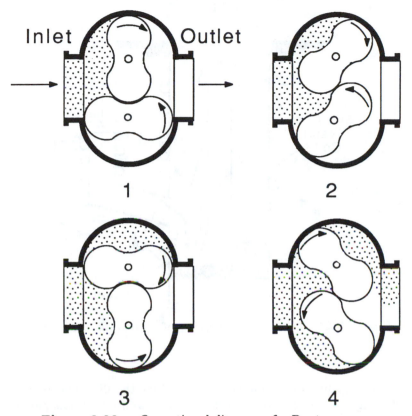

Figure 2.20.   *Operational diagram of a Roots pump.*

(The higher speed Roots pumps from Leybold-Heraeus are available on special order only.) The actual speed of the diffusion pumps in operation will likely be about 25 to 30% less than shown, depending on what sort of baffling scheme is used to prevent backstreaming of the pump oil. Which brings up the principal problem with diffusion pumps—oil. More than one group of experimenters at Cornell have found themselves clustered around their cryostat wondering just how to get the pump oil out of the mixing chamber after their diffusion pump sprang a leak (high pressure Freon seems to work all right, but the sinter has to be replaced). Even if nothing so catastrophic happens, day to day operation can lead to macroscopic amounts of oil in awkward places, especially with the higher speed pumps.

Whether a diffusion pump or a Roots pump is appropriate for a particular application depends on, among other things, the required pumping speed and throughput, Q. Figure 2.23 shows the pumping speed and throughput, as functions of inlet pressure, of both a diffusion pump and a Roots pump of ap-

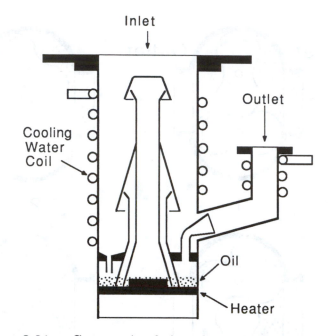

Figure 2.21.   *Cross sectional view of a small diffusion pump.*

proximately the same maximum speed. Note that the diffusion pump doesn't even get going until the pressure is below .01 torr while the Roots pump has its maximum speed at .1 torr. The real difference between the two is not in their speed, however, but in their throughput, which is after all, the relevant parameter for a fridge. The maximum Q of the diffusion pump is 1.5 torr-liters/sec, while that of the Roots pump is two orders of magnitude (or so) larger: 230 torr-liters/sec. (The largest diffusion pumps on the market do not even approach the throughput of this medium-sized Roots pump.) Throughput, however, is not the only consideration. Roots pumps put out pretty strong vibrations, which may be a problem if one is looking to achieve really low temperatures.

As an example, consider the problem of selecting a pump for the ^{3}He fridge discussed in the last section ($Q_m = 1.8 \times 10^{-5}$ gm/sec, pressure at the top of the cryostat $= 10^{-3}$ torr). At room temperature, the required volumetric flow rate (see Equation 2.7) is .5 torr-liters/sec. If the plumbing that runs from the fridge to the pumps is fat enough, the pressure at the pump inlet will also be about 10^{-3} torr. In that case, the required pumping speed (calculated according to Equation 2.26) is 500 l/s. Varian sells an 800 l/s pump with an ample 1.5 torr-l/s throughput for $1300. An appropriate backing pump costs about $1500. A glance at Figure 2.23 shows that a Roots pump is totally inappropriate for this particular application. Since both the pump inlet pressure and volumetric flow rate are fixed by the operating point of the fridge, we would have to purchase

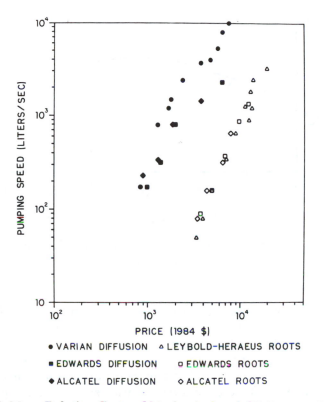

Figure 2.22.   *Relative Costs of Mechanical and Oil Booster Pumps.*

a Roots pump whose maximum speed is about 2500 l/s in order to achieve the required speed of 500 l/s given the small (for a Roots pump) inlet pressure of .001 torr. It is important to note that things are quite different for a dilution refrigerator. In that case, the pump inlet pressure will be between .01 and .1 torr. A second glance at Figure 2.23 reveals that a Roots pump will be in full form, while a diffusion pump will be unable to cope with that kind of an inlet pressure, which makes it the preferable choice.

It is worth mentioning two other kinds of pumps. Turbomolecular pumps work essentially the way a fan does. A 1500 liter per second turbo pump from Edwards rotates at 21,000 rpm, has an ultimate pressure somewhere around 10^{-10} torr, and can be started at atmospheric pressure. It also costs $16,700, and its backing pump another $3400. At the opposite end of the economic scale is the activated charcoal adsorption pump. Non-circulating dilution and ^{3}He refrigerators can be constructed using a small bag of activated charcoal suspended in the ^{3}He pumping line. As the charcoal is lowered into the colder regions of the cryostat, relatively large pumping rates can be achieved without any vibration

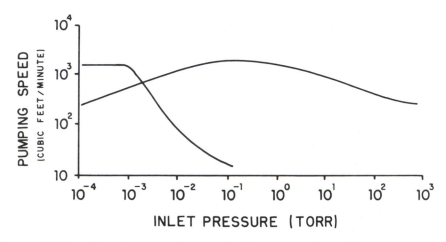

Figure 2.23.   *Pumping speed versus inlet pressure for a Roots pump and a diffusion pump of approximately the same maximum speed.*

whatsoever. The disadvantage of such an arrangement is that once all the ³He has been adsorbed into the charcoal, the bag must be raised far enough in the tube so that the ³He condenses onto the 1 degree pot and flows back down. On the other hand, the pump costs essentially nothing, and is completely free from vibrations. See Lounasmaa (1974) for more details on such single shot designs. It may even be possible to build a recirculating system using a pair of charcoal bags. (See Mikheev, 1984.)

Once you have settled on the appropriate system for your fridge, but before you actually go ahead and blow your research budget for the next n years on 10^4 or so dollars of pumping equipment, you should shop around. The information on manufacturers and their prices presented here is not intended to be complete. In addition, be aware that some manufacturers offer various educational and/or volume discount schemes. Their sales staff, however, tend to be a bit reticent about this sort of thing, so press hard.

2.4.5   Protection

After you've plunked down all that money for your pumps, you'd like to make sure they stay around for a while. This section discusses a few of the common techniques available to protect your pumps from your system (and your system from your pumps). Three of the major causes of pump death are cooling water loss, power failure and return line blockages. If it is at all possible that your cooling water supply will fail while the pumps are running and nobody's around, it's a good idea to have some kind of automatic shut-off system linked to the water flow. The same sort of idea applies in the case of a power loss. If all the pumps lose power, it's not a problem. But if a backing pump should fail, you'd like to have some sort of pressure switch sensing the pressure between it and the Roots or diffusion pump. It's also a good idea to have at least a gauge, if not a similar switch on the return line side of the backing pump. A line blockage can cause the pressure to get high enough to blow the seals in the rotary pump, causing a lot of air to go where you don't want it and a lot of ^{3}He to go where you'll never get it back. A good precaution against this sort of catastrophe is an automatic pressure switch that will vent the return line into a holding tank. It is probably a good idea to monitor the pressure inside the tank as well. There's no sense in trying to stuff the entire room into a few liters of storage space.

The precautions to take against line blockages really come under the category of protecting your system from your pumps. There is almost always some small amount of air that leaks through the rotary pump and into the return line, not to mention a liberal amount of pump oil. If you don't get rid of it somehow, it will freeze out in your fridge and block the ^{3}He return path. The easiest way to get rid of the air is with a nitrogen cold trap. It's a good idea, however, to try and take out most of the oil before the trap. Most pump manufacturers will sell you various filters for this purpose, but a section of glass tubing with screens at either end and loosely packed cotton or glass wool in between works pretty well. The best kind of oil filter, however, is a coalescing one. These filters use some kind of polymer membrane to accumulate the fine mist that comes out of the pump into large drops, which then remain in the filter. Since the Zeolite in cold traps is in chunk form, it actually lets a fair amount of the mist through. A relatively wide selection of coalescing filters is available from Balston, Inc.

In spite of all filters and LN_2 cold traps, it is still possible that enough junk will get through to cause a block. The best protection against this eventuality is a 4 K cold trap. A relatively simple design for such a trap is a pair of concentric tubes that dip down into the helium bath (or into a separate helium dewar). The mash enters the trap through the inner tube and is forced back up next to the cold outer wall before returning to room temp. Then, if a block occurs, it will be in the trap (or so one hopes, anyway), which can just be valved off and removed, pumped out and put back on. (It will take several valves to make this work, but it means you don't have to warm up the whole cryostat.) Oxford Instruments now offers this sort of a trap as a standard feature on some of their dilution refrigerators.

2.4.6 Further Design Examples, References, and Manufacturers

One aspect of refrigerator design so far not dealt with is the one-degree pot, which in a typical fridge will consume a few hundred cm^3 per hour. How critical is the design of this pumping line to the overall performance of the cryostat? Well, as it turns out, not very. The pressures involved in this case are on the order of a few torr. A quick glance at Figure 2.12 shows that the flow in this case is completely viscous. If the room temperature section of the pumping line is of approximately uniform cross-section, Equation 2.9 can be re-written as

$$Q = S^2 \frac{8\eta l}{\pi a^4} \left[-1 + \sqrt{1 + \left(\frac{\pi a^4}{8\eta l} \right)^2 \frac{P_1^2}{S^2}} \right] \tag{2.28}$$

where Equation 2.27 has been used to substitute for P_2, and S is the speed of the pump. Since the messy term inside the square root will be large compared to 1 for all but the wimpiest of pumps (and *tiniest* of tubes), Equation 2.28 reduces to $Q \simeq SP_1$, independent of the radius of the tube. A similar argument holds for the colder sections of the pumping line as well.

As a final example, consider the problem not of emptying the 1 K pot, but of keeping it full. On most refrigerators this is accomplished by extending a fill line (often vacuum jacketed) from the 4 K plate down along the side of the vacuum can. A length of capillary tubing between the fill line and the 1 K pot provides the impedance over which the liquid pressure drops from an atmosphere to a few torr. The trick is to design this impedance so the pot neither overfills nor runs dry.

For definiteness, assume that we have a 1 K pot with a boil-off rate of 200 cc/hr ($dV/dt = 3.3$ cm^3/sec). Since the flow is clearly viscous, it will be fairly well described by Equation 2.9. Since the density of the liquid varies only weakly with pressure, the throughput is just given by $Q = P_A(dV/dt)$, where P_A is atmospheric pressure. Approximating $(P_1^2 - P_2^2)$ as P_A^2, and taking the viscosity of the ^{4}He normal liquid to be about 35 micropoise gives $a^4/l \simeq 6 \times 10^{-10}$cm^3. If one uses .025 cm (.010 inch) I.D. capillary, then the impedance should be about 43 cm long.

Once you've made the impedance, you might want to test it at room temperature. This can be easily done by flowing helium gas in one end and attaching the other to a rotary pump. Putting pressure gauges in the obvious places will tell you P_1 and P_2. You can measure Q by running a rubber hose from the outlet of the pump to a basin full of water with an overturned beaker in it. The time it takes to expel the water from the beaker gives a direct measure of the throughput.

Of course, if you really want to get fancy, you can use a needle valve as the impedance, and run a control rod up through the top plate of the cryostat. This nifty feature is standard on Oxford Instruments' refrigerators, and allows you to close off the fill line completely and "one shot" the pot (useful when condensing in the mash).

So much for examples. There are several rather nice reference books that contain relevant information on gas flow and pump designs. Kennard's *Kinetic Theory of Gases* (McGraw-Hill, 1938) is a little old fashioned, but very complete. *Scientific Foundations of Vacuum Technique* by Saul Dushman (John Wiley and Sons, 1962) is an extremely useful text. It contains a brief outline of kinetic theory, but rather extensive discussions of transition flow, flow through short tubes and orifices, as well as sections on pumps, manometers, diffusion of gases through metals, etc. *Vacuum Technology* by Alexander Roth (North-Holland, 1976) covers much the same ground, but is somewhat more mathmatical in approach. When it comes to actually purchasing a pump, however, there is no substitute for long hours spent poring over catalogs and technical data, all of which can be obtained directly from the manufacturers:

- Alcatel Vacuum Products (rotary and Roots pumps, the local choice for sealed pumps), 40 Pond Park Road, South Shore Park, Hingham, MA 02043, 617-749-8710.
- Balston, Inc. (oil filters), 703 Massachusetts Avenue, P.O. Box C, Lexington, MA 02173, 617-861-7240.
- Edwards High Vacuum, Inc. (rotary, Roots, and diffusion pumps, the local choice for Roots blowers and small rotary pumps), 3279 Grand Island Boulevard, Grand Island, NY 14072, 716-773-7552.
- Leybold-Heraeus (rotary and Roots pumps), 5700 Mellon Road, Export, PA 15632, 412-327-5700.
- Varian Associates, Inc. (rotary and diffusion pumps), 645 Clyde Avenue, Mountainview, CA 94043, 415-964-2790.

2.5   Magnetic Cooling

by Henry E. Fischer

All cooling processes involve the idea of a refrigerant which is in thermal contact with the system to be cooled, and whose temperature can be lowered by appropriately changing the relevant thermodynamic parameters. The usual method is to take advantage of the entropy or enthalpy being a function not only of the temperature, but also of another parameter which may be changed externally (with the entropy or enthalpy held constant) to produce cooling in the refrigerant. This other parameter can be called the manipulable cooling parameter (mcp). Some well known examples are:

- Isentropic (reversible adiabatic) expansion of a gas. Here entropy is a function of temperature and volume, and an adiabatic increase in volume produces a decrease in temperature.
- Isenthalpic expansion (throttling) of a gas. Here an isenthalpic decrease in pressure causes a decrease in temperature.
- Adiabatic demagnetization of a paramagnet. Here the manipulable cooling parameter is the applied magnetic field H. A decrease in H at constant entropy produces a decrease in T.

The examples above actually involve two steps, so that we end up with the mcp at some nice livable value. First the mcp is varied at constant temperature (in contact with a heat bath) so as to remove entropy (or enthalpy) from the refrigerant. Then the mcp is varied in the other direction with the entropy (enthalpy) constant so as to produce cooling. This process is represented most conveniently on an entropy versus temperature (TS) diagram, as in Figure 2.24.

Once a gas has been liquified by using either of the first two processes above, its temperature can be further reduced by pumping on its vapor. This method is easy to do and has been known for a long time, but it is only effective for temperatures above about 1 K with liquid ^{4}He and .3 K with liquid ^{3}He. In order to get below that (i.e., before dilution refrigerators were available), it was necessary to find some system which remains partially disordered at temperatures of about 1 K, so that we can control the entropy by changing a suitable mcp. The idea of using the assembly of magnetic dipoles in a substance that remains paramagnetic at temperatures of about 1 K was conceived of independently by Debye in 1926 and Giauque in 1927.

2.5.1   Why it Works

For a paramagnetic atom or ion of total angular momentum J, its magnetic moment may point in any of $2J+1$ possible orientations with respect to the local magnetic field. The probability distribution of these $2J+1$ states is proportional

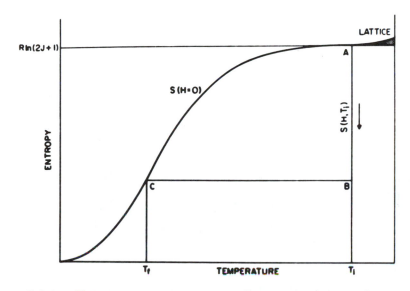

Figure 2.24. *Entropy versus temperature for magnetic cooling. [Hudson, R. P. Principles & Applications of Magnetic Cooling, (North Holland, Amsterdam), 1972.]*

to the Boltzmann factor

$$e^{-mU/kT} \quad \text{for} \quad U = g\beta B \qquad (2.29)$$

where U is the energy difference between adjacent levels, m is the level number, g is the Lande splitting factor, β is the Bohr magneton, and B is the local magnetic field. For $U \ll kT$ (i.e. B is small), the $2J+1$ levels are all about equally populated. But if a large external field B_i is applied at constant temperature T_i, then the distribution shifts in accord with Equation 2.29. Hence the system has become partially ordered —that is, its entropy has been reduced. Now if the field B_i is reduced to B_f isentropically, the degree of order must remain constant, and so the two distributions must be equal. Therefore,

$$\frac{g\beta B_i}{kT_i} = \frac{g\beta B_f}{kT_f} \quad \text{or} \quad T_f = B_f T_i/B_i. \qquad (2.30)$$

The process is depicted graphically in Figure 2.24. But B_f cannot be reduced all the way to zero, since it is limited by B_{int} —the field due to the mutual interaction of the dipoles and other interactions, such as that between the magnetic moments of the nucleus and the electrons. Equation 2.30 can also be derived using simple thermodynamic relations, and in general we find that for a magnetic system at constant entropy

$$B/T = \text{constant} \quad \text{or} \quad S(B,T) = S(B/T). \qquad (2.31)$$

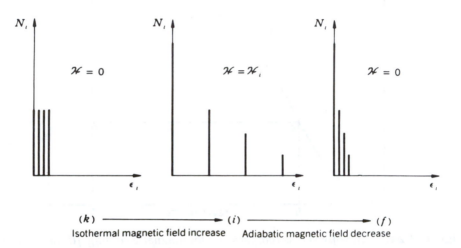

Figure 2.25.   *Energy level distributions during magnetic cooling.*
[Hudson, R. P. Principles & Applications of Magnetic Cooling,
(North Holland, Amsterdam), 1972.]

If the field B is broken up into an applied external field and a residual interaction field, then the relation is

$$(B_{app}^2 + B_{int}^2)/T^2 = \text{constant}. \qquad (2.32)$$

2.5.2   The Basic Idea

There are 3 types of magnetic cooling refrigerants:

• Paramagnetic salts
• Nuclear demagnetization materials (e.g., copper)
• Hyperfine enhanced (hfe) nuclear demagnetization materials (e.g., $PrNi_5$)

The choice of refrigerant depends on the temperature range desired, the amount of material (i.e., heat capacity) to be cooled to that temperature range, and the amount of time spent there. These three types of refrigerants and magnetic cooling systems will be discussed separately in later sections.

A schematic of the basic experimental setup for magnetic cooling is shown in Figure 2.26. First the thermal switch is turned on as the cooling material is magnetized in order to remove the heat of magnetization, and then the switch

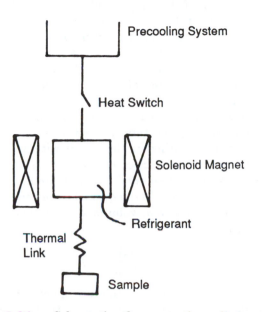

Figure 2.26. *Schematic of a magnetic cooling system.*

is turned off for demagnetization. The cooling material may be any of the three types of refrigerants listed above, and the precooling stage may be a liquid ^{4}He bath, a ^{3}He pot, a dilution refrigerator, or another demagnetization stage (Mueller *et al.* 1980). The magnet is superconducting, and these can be either purchased or constructed to provide the desired maximum field strength (up to about 9 tesla) and cross sectional area.

Magnetic cooling systems are usually operated as single shot devices, where the demagnetization is carried out to almost zero field, and then the sample and refrigerant are allowed to warm back up through heat leaks. Or, the sample may be temperature controlled by a small heater, using the refrigerant as a heat sink. It is also possible to start out with an initially larger field, reaching the desired temperature by partial demagnetization, and then to temperature control by slowly demagnetizing just enough to counteract the heat leaks. The off balance signal of the thermometry bridge can be fed into an active integrator, which can then give a ramp voltage to the programmable power supply controlling the magnet current. The advantage of this method is that no extra heat is being put into the sample. However, the magnet's persistent current switch must be kept open so that its current can be controlled by the power supply, and the necessary heat applied to the switch considerably increases bath boil off.

If one desires to spend an indefinite amount of time at some low temperature (such is almost always the case) and a dilution refrigerator is not suitable, it is possible to use a cycling demagnetization system. Part of the cycle then involves the magnetization of the sample and removal of the heat of magnetization. During the cold part of the cycle the temperature can be held constant by temperature controlled demagnetization, until the field is reduced to zero and the cycle repeats. The fraction of "cold time" is the duty cycle. Castles (1985) describes a system for keeping three infrared bolometers at .1 K with a 90% duty cycle, using the paramagnetic salt Ferric Ammonium Sulfate and a space pumped helium bath.

The rate of demagnetization must be fast enough so the heat leak doesn't warm things up too much by the time you get down there, and slow enough to minimize eddy current heating in the metal within the refrigerant. The paramagnetic salts are non-metallic, of course, but usually include embedded copper wires or plates to aid in heat conduction. A simple calculation from Faraday's Law gives the power dissipation due to eddy currents for a metal cylinder (axis parallel to B) of resistivity ρ, volume V, and cross sectional area (area perpendicular to B) A_p as

$$P_{\text{eddy}} = \frac{A_p V}{8\pi\rho}(dB/dt)^2. \tag{2.33}$$

Hence for a given demagnetization rate, the rate of heat production in the refrigerant is proportional to its cross sectional area. Note that this area is the area of a continuous piece of metal, and so if the metal is in the form of thin insulated wires parallel to the magnetic field, it will have less eddy current heating than a solid piece of metal of the same volume and A_p. The general formula is

$$P_{\text{eddy}} = \frac{1}{4\rho}(dB/dT)^2 \int dx \int_{A_p} r^3 dr\, d\theta \tag{2.34}$$

where x is the direction of the magnetic field, and the integration over A_p is done as a sum of integrations over continuous pieces of metal. Another geometry appropriate for paramagnetic salts is to alternate layers of salt disks and copper disks such that the lines of B lie in the disks, and to have the Cu disks connected together on one side.

2.5.3 Heat Switches

The two main types of heat (or thermal) switches are superconducting heat switches and gas gap heat switches. Superconducting heat switches work well for various temperature ranges below 1 K, depending on the superconductor. A superconducting heat switch relies on the fact that only phonons contribute to thermal conductivity in a superconductor (not electrons), and their contribution goes as T^3 at low temperatures. So the switch is off as long as it is superconducting, and can be turned on by the application of a magnetic field B, where $B > B_c$. The switching ratio is given by

$$\frac{K_n}{K_s} \propto \frac{\gamma T}{AT^3} = \frac{C}{T^2} \quad \text{typically} \quad C \approx 100K^2 \tag{2.35}$$

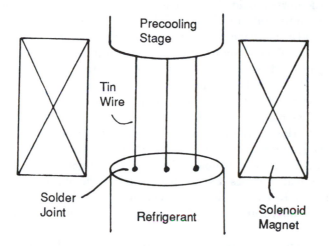

Figure 2.27. *A superconducting tin heat switch.*

where C is a parameter which depends on the superconducting material's residual (normal state) resistance, Debye temperature, and its average crystallite grain size (White 1979, pp. 152–153). Purer materials in general have better switching ratios, because of a lower residual resistance. Note that since a superconducting heat switch is off when the magnetic field is off, it would be possible to simply place the switch in the fringes of the main magnetic field, so that it stops conducting heat after demagnetization, but usually it is better to have the extra control offered by a separate little superconducting magnet just for the heat switch. A schematic of a superconducting heat switch appears in Figure 2.27. The heat switch's magnet should be thermally anchored to the precooling stage, of course.

Two popular materials for superconducting heat switches are tin and aluminum. For temperatures below 0.1 K, switching ratios of 10^3 or higher are possible. Tin heat switches have the advantage of being relatively easy to construct, often consisting of a few tin wires of millimeter size which are soldered to metal surfaces of the precooling stage and the cooling material using appropriate flux. The superconducting transition temperature for tin is about 3.7 K, and the critical field (for $T \ll T_c$) is 300 gauss. A disadvantage of using tin is that it may undergo a phase transition during cool down, such that at temperatures even warmer than that of liquid nitrogen it turns into "gray tin," a brittle substance which may or may not fall apart before it is completely cooled. Fortunately, ordinary white tin is a robust metastable phase, so that unless there are seeds of the grey tin phase in contact with the heat switch, or the apparatus is kept near the transition temperature of 240 K for long periods of time, the transformation is unlikely to occur.

Aluminum heat switches do not have this problem, but they are harder to make, since the normal oxide layer on aluminum prevents good thermal contact. Pieces of aluminum foil are often used (since the narrow cross section can be aligned parallel to the magnetic field to cut down on eddy current heating when the field is turned off) and they can be gold electroplated on both ends, and then secured by gold electroplated clamps mounted on the precooling stage and the cooling material. It can be a pain, but it will last much longer than a tin heat switch. Two other advantages of using aluminum are that it requires a smaller critical field, 100 gauss, and it has a higher Debye temperature, so that the phonon contribution in the off state is smaller. However, T_c for aluminum is a factor of three smaller than for tin, being about 1.2 K.

Equation 2.35 gives an estimate of the limiting functional temperature of superconducting heat switches. Even a heat switch made of lead, which remains superconducting up to 7 K, would have a switching ratio of only about 40 at 1.5 K. And so above temperatures of about 1 K, superconducting heat switches become ineffective.

For higher temperatures, such as those of a pumped helium bath (used as a precooling stage for a paramagnetic salt), a gas gap heat switch works well. For a gas filled gap of width d between two plates, and with a pressure P_H such that the mean free path $L \ll d$, the thermal conductivity of the gas is independent of the pressure, and varies almost linearly with temperature. But for $L \gg d$, the thermal conductivity goes as

$$K \propto P(1/MT)^{1/2} \qquad M = \text{the molecular mass.} \qquad (2.36)$$

Hence for a given operating temperature, the thermal conductance can be varied from a maximum value to a lower value which scales as the pressure or density. Figure 2.28 shows a basic schematic of a gas heat switch. The exchange gas is usually helium or hydrogen, and the pump consists of an adsorbing medium such as charcoal or zeolite. By heating the adsorber from 4 K to say 20 K, the adsorbed gas is driven off into the gap. For $d > 10^{-2}$mm a pressure of 1 torr is sufficient to reach the maximum conductance at 4 K, and the equivalent volume of this amount of gas can easily be stored in a small volume of adsorbing medium. Allowing the adsorber to cool to bath temperature reduces the pressure to the order of a millitorr or less, for which the thermal conductance of the gas is often smaller than that due to radiation and the isolation supports (usually stainless steel). In particular, by using helium adsorbed on zeolite at 1.5 K, with heating to 20 K, Castles (1985) has obtained a pressure range of 10^{-6} to 10^{-1} torr, and a switching ratio of 10,000 for his gas heat switch.

2.5.4   Adiabatic Demagnetization of Paramagnetic Salts

The paramagnetic material is in the form of a salt of a magnetic ion, which can be used either as a confined powder, or can be pressed in order to form a compact "pill." Usually some vacuum grease or epoxy is mixed with the powder to make the pill stronger, and also some foil or copper wires are mixed in to increase the thermal conductivity (but keeping in mind eddy current heating).

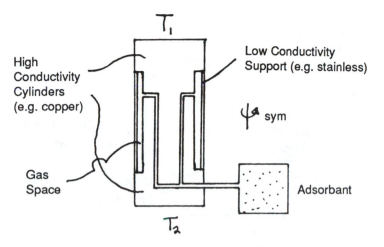

Figure 2.28.   *A gas heat switch.*

White (1979, p. 242) gives values for thermal boundary resistances between various salts and metals. Suitable magnetic ions are found in the transition and rare earth elements, where the atoms have a residual magnetic moment due to incompletely filled electron shells—the 3d shell in the case of the transition elements (e.g., Mn, Fe, Co, Ni) and the 4f for the rare earths (Ce, Gd, Nd).

Looking back at Figure 2.24, we see that for a salt to function well as a refrigerant, it should have the following properties:

- The zero field entropy S should be about $nR\ln(2J+1)$ at the starting temperature T_i, such that $S_{\text{magnetic}} \gg S_{\text{lattice}}$.
- The fraction of magnetic entropy that can be removed at T_i should be large, and can be measured from the relation $\Delta Q_{\text{magnetic}} = T_i \Delta S_{\text{magnetic}}$.
- The zero (applied) field heat capacity of the salt in the final temperature range should be large.

The first requirement is satisfied by most salts at 1 K, except for cerium magnesium nitrate (CMN), whose lattice entropy only becomes small enough for $T_i \leq$ 0.3 K. Hence magnetic cooling for CMN requires the precool stage to be a ^{3}He pot or a crude dilution refrigerator, whereas a pumped ^{4}He bath is sufficient for the others. The second requirement is satisfied for the initial field $B_i = 0.5$ or 1.0 telsa, easily produced by a small superconducting magnet.

The analysis of the third requirement is a little more complicated, and what follows is an attempt to present it in simple terms. The magnetic specific heat C_m for a given magnetic field B and temperature T is best thought of as a function of the ratio B/T. If the magnetic field is too large, that is $g\beta B \gg kT$, then almost all of the dipoles are in the ground state. Since it is unlikely

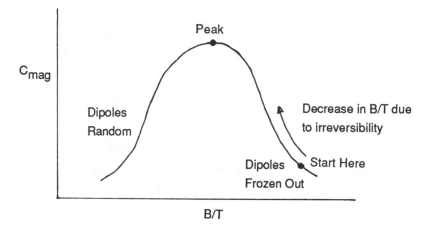

Figure 2.29. *Magnetic specific heat as a function of B/T.*

that a dipole will acquire enough thermal energy to make a transition to a higher state, the dipoles are effectively "frozen out," and can't make much of a contribution to the heat capacity. On the other hand, if $g\beta B \ll kT$, then the energy level distribution is uniform and saturated—there are no available higher energy states, and again C_m is small. Between these two extremes C_m has a maximum. This is depicted qualitatively in Figure 2.29. Note that for zero applied field and a constant B_{int}, this graph is similar to that of a Schottky specific heat hump for a two level system (taking the reciprocal of the horizontal axis).

Ideally, if our initial condition B_i/T_i gives us a maximum in C_m, then our specific heat should remain at that maximum value as we demagnetize adiabatically, since $(B_{\text{app}}^2 + B_{\text{int}}^2)/T^2$ is constant at constant entropy (note that $B_i \gg B_{\text{int}}$). However, several processes are at work to increase the entropy of our refrigerant as it cools down.

- Eddy current heating
- External heat leaks
- Heat flow from the sample to be cooled

Since a larger B/T means more order and less entropy, then an increase in entropy means that B/T decreases a little on cooling, and this pushes us to the left in Figure 2.29. Hence it is best to start out with a large enough value of B_i/T_i so that we are at the right of the hump, and the dipoles are basically frozen out. Then after demagnetization, as the salt begins to warm up, it passes through the maximum in heat capacity. And so in the case of demagnetization of paramagnetic salts, it does not make sense to demagnetize only partially in order to get a larger heat capacity. However this is not the case in nuclear demagnetization, as will be shown later.

It's important to emphasize that the relevant parameter for cooling capacity in magnetic cooling is the amount of entropy, not heat, that can be removed during magnetization. After cooling down, the refrigerant will readsorb the same amount of entropy from the sample and external sources as it warms up, in addition to the entropy from irreversibility. The amount of heat absorbed by the refrigerant during warm up at a constant B is equal to $Q = \int T(S)dS$, which is just the area to the left of the $B = $ const curves on a TS diagram. Then for a given initial entropy reduction, the refrigerant has the trade off of either absorbing a lot of heat from the sample at a high temperature (if the demagnetization is not carried out all the way, or if temperature controlling is done by demagnetization) or less heat at a lower temperature. In both cases the refrigerant passes through the same maximum in heat capacity, it is just that the width of the hump in degrees Kelvin is larger for larger B, so that $\Delta Q = \int C_H(T)dT$ is larger.

The entropy increase from irreversibility points out another motivation for demagnetizing slowly. For a given small amount of heat flowing from the sample to the refrigerant, the total change in the entropy of the salt-sample system is given by

$$\Delta S = \Delta Q(1/T_{\text{salt}} - 1/T_{\text{sample}}). \qquad (2.37)$$

Therefore it is desirable to keep the temperature difference between the two as small as possible by demagnetizing slowly, so as to conserve entropy. Of course this is easier if $C_{\text{salt}} \gg C_{\text{sample}}$, and if they are thermally well coupled to each other.

And so, in general, efficient magnetic cooling means taking out as large an amount of entropy as possible during magnetization, and then keeping things as reversible as possible during demagnetization and refrigeration of the sample. Parpia *et al.* (1984), describes an experimental method for finding the optimum demagnetization rate for his PrNi$_5$ refrigerator to minimize irreversibility. Since a steady decrease in B/T during demagnetization reflects an increase in entropy, the degree of irreversibility can be parameterized by the "zero temperature" magnetic field intercept on a plot of T vs B.

Another consideration in selecting a paramagnetic salt is the desired final temperature T_f attainable. As shown by Equation 2.32, the low temperature limit arises because of the internal magnetic field B_{int}, and is approximately equal to the Neel temperature, at which spontaneous magnetic ordering occurs. To achieve a small B_{int} and hence a small T_f, the salts are often significantly hydrated, to increase the dipole-dipole spacing and hence reduce the field due to the dipole-dipole interaction, as long as the lattice entropy does not become comparable to the magnetic entropy. An entropy versus temperature diagram (in zero applied field) for some of the common salts used appears in Figure 2.30. The limiting temperature is about where the curves start to bottom out. Table 2.4 lists some relevant parameters of different salts, and was taken from a more complete listing in Hudson (1972, pp. 110–111). The constant b describes the temperature variation of C_m, but only for the high temperature tail of the

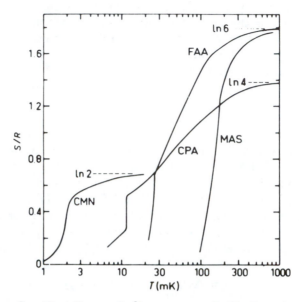

Figure 2.30. S vs T at $B_{app} = 0$. [Lounasmaa, O. V., Experimental Principles
& Methods Below 1K, (Academic Press, NY), 1974.]

Table 2.4 Properties of Common Paramagnetic Salts [White, GK,
Experimental Techniques in Low Temperature Physics (Oxford University
Press, Oxford), 1979.]

Salt		$T_1(K)$	$T_N(K)$	$\left(\dfrac{c}{\text{emu/g-ion}}\right)$	b
$Cr(NH_3CH_3)(SO_4)_2.12H_2O$	CMA	0·21	0·016	1·83	0·018
$CrK(SO_4)_2.12H_2O$	CPA	0·20	0·009	1·84	0·017
$CuK_2(SO_4)_2.6H_2O$		0·099	0·05	0·445	$5·7 \times 10^{-4}$
$FeNH_4(SO_4)_2.12H_2O$	FAA	0·090	0·026	4·39	0·013
$Gd_2(SO_4)_3.8H_2O$	GS	0·28	0·182	7·80	0·32, 0·37
$Mn(NH_4)_2.(SO_4)_2.6H_2O$	MAS	0·165	0·173	4·38	0·032
$Ce_2Mg_3(NO_3)_{12}.24H_2O$	CMN	~0·01	0·001₅	$C_{\parallel} \sim 0$ $C_{\perp} = 0·317$	$6·6 \times 10^{-6}$

Schottky peak, i.e., for $T_f \gg T_N$. The Curie constant gives an estimate of the
relative sizes of the peaks in heat capacity. A plot of the zero field specific heats
of some of these salts appears below. Note that the peaks occur at temperatures
above T_N.

One of the more popular salts for "low" temperature work has been cerium
magnesium nitrate (CMN), and a plot of S vs T for several B appears in Fig-

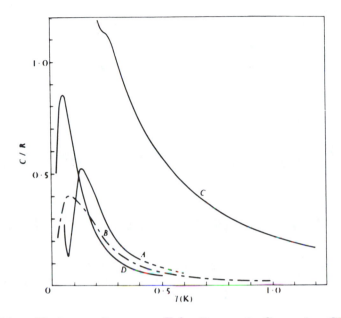

Figure 2.31. *Heat capacity versus T for $B_{app} = 0$. Curve A – CMA, Curve B – CPA, Curve C – GS, Curve D – FAA. [Wright, G. K., 1979 Experimental Techniques in Low-Temperature Physics (Oxford University Press, Oxford.]*

However, if one does not wish to get that cold, say only about .1 K, then CMN is a bad choice, because its peak in heat capacity (for zero field) is much lower than that. A better choice for complete demagnetization would be chromium potassium alum (CPA) or ferric ammonium alum (FAA—also called ferric ammonium sulphate), which have their peak in C around 0.1 K as shown in Figure 2.31.

2.5.5 Nuclear Demagnetization

The idea of using nuclear magnetic moments instead of electronic ones for adiabatic demagnetization was first proposed in 1934. Nuclear demagnetization experiments fall into two categories: either the cooled nuclear spin system itself is of interest, or the nuclear spins are used to refrigerate another system (e.g., the electrons or another attached sample such as ^{3}He). The first is called simply nuclear cooling, and the second, more common application, nuclear refrigeration. The advent of efficient off-the-shelf dilution refrigerators which can keep a constant temperature below 5 mK has made demagnetization of paramagnetic salts obsolete for work in this temperature range, except where a dilution refrigerator would be mechanically inappropriate, such as in a space shuttle launch (Castles 1985).

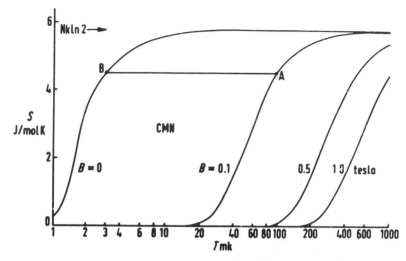

Figure 2.32. *Entropy of CMN along several lines of constant B. [Betts, D. S., Refrigeration & Thermometry Below 1K, (Sussex Univ. Press, Brighton), 1976.]*

Recall that the limiting temperature which can be reached by demagnetizing a system of magnetic dipoles is limited by the residual magnetic field, which is in turn due to the interaction between the dipoles. Since nuclear spins have magnetic moments which are 2000 times smaller than those of electronic spins (i.e., the nuclear magneton compared to the Bohr magneton), then the dipole-dipole interactions are reduced, and temperatures can be reached which are about 1000 times smaller than those reachable with electronic spins. Spontaneous ordering of the nuclear spins occurs around 1 μK, which is about the limiting temperature. But this is only the temperature of the nuclear spins; in order for the nuclear refrigerant to cool anything else, the electronic system of the refrigerant must be cooled by the nuclear spins.

For the operation of nuclear refrigerators, the magnetic entropy can be approximated quite well by the expression

$$S = nR\ln(2I+1) - \frac{n\lambda(B_{\text{app}}^2 + B_{\text{int}}^2)}{2\mu_0 T^2} \tag{2.38}$$

and the heat capacity by

$$C = \frac{n\lambda(B_{\text{app}}^2 + B_{\text{int}}^2)}{\mu_0 T^2} \tag{2.39}$$

where

$$\lambda = N_0 I(I+1)\mu_0\mu_n^2 g_n^2/3k \tag{2.40}$$

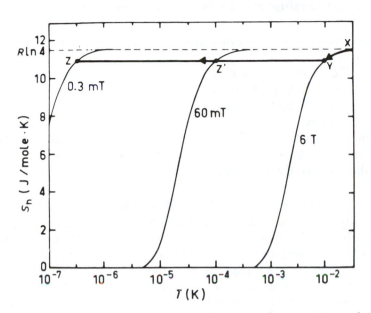

Figure 2.33. *Nuclear demagnetization of copper ($B_{int} = 0.3$ mT). [Lounas-maa, O. V., Exp. Princ. & Methods, (Academic Press, NY), 1974.]*

is the molar Curie constant, n the number of moles, I the total nuclear angular momentum number, μ_n the nuclear magneton, g_n the nuclear Lande factor, k Boltzmann's constant, and N_0 Avogadro's number. The same formula is valid for paramagnetic systems (for small values of B/T), replacing I with J and μ_n with β. Since μ_n is about 2000 times smaller than β, then in order to get the same percentage of entropy reduction, we must have starting values of B_i/T_i which are 2000 times higher than those for the paramagnetic systems. Using copper as an example, for $T_i = 10$ mK, and $B_i = 6$ T, the entropy reduction is only 5% (see Figure 2.33), less than that for paramagnetic salts with initial conditions of $T_i = 1$ K and $B_i = 1$ T, but still good enough for successful experiments. These starting conditions can be obtained with a dilution refrigerator for a precooling stage, and a high field superconducting magnet.

After the temperature of the nuclear spins has been reduced, it is necessary to cool the conduction electrons, since they are the medium of heat transfer from the refrigerant to the sample (the thermal conductivity of the lattice goes as T^3 and is negligible). The rate at which equilibrium is established between nuclear spins and conduction electrons is proportional to $1/\tau_1$, where τ_1 is the nuclear spin-lattice relaxation time, and is in turn proportional to the Korringa constant κ.

A good nuclear refrigerant should have the following properties:

- a large Curie constant λ in order to get a large entropy reduction and a large heat capacity.
- A small Korringa constant κ, in order to get good refrigeration of the conduction electrons from the nuclear spins.
- High thermal conductivity, so that there is thermal equilibrium over macroscopic distances.
- The final magnetic field B_f must be greater than the critical field B_c of the refrigerant, because in the superconducting phase the electrons cannot contribute to the thermal conductivity.

It is also desirable to have a low electrical conductivity $\sigma = 1/\rho$, in order to minimize eddy current heating, but the Wiedemann- Franz Law prevents us from having that in addition to a large thermal conductivity. Eddy current heating is minimized by using very thin insulated wires for the refrigerant.

Table 3.5 in section 3.2 (Special Materials) lists some properties of some metals which can be used. An important point to note is that in order to obtain the lowest conduction electron temperature with a nuclear refrigerant, the demagnetization should not be carried out to $B_f = 0$, but should stop at some higher, optimal value $B_f(\text{opt})$. This comes about because of the very small B_{int} of these refrigerants. The heat capacity of the nuclear spins is given by Equation 2.39, and would remain constant during demagnetization if it was reversible, but because of external heat leaks into the conduction electrons, which are being cooled by the spins, T_f is going to be appreciably finite, and hence the heat capacity is going to be very small unless $B_f > B_{\text{int}}$. The faster that the spins are able to cool the electrons during demagnetization (proportional to $1/\kappa$), the greater the reversibility since $T_{elec} - T_{spin}$ is smaller (cf. Equation 2.37), and the lower $B_f(\text{opt})$ can be. Lounasmaa (1974) gives a detailed derivation, and finds that for $B_f \gg B_{\text{int}}$, an approximate formula is

$$B_f(\text{opt}) = (\mu_0 \kappa \dot{Q}/n\lambda) \tag{2.41}$$

where $\dot{Q}$ is the external heat leak into the conduction electrons.

2.5.6   Hyperfine Enhanced Nuclear Refrigeration

The interaction of the magnetic moments of the nucleus with the electrons is called the hyperfine interaction. Normally the magnetic field at the nucleus produced by the electrons is very small, but there is an exception. In singlet ground state ions, with high Van Vleck susceptibilities, large hyperfine fields can be induced by moderate external fields. What happens is that in zero applied field the ions are in nonmagnetic singlet ground states, but in an applied field the wave function of the ground state changes and acquires a magnetic moment. This increases the net magnetic field at the nucleus by an enhancement factor α,

$$\alpha = 1 + K \tag{2.42}$$

where K is the Knight shift. Hence we can use the same formulas that we have been using for nuclear refrigeration, replacing B with αB. It is common for α to be on the order of 20 or 100. What we have then is a nuclear refrigerant with an apparent nuclear magnetic moment that is one or two magnitudes larger than the nuclear magneton, and one or two magnitudes smaller than the Bohr magneton. In order to get big α's, we need a small energy separation between the singlet ground state and the first excited states, which leads to a high Van Vleck susceptibility and a large hyperfine coupling constant. The best candidates are rare earth ions with integral values of the total electronic angular momentum J. Pr^{3+} and Tm^{3+} in particular have the lowest spin angular momentum $S = 1$, which reduces exchange interactions and allows lower temperatures before spontaneous ordering of the electronic dipoles sets in. Typical limiting temperatures for hyperfine enhanced (hfe) cooling materials are on the order of a couple millikelvin.

The motivation behind using hfe nuclear magnetic cooling is to combine the most useful features of insulating paramagnetic salts and metallic nuclear refrigerants. The enhanced B field at the nucleus makes possible a higher starting temperature T_i, a smaller starting field B_i, and also a larger percentage of entropy reduction than in ordinary nuclear demagnetization, so that a smaller amount of refrigerant is needed. The temperatures attainable with hfe cooling materials are lower than those of most paramagnetic salts, and since they are metallic, their thermal conductivity is greater, and they can be soldered to for good thermal contact. So far the best hfe refrigerant found is praseodymium nickel ($PrNi_5$).

As alluded to earlier, the effective cooling power of a demagnetization stage, and therefore its lowest attainable temperature for a given load, depends on how quickly heat can be absorbed by the refrigerant. Because the thermal conductivity decreases with decreasing temperature, the geometry of the refrigerant is an important consideration. Copper, which is used for the lowest temperatures has about the best thermal conductivity possible, so there is little you can or need to do for improving its thermal performance. For $PrNi_5$, the thermal conductivity can vary by almost an order of magnitude due to lattice imperfections. A good sample will have a RRR of over 15. Even the best $PrNi_5$ will not work as a single solid piece of refrigerant—it must be thermally anchored to a very good conductor so that there is at most a short thermal path to any part of the refrigerant. A good discussion of one way this is done can be found in Mueller *et al.* (1980). The basic idea is to form the $PrNi_5$ into rods of diameter on the order of 6 mm. These rods are then thermally anchored along their entire length to high purity copper, which is anchored to the rest of the $PrNi_5$ and the stage. The $PrNi_5$ and copper can be joined together with various solders; an advantage of using cadmium is that it has a relatively low critical field, $B_c = 30$ gauss, so that nearly

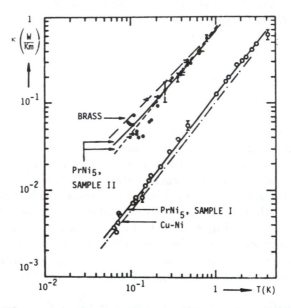

Figure 2.34.   *Thermal conductivity of two samples of PrNi₅. [Meiyer, H. C.,*
G. J. C. Bots and H. Postma, Physica 107b, 607, 1981.]

complete demagnetization is possible without it going superconducting. A more
recent paper by Meijer *et al.* (1981) suggests that the PrNi₅ can be nicely electroplated with copper, and then presumably annealed. The copper can then be
machined. This method may not produce quite as pure a copper, so some hybrid
technique may be optimal. As another consideration, it has been observed that
the PrNi₅ tends to crack upon thermal cycling, making it even more important
to thermally anchor the refrigerant along its entire length. What happens is that
the PrNi₅ absorbs the helium exchange gas during the initial cool down, and
then it cracks apart when the stage is warmed back up. Neon is a better bet for
an exchange gas. Here again is a reason for modular design, because the PrNi₅
may need to be repaired or replaced periodically. Figure 2.34 shows the thermal
conductivity of two samples of PrNi₅.

A group at Texas A&M (Parpia *et al.* 1984) has developed a nuclear refrigerator which consists of a single hybrid stage of 0.27 moles PrNi₅ and 2.8 moles
copper to cool a sample of ^{3}He to 350 μK. The copper is in the form of a thermal
link of wires to the main heat exchanger containing the ^{3}He sample. The whole
nuclear stage was precooled by a commercial dilution refrigerator to a temperature of 8 mK in an applied field of 7.8 T, and the final field $B_f = 300$ gauss.
Temperatures below 1.1 mK were maintained for periods exceeding two weeks.
A schematic of the cryostat appears in Figure 2.35.

The advantages of using PrNi₅ over copper alone for this temperature range

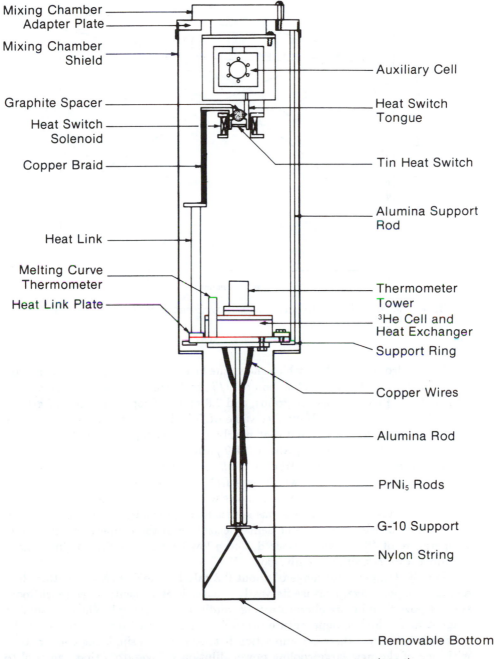

Figure 2.35. *PrNi$_5$ refrigerator of Parpia et al. (1984).*

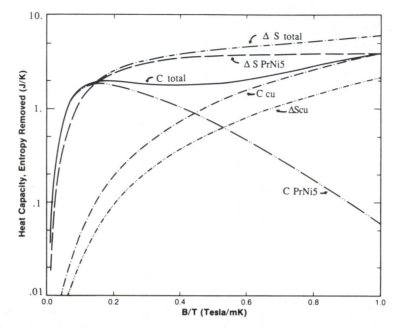

Figure 2.36.   *Heat capacity and removed entropy of copper and PrNi₅.*

can be seen from Figure 2.36 which shows the heat capacity and removed entropy of the copper and PrNi5 as a function of B/T for Parpia's refrigerator. Note that 0.27 moles of PrNi5 means 116 grams, and 2.8 moles of copper means 177 grams. It can be seen that a greatly smaller value of B/T is needed to achieve the peak in heat capactiy for PrNi5, and that nearly all the entropy of the PrNi5 can be removed with B/T being only about 0.1 T/mK.

In Jülich, West Germany, Mueller *et al.* (1980) have built a two stage nuclear demagnetization refrigerator which uses PrNi5 in the first stage and copper in the second. For the PrNi5 stage, 4.29 moles were cooled to 0.19 mK starting from $T_i = 25$ mK and $B_i = 6$ T. This was sufficient to precool 10 moles of copper to 5 mK under a field of 8 T. After demagnetization of the copper, and electronic temperature of 48 μK was measured, and a low heat leak of 0.1 nW/mole gave a temperature rise of only 2 μK/day.

For the temperature range of about 0.3 mK to 2 mK, PrNi5 is better than a single copper stage, because its vastly greater heat capacity gives much more staying power, and more liberal starting conditions are possible—hence a smaller magnet and a higher temperature precool. But for lower temperatures, in the microkelvin regime, there is competition between using a single big copper stage with one of the new large cooling power dilution refrigerators that can cool to a temperature of 5 mK, or using a double stage system similiar to the Jülich group, where an upper stage of PrNi5 is used to precool the copper stage. The advantages of the former are that only one magnet is required, so that the design is simpler, and that a bigger copper stage may be used, since the precooling is done by a dilution refrigerator. But the disadvantage is that the precooling of the copper is done at 2 mK, where the cooling power of the fridge is smaller, and where the Kapitza resistance between the liquid helium in the mixing chamber of the dilution refrigerator and the copper is much greater than it would be for precooling PrNi$_5$ at 10 mK (it goes as $1/T^3$ in that temperature range). A heat exchanger can be used in the dilution refrigerator (see Section 3.5) but it may still take a while to cool all that copper. Of course, for the double stage design, the precooling of the copper can be done through a nice soldered connection to the PrNi$_5$.

Another cooling method worth mentioning is that of electric cooling, the electric analogue of cooling by adiabatic demagnetization. It employs the electrocaloric effect, where an electric field applied isothermally to a crystal (such as a doped alkali halide) produces a reduction in entropy. Hence a subsequent adiabatic decrease in the electric field leads to a decrease in temperature. See Taylor (1968), Radebaugh (1978), and Pohl *et al.* (1969) for theoretical and experimental details.

2.5.7   Recommended Reading

Books

- Betts, D. S., 1976, *Refrigeration and Thermometry Below 1K* (Sussex Univ. Press, Brighton).
- Hudson, R. P., 1972, *Principles and Application of Magnetic Cooling* (North Holland, Amsterdam).
- Lounasmaa, O. V., 1974, *Experimental Principles and Methods Below 1K* (Academic Press, NY).
- White, G. K., 1979, *Experimental Techniques in Low- Temperature Physics* (Oxford University Press, Oxford).

Preprints and Journal Articles

- Andres, K., "Hyperfine Enhanced Nuclear Magnetic Cooling," 1978, *Cryogenics* **18**, 8.
- Babcock, J., L. Kiely, T. Manley, and W. Wehmann 1979, *Phys. Rev. Lett.* **43**, 380.
- Castles, S., 1985, "Design of an Adiabatic Demagnetization Refrigerator for Studies in Astrophysics," (unpublished). Address: Cryogenics, Propulsion and Fluid Systems Branch, Space Technology Division, Goddard Space Flight Center, Greenbelt, Maryland.
- Kubota, Y., H. R. Folle, Ch. Buchal, R. M. Mueller, and F. Pobell, "Nuclear Magnetic Ordering in PrNi$_5$ at 0.4 mK," 1980, *Phys. Rev. Lett.* **45**, 1812.

- Meijer, H. C., G. J. C. Bots, and H. Postma, "The Thermal Conductivity of $PrNi_5$," 1981, *Physica* **107b**, 607.
- Mueller, R. M., Chr. Buchal, H. R. Folle, M. Kubota, and F. Pobell 1980, *Cryogenics* **20**, 395.
- Parpia, J. M., W. P. Kirk, P. S. Kobiela, T. L. Rhodes, Z. Olejniczak, and G. N. Parker, "Optimization Procedure for the Cooling of Liquid ^{3}He by Adiabatic Demagnetization of Praseodymium Nickel," 1984, *Texas A & M Technical Report*.
- Peshkov, V. P. and A. Ya. Parshin, "Superconducting Thermal Switches," 1965, *JETP* **21**, 258.
- Pohl, R. O., V. L. Taylor, and W. M. Goubau, "Electrocaloric Effect in Doped Alkali Halides," 1969, *Phys. Rev.* **178**, 431.
- Radebaugh, R., "Electrocaloric Refrigeration for the 4-20 K Temperature Range," 1978, *NBS Special Publication* **508**, 93.
- Taylor, V. L., 1968, "A Study of the Electrocaloric Effect in KCl:Li," thesis, Cornell University.
- Tward, E., "Gas Heat Switches," 1981, *NBS Special Publication* **607**, 178.

Chapter 3

Cryogenic Design Aids

Eric N. Smith, Jeffrey E. VanCleve, Roman Movshovich,
Robert S. Germain, and Eric T. Swartz

3.1 Recipes

by Eric N. Smith

This section will deal with various techniques that are useful in the construction of cryogenic apparatus. A substantial fraction of the space will be devoted to various methods of putting together the tubing, plates, chambers, etc. typically found in a cryostat, and the remainder to more specialized procedures such as sintering, electroplating, and heat treating of materials. As a general comment on procedures for making cryostats, there are two principles which should be followed whenever possible. First, methods used must have exceedingly high reliability. A cryostat may have several hundred joints which must all be vacuum tight. 99% reliability in each joint is clearly not good enough. Further, a highly modular construction method is very desirable. This permits easy repairs if problems should develop, and more readily permits modifications if the needs of an experiment change with time. It is often a temptation to cut corners on the modularity, because making lots of little flanges is more time consuming than just soldering a bunch of tubes all through the same plate. Also, many older pieces of apparatus in many labs provide poor examples. Ease of assembly repays much early machining time, and any necessary repairs will more than repay the rest.

Cryostats usually consist mainly of cans filled with vacuum, helium, or experimental apparatus connected to the rest of the world by thin-walled tubing and electrical wiring. The tubing is usually thin-walled stainless steel or cupronickel (70-30), chosen because of low thermal conductivity and high strength. When dealing with pressure differences on the order of 1 atmosphere to be supported by the tubing, a wall thickness of 1% of the tube diameter is safe against collapse. For larger pressure differences, or for larger tubes where weight, heat or space considerations may be more important, Figure 3.1, adapted from similar figures in Hoare, Jackson, and Kurti, Experimental Cryophysics, p.121 (Butterworths, London, 1961), may be used to give a more exact estimate. Many European

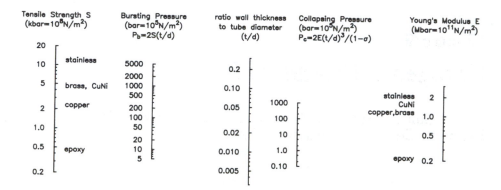

Figure 3.1.   *Nomogram for determining bursting pressure (left side) or collapsing pressure for tubes of various materials.*

cryostats are made with tubing which contains small corrugations every few inches of length, which provides very good stiffening against collapse, and permits much thinner walls. I have not found any US manufacturers of such tubing, however. G-10 fiberglass tubing is also often used for the necks of dewars, etc., and is a useful material available from a number of manufacturers. At higher temperatures (near room temp.), it permits the diffusion of helium, and should preferably have a thin lamination of .001" stainless foil attached if it is to separate helium from vacuum. Thin walled stainless tube is generally available. For most purposes 304 or stainless steel 316 alloys are most suitable, though 321 gives somewhat stronger and more corrosion-resistant welds. The 321 is unsuitable for hydrogen brazing (the titanium reacts with any water vapor present), and seems a bit harder to soft solder. Cupronickel tubing is easier to soft solder than stainless, and is much more flexible for small interconnecting tubing–we prefer to use it with an "as-drawn" temper. Two suppliers for larger quantities are Uniform Tubes, Inc., 7th Ave, Collegeville, PA 19426, and Superior Tubes, 1938 Germantown Ave., Norristown, PA 19426, but they have a substantial setup charge and several weeks delivery times. Until recently, for quantities less than a hundred feet or so you were better off with Oxford Instruments, 3A Alfred Circle, Bedford, MA 01730, or Biomagnetic Technology, Inc., 4174 Sorrento Valley Blvd., San Diego, CA 92121 (formerly SHE), each of whom kept a fair selection of sizes in stock. Unfortunately, this service has been disrupted and it can now be quite difficult to obtain small quantities of capillary tubing very quickly.

Another question is the length of tubes. Don't forget that tubes shrink by 3 to 4 mm per meter of length when they're cooled to helium temperature. That includes the tubes your dewar is made of, and on occasion you may want to put warm apparatus into a precooled dewar, so make the apparatus short enough to fit this worst case. Also anticipate this contraction if you are planning to make an optical cryostat, or your windows may end up in the wrong place. Often in the lower part of a cryostat, a pumping tube may double as one of the support posts. Try to arrange things so that the other support posts have similar thermal contractions, so that things don't bend off to the side, or solder joints receive unnecessary stresses.

3.1.1   Methods for Joining Tubes to Flanges, Plates, Other Tubes, etc.

Welding One of the most satisfactory methods for joining two pieces of stainless steel or two pieces of aluminum is welding in an inert gas atmosphere, commonly known as TIG welding or heliarc welding. The advantage is that the material is homogeneous, and less subject to stresses from differential thermal contraction. No corrosive fluxes are needed, and a skilled welder can produce highly reliable joints. However, particularly with aluminum, where much of the material gets quite hot during the welding process, distortion of the material may occur, and it is advisable to allow extra material for further machining after the weld if close tolerances are necessary. When a thin-walled tube is to be joined to a heavier plate, it helps to have a weld-relief groove (looking essentially like an O-ring groove) machined into the plate, typically 2-3 mm deep and of a comparable width. A shoulder with a width of 1-2 mm for tubes of wall thickness .25 to 1.5 mm should be left for actually forming the weld bead, so that the tube and flange melt together at a comparable rate. On thinner walls, .25 mm and down, it helps to also have a snugly fitting ring on the inside of a mm or two in thickness to help keep from melting the very thin wall too fast. Tolerances should be very tight (.001" on the diameter) to avoid having the two edges melt away from each other rather than smoothly joining (though a sufficiently good welder can manage to salvage some incredibly sloppy machining if his pride is sufficiently challenged). Pieces should be free of oil, cleaned with acetone or trichlor, allowed to dry, rinsed with alcohol and water, and again allowed to dry. Don't use Q-tips or Kimwipes for cleaning, as bits of fiber may catch on burrs and get involved in weld, causing pinholes as they disintegrate in the heat.

Under some conditions it is possible to weld dissimilar materials, such as copper to stainless steel. On such things, if the welder hasn't done it before, have him practice on pieces that don't take too much machining first. It is possible to weld copper, or silver, much as with aluminum. Structural copper, however, has quite a lot of impurities and is unlikely to result in a leaktight joint, even though it works fine in the even-dirtier aluminum. Alloys like brass with a high-vapor-pressure constituent can't be welded.

Vacuum Brazing and Brazing in a Reducing (H_2) Atmosphere This technique works very well on materials which can be raised to 800-1000° C in a vacuum. One can achieve high-strength joints with no flux residues to contend with. Leave .001-.002" clearance on the diameter. This leaves enough space for

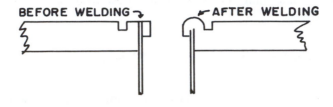

Figure 3.2. *Schematic cross section of a weld joint.*

Table 3.1.
Some Brazing Materials

stainless-stainless	NiOro	82% Au	melts at 950° C
stainless-copper		18% Ni	usually braze at 1000° C
	(GTE-WESGO, 477 Harbor Blvd., Belmont, CA 94002)		
copper-copper	BT	72% Ag	melts at 779° C
copper-silver		28% Cu	usually take up to 800° C
copper-cupronickel			(copper-silver eutectic)
	(Lucas Millhaupt, Inc., 560 S. Penn Ave., Cudahy, WI 53110)		

the alloy to enter, but is sufficiently tight so that capillary action holds the material in place rather than simply letting it all run through. There is a limit on the size of pieces which can be done this way which depends on the ovens available. Some combinations of materials which may be used are given in Table 3.1. If the copper is done in an H_2 atmosphere, it must be OFHC copper, or it is likely to come out porous. The BT braze is nice if you want to avoid having a superconductor in the vicinity, for either thermal conductivity or magnetic reasons.

There exist many other brazing alloys. Be careful to ensure that the one you use is compatible with the materials. People in our group once tried a mostly gold and a little copper alloy that happened to be available to try to join some delicate copper pieces in a non-superconducting fashion. The solder being on the copper-poor side of the eutectic quickly proceeded to consume and destroy the delicate copper pieces being attached together.

Hard Solder or Silver Solder This is another high temperature method producing high strength joints. It is useful on copper, stainless steel, brass and cupronickel, but not aluminum. More surprisingly, neither this nor vacuum brazing works on beryllium copper (one gets structurally sound, but leaky, joints). The usual alloys, typified by Easy-flo 45, Handy and Harman, 850 Third Ave., New York, NY 10022, which is 45% Ag, 15% Cu, 16% Zn, 24% Cd, work nicely but rely on Cd to lower the melting point and enhance flow. At Cornell, we are currently not allowed to use such alloys. Eutectic/Castolin, 40-40 172nd St., Flushing, N.Y. 11358 makes Eutecrod 1801, an alloy with about half the Cd content, melting at 605° C, that is available in the Clark stockroom. I prefer their 1800 alloy which is Cd free, melts at about 620° C, and appears to flow a little more freely. Eutectic/Castolin makes the same stuff under the name braze 560 with formula 56% Ag, 22% Cu, 17% Zn, 5% Sn, and at about half the price last time we purchased some. The Eutectic 1801B flux works very nicely for any of these alloys, and should be applied liberally on and around the joint. Unfortunately Eutectic/Castolin no longer sells 1801B. Instead, we use Handyflux for copper and brass, and the more corrosive Handyflux Type B-1 for stainless steel. Residues may be removed afterwards by immersion in boiling water (this is slow), or in Eutectic 1005 Exflux which works much faster. Heat is applied with a gas-oxygen, hydrogen-oxygen, hydrogen-air (some people feel that the hydrocarbon flame on stainless incorporates excessive carbon into the hot metal and ruins the metal properties, though I haven't experienced such problems myself, and find the nearly invisible hydrogen flame a bit of a pain to work with), or acetylene-air flame. The latter is quite useful when doing thin-walled stainless, as the flame is a bit cooler, and less likely to burn holes in the tubing if left too long in one place.

Soft Solders The low temperature group has generally used 60% Sn-40% Pb solder, which melts at about 190° C. Generally joints should be made with a clearance of .002-.003" on the diameter between the mating parts. The joint should allow a section of thin-walled overlap between the pieces being joined, as shown in the sketch below, to prevent excess stresses produced locally by screws on a flange, or by thermal contraction of dissimilar materials, from being taken up entirely by the weaker solder joint rather than the metal of the pieces being joined. Ideally, the joint should have an overlap length of about 1 cm of .5-1 mm thick wall. It is desirable to pre-tin the surfaces before assembly and final soldering because solder will not reliably run in to completely fill the gaps with weaker and less corrosive fluxes, or with stainless. Often simply soldering from the outside followed by pulling the pieces apart with a bit of twisting is sufficient to wet everything, but sometimes a soldering iron is very helpful, as it also provides some abrasive action. For small joints and thin-walled tubing a soldering iron is often an adequate heat source, and a hot plate works for many intermediate-size joints. Otherwise a torch works fine, using any of the brazing flames, or a

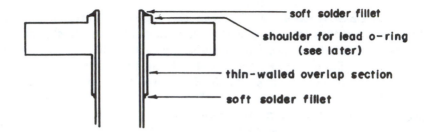

Figure 3.3. *Cross section of a soft solder joint between a tube and a flange.*

portable propane torch. For soldering stainless, a relatively viscous ZnCl in HCl flux (eg., Stay-Clean by J.W.Harris Co., Inc., 10930 Deerfield Rd., Cincinnati, OH 45242) is necessary. However residues of this are deadly to cryostats, and will corrode through thinwalled tubing on the weeks-years timescale, so thorough washing is crucial. (Also protection of other tubing in the vicinity is necessary if soldering is done in situ, since a mist of hot flux tends to be produced in the process of making the joint.) For other materials, it is desirable to use a milder organic flux such as Supersafe #30, made by Superior Flux and Manufacturing Co., 95 Alpha Drive, Cleveland, OH 44143, which is less likely to cause long term damage, but is trickier to use. (It will not tolerate overheating.) Although ideal for electrical work, rosin flux and rosin-core solder are risky things to use on vacuum joints. The rosin has a tendency to leave little threads of flux in the joint, which may open up as leaks on thermal cycling, or simply with time.

Eutectic 157 solder is a Ag-Sn eutectic with a melting temperature of about 218° C. It is similar in use to SnPb solders, using 157 flux from Eutectic, and seems to wet rather better than conventional soft solder. Most joints may be flowed in from the outside, but it is still well to pull them apart to inspect for complete wetting before making the final assembly, as particularly some stainless (321 especially) seems to be unreliable about wetting. If 157 and SnPb solder are both being used on a cryostat, it is useful to keep careful records of which is used where for future changes, since it is difficult or impossible to distinguish the joints visually.

Low Temperature Solders From a mechanical standpoint, most of these solders are greatly inferior to soft solder or hard solder. Indalloy 13, available from Indium Corp. of America, PO Box 269-A, Utica, NY 13503, is an alloy with some mechanical strength, melting a bit over 100° C. Wood's metal (50% Bi, 25% Pb, 12.5% Sn, 12.5% Cd) melts at about 70° C, and Cerrolow 117 (44.7% Bi, 22.6% Pb, 8.3% Sn, 5.3% Cd, 19.1% In) available from Cerro Metal Products, Division of the Marmon Group, PO Box 388, Bellefont, PA 16823, melts at 47° C. The last two alloys require quite active fluxes to work successfully, and cryostats using these materials for the sealing of vacuum cans, for example, can usually be distinguished from a distance by the distinctive greenish hue of corroding copper and brass.

Epoxy Joints and Seals These can be made at room temperature, and can provide electrical insulation. They are very useful for electrical feedthroughs. The epoxy can also be itself used as a structural member. Favorite epoxies in our group are:

- Stycast 1266 available from Emerson and Cuming, Canton, MA. 02021, which is strong, transparent, low viscosity in fluid state, easily machinable, and which bonds relatively well to most metals. It has the disadvantage of a rather large thermal expansion coefficient (1.5% change in linear dimension on cooling to 4 K), and can only be made in rather small batches (< 100 g), as the exothermic reaction involved in curing causes thermal runaway in larger castings. 1266 comes initially as 2 parts of clear liquid, but part A sometimes becomes a pasty white material on storage. It may be restored to original form by heating to 70-80° C for a few minutes.

- Stycast 1269, same manufacturer, has similar mechanical properties, and lower dielectric loss. It may be made in large batches, but requires a cure at elevated temperatures, which is not convenient in many circumstances.

- Stycast 2850FT and GT, again same manufacturer, are a black epoxy resin filled with silica powder to give a lower thermal expansion coefficient, matched roughly to copper (FT) or brass (GT). The resulting material is very strong, adheres well to metals (better than 1266 or 1269), and tolerates brief exposure up to 200° C for soft soldering nearby. Unfortunately, it is essentially unmachinable, has a non-negligible magnetic susceptibility and a temperature-dependent dielectric constant at low temperatures, and at room temperature is vastly more viscous prior to curing. (Using Catalyst 11 for elevated cure at 100-125° C gives very acceptable properties to the liquid form if the casting is done with the resin preheated to 80° C or so.) Stycast 2850 is very much more reliable for making seals to niobium than is Stycast 1266. 2850 appears to adhere better to metals in general than does 1266, and on larger diameter joints between an epoxy chamber and a metal part, it can be utilized effectively as an intermediate glue, in a layer 1 or 2 mm thick.

All of the above epoxies may be removed with the strippers Ecostrip 93 or 94, and the 1266 and 1269 can be cut away with the judicious use of a good hot soldering iron. All of these epoxies will be attacked by various organic solvents over a period of hours to weeks, examples being acetone, which acts slowly, and is fine for short-term cleaning off of pieces, and methylene chloride, which acts considerably faster. They are only weakly attached by concentrated acids, and appear impervious to strong bases. A useful fabrication technique is to create a mold from aluminum, which can be machined with a very smooth surface to

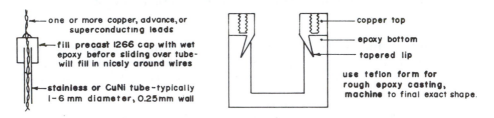

Figure 3.4.   *Epoxy to metal joints.*

high precision, cast epoxy in or around it, then later etch away the aluminum with concentrated NaOH. Very precise and delicate structures have been made in this way. Multiple pieces of epoxy may be glued together with fresh epoxy, but the joint is much stronger if the surfaces to be joined have been machined, or at least roughened with sandpaper. When making joints between metal pieces and epoxy, two principal design features are important. The metal should be tapered down to a very thin wall thickness (.25 mm or so) to allow it to yield slightly upon thermal contraction, and the epoxy should be on the outside of the joint, so that it contracts around the metal rather than pulling away from it. Shown are sketches of a typical design for a feedthrough for some electrical leads, and a joint between a copper sample chamber top and an epoxy bottom.

Epoxy may also be used to glue together two metal pieces, and is reasonably reliable, though not quite as much so as solder, if the two pieces are made of the same metal and the clearance is quite small (.001" or so), even if the metal pieces are of quite substantial size.

- Another epoxy finding frequent use in our group because of speed of curing is a 5 minute epoxy made by Hardman, Inc., Belleville, N.J. 07109. This has mechanical properties rather inferior to those of 1266, but does have a great convenience advantage in non-critical applications (or even sealing between parts of imperfectly matching forms, or on corners of joints being made with 1266 as the really functional epoxy).

- In the same vein of higher speed, Eastman 910 or any of the myriad other cyanoacrylate adhesives by other manufacturers is very convenient, very fast, and strong in thin layers between clean surfaces. It cycles reasonably well to low temperatures.

- Sometimes it would be convenient to have a conductive epoxy. Silver filled epoxies are rather expensive, but some of them work quite nicely. A couple that we have had a bit of experience with in the low-temperature group are ECR4100 by Formulated Resins Inc., PO Box 508, Greenville, RI 02828, which comes in convenient 2 g bubble packs for individual use applications, and Epo-Tek 417 by Epoxy Technology, 14 Fortune Drive, PO Box 567,

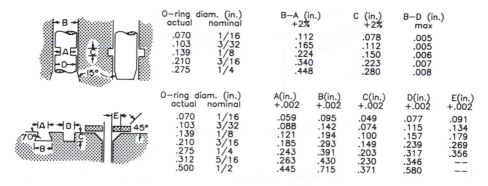

O-ring diam. (in.) actual	nominal	B−A (in.) +2%	C (in.) +2%	B−D (in.) max
.070	1/16	.112	.078	.005
.103	3/32	.165	.112	.005
.139	1/8	.224	.150	.006
.210	3/16	.340	.223	.007
.275	1/4	.448	.280	.008

O-ring diam. (in.) actual	nominal	A(in.) +.002	B(in.) +.002	C(in.) +.002	D(in.) +.002	E(in.) +.002
.070	1/16	.059	.095	.049	.077	.091
.103	3/32	.088	.142	.074	.115	.134
.139	1/8	.121	.194	.100	.157	.179
.210	3/16	.185	.293	.149	.239	.269
.275	1/4	.243	.391	.203	.317	.356
.312	5/16	.263	.430	.230	.346	--
.500	1/2	.445	.715	.371	.580	--

Figure 3.5.   *Dimensions for rubber O-ring seals.*

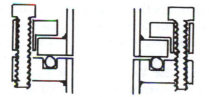

Figure 3.6.   *Flange with rotable orientation. The separately rotable bolt circle can also be used with metal O-rings.*

Billerica, MA 01821, which is more expensive and requires elevated temperatures to achieve curing in a reasonable time, but has slightly higher electrical conductivity. Silver conductive paint intended for room-temperature printed circuits is OK for use on some surfaces, but not nearly as strong as the conductive epoxies, and it sometimes flakes off upon thermal cycling.

Demountable Flanges of Various Sorts In keeping with the idea of modularity mentioned at the outset, the previously mentioned solder and weld joints will mostly be made to small flanges which will be bolted with some type of vacuum-tight gasket to another mating flange. For instance, half a dozen stainless tubes coming into the top of a vacuum can at 4.2 K might all have small flanges on their ends, be bolted to the top plate of the vacuum can, and be continued below with an additional set of flanged tubes, each joint being sealed with a metal gasket or "O-ring". A number of different types of O-ring seals will be discussed.

Rubber O-rings These are very common and useful at room temperature,

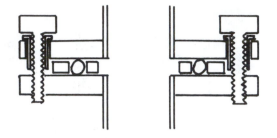

Figure 3.7.   *Flange providing electrical isolation.*

although they don't work at low temperatures where the rubber is no longer elastomeric. One finds many different prescriptions for groove dimensions, often oriented toward hydraulic seals at high pressures. Dimensions are not supercritical, but a set which works well for vacuum work is shown below. Usually the I.D. of the groove is chosen as the nominal I.D. of the O-ring, while the O.D. of the groove is rather wider than the O-ring, which will expand outward as the flanges are bolted together. Neoprene O-rings are OK for most purposes, though Viton is better at higher temperatures, retains elasticity for a few more years, and costs a bit more. Locally, the physics stockroom carries neoprene and the chemistry machine shop is the source for Viton (which is also more chemically resistant). The physics stockroom also carries neoprene cord in the standard O-ring sizes, which may be cut squarely at the ends with a razor blade and joined with Eastman 910, and used for larger or odd size O-ring seals. Parker Seals, O-ring Div., 2360 Palumbo Drive, Lexington, KY 40509 is a source for essentially any size or material O-ring. With rubber O-rings as with the metal O-rings to be discussed later, the flanges should not be made too thin or the screws placed too far apart, or the metal will warp badly when the screws are tightened–1/4" thick for 1/16" O-rings with screws every inch or so of perimeter is OK, 1/2" thick with screws every 2-3" for a 3/16" O-ring. It is sometimes convenient to make the bolt circle on a separate piece of material to allow arbitrary orientation of one piece relative to the other. If it is possible, put the O-ring into the gravitationally stable side of the joint for ease of initial assembly. If it is desirable to have an electrically insulated coupling, it is possible to replace the O-ring groove machined into one flange by a plexiglas O-ring support pinched between the two flanges, with small insulating sleeves to prevent conduction through the screws (Delrin, nylon, or plexiglas). An analogous construction can also be used if an O-ring groove is inconvenient to put into either flange, or has been inadvertently left out during construction.

Lead and Indium O-rings In this case, one simply uses thin lead or indium wire wrapped in a circle with the ends twisted together. There is no need for an O-ring groove, though many commercial items have been constructed using essentially conventional O-ring grooves to be filled with indium rather than rubber.

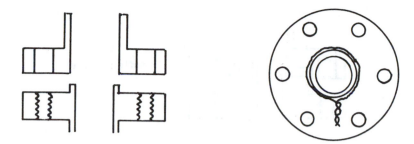

Figure 3.8.   *Lead O-ring flange in cross section, and view of the lower flange from above, showing how the lead wire is twisted before the flange is assembled and tightened.*

Upon recycling these, it is not necessary to replace the entire joint, but simply to add a small additional amount of indium (perhaps about the same loop of .020" diameter wire that we normally use for our O-rings. The form which we conventionally use for such a joint involves making a small shoulder on one side of the flange which fits snugly (preferably about .001" clearance on diameter) into a hole in the other flange. The O-ring wire is wrapped around this shoulder (which should be at least .050" tall for convenient assembly) and the flanges are compressed together, usually with socket head cap screws with an allen key, though regular slot-head screws are possible with indium which requires somewhat less force. More screws are necessary than with a rubber O-ring–about one screw for every 3/4" of perimeter. Flanges should be thick enough to avoid distortion(1/4" for small diameter flanges, 3/8-1/2" for flanges of several inches in diameter. Generally 2-56 screws are about as small as advisable for small flanges, with 4-40 or 6-32 being easier to tap and less easy to stretch or snap. People have used 0-80 screws successfully (metric sizes of M1.5, M3, M4, M1 respectively). A radial clearance of 1-1.5 mm should be left between the shoulder and the inside edge of the screw holes. The lead or indium wire will typically squeeze down to .003" thickness, making the .015" Pb or .020" In wire spread to .040-.060" wide. However, the screws should not be placed too far from the O-ring because bending or flexing of the metal flange will reduce the pressure on the wire. Especially on larger diameter seals, such as on the vacuum can for a cryostat, it should be remembered that it is much easier to mount the O-ring if the shoulder is on the lower piece as shown in Figure 3.8, so that gravity is an aid to assembly.

On indium wire it is sufficient to simply overlap the ends rather than giving two or three twists as we usually do with the stronger lead wire. The indium O-rings tend to persistently adhere in a fragmentary fashion to the flanges after separation. A light coat of silicone vacuum grease on the wire before assembly makes the used O-ring peel off much more conveniently. We prefer to use lead wire on flanges of harder materials such as brass, stainless, or beryllium copper, and indium on softer materials such as copper, which tends to distort on repeated

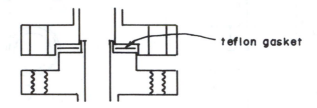

Figure 3.9.   *Flange using a teflon gasket, with shoulders arranged to eliminate creep of the gasket material with time.*

use of lead. Usually our "lead" wire actually contains about 4% Sb to make it draw more easily (this is typical for fuse wire).

Gold O-rings A defect of both lead and indium O-rings is that they are super-conducting. In magnetic fields <Hc this can produce field gradients due to the diamagnetic screening, and trapped fields may persist after temporary exposure to fields >Hc. Thus, in some experiments other materials are needed. .010" gold wire can be cut to the right length and flame welded into a circular loop using a miniature gas-oxygen torch. With care and practice a loop of essentially uniform cross section may be formed. The gold requires somewhat harder surfaces for the flanges, even after the wire has been softened by flame annealing. OFHC copper will deform about as much as the gold wire, and is good for at best one seal if used as the flange material.

Teflon O-rings Various plastics, such as Teflon, Mylar, and Kapton (all manu-factured by duPont) can be used as sealing gaskets at low temperatures. However, they have a tendency to flow with time, and the structure of the flange must be designed with shoulders on each side of the gasket material to inhibit this ten-dency, as shown in Figure 3.9. A convenient technique for making the gaskets of the thin (.002-.005" typically) plastic films is to glue a piece of the material to the smooth end of a brass or aluminum rod with Eastman 910, put in the lathe, and use a pointed cutting tool to scribe concentric circles of the appropriate diameters in the plastic. Then soak in acetone for a few minutes to dissolve out the superglue and extract the gasket.

Conical Taper Joints Sealed with Grease or a Glycerine Soap Mixture
Two pieces with a matching conical taper are machined from the same material–either metal or epoxy, and pressed together with a screw thread arrangement after greasing the metal surfaces. The virtue of the conical surfaces rather than simply flat mating surfaces is that they are somewhat self aligning, and give a substantial surface of overlap with limited space required on the diameter. In low pressure applications, it is not even necessary to use a screw to tighten down the

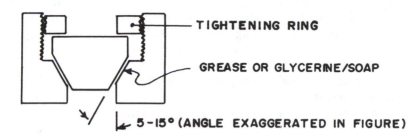

Figure 3.10. *Tapered seal using grease or a soap and glycerine mixture as a sealant.*

seal. A vacuum can may be held in place by simply pumping out the interior. At low temperatures, the form of seal with the threaded retainer can tolerate pressures of at least 40 atmospheres, but pressure differentials of no more than an atmosphere or two are allowed at room temperature, where the grease is still somewhat fluid. Dow-Corning silicone vacuum grease works well, but Apiezon L, M, N, etc. fragment upon cooling, and will cause leaks.

PVC Tubing Joints Two small diameter metal tubes of the same O.D. may be joined together by sliding PVC spaghetti over the two juxtaposed ends. The seal appears to be fairly reliable at low temperatures, but has the disadvantage that He diffuses through the tube relatively rapidly at room temperature making leak testing very difficult. It is desirable to keep the unsupported length of tube short, have no sharp bends in the PVC (or teflon) spaghetti, and to have the spaghetti a close match to the metal tubing diameter (so that it has to stretch a few mils to slide over the metal tubes). Some people prefer to lightly grease the ends of the tubing with silicone grease before sliding on the tubing.

Pipe Threads At room temperature, one common form of connection is a tapered thread called pipe thread. It is commonly found on thermocouple gauges, bourdon gauges, small valves, etc. The sealing principle is that some deformabel material is squeezed between the threads and maintained essentially immobile and impermeable to gases or liquids as the tapered threads jam together. Traditionally these have been used with various thread sealing compounds which are painted on as a liquid which partially hardens with age, but retains some plasticity. Often seen are Glyptal paint, made by General Electric and LeakLock joint sealing compound by Highside Chemicals, Inc., 10 Colfax Ave., Clifton, NJ (familiarly known as blue goop). Stycast 1266 is another local favorite on these joints. Teflon thread seal tape also works quite well on these joints. These seals are only for room temperature use, and don't cycle well to low temperatures.

Swage Type Fittings A common type of seal at room temperature is formed by compressing a small brass, stainless, or nylon ferrule simultaneously around a tube and against a tapered surface in a surrounding fitting. Examples of this are high pressure fittings in stainless steel from HIP Inc., in Erie, PA, and Swagelock fittings for copper and plastic tubing. We have used these at room temperature with moderate success. Vibration, flexing of joints, and simply long time scale relaxation of the metal appear to present some reliability problems on the time scale of a few years. Although these joints on the HIP tubing are probably OK at low temperatures, it is so easy and reliable to soft-solder small tubes, with much less space required, that it is more sensible do that in the cryostat.

Copper Gasket Seals Typified by the Varian Conflat flange, these seals are formed by cutting into a soft copper anulus with a stainless steel knife edge. They are used extensively in UHV systems, are highly reliable, and also work just fine at low temperatures. They are, however, very expensive and quite bulky, so their cryogenic use tends to be limited to people with big budgets and big cryostats.

3.1.2 Making Thermal Contact at Low Temperatures

Making thermal contact at low temperatures involves both establishing strong thermal links between a refrigerator and a sample, but also heat sinking of wiring, fill lines, structural members, etc. at various intermediate temperature levels in a cryostat to reduce unwanted heat leaks into the sample.

Welds Often copper or silver wires, rods, or plates are used to thermally connect parts of the apparatus. The most reliable way to join these with low thermal resistance is to weld them together. Small wires can be welded in a gas-oxygen flame with a small torch, using brazing flux to limit oxidation. Somewhat nicer is TIG welding, which can be done on pieces ranging from small wires to large plates in size. This can even be done in situ on a cryostat. Electron beam welding works very nicely if the piece can be put in a vacuum chamber, and if you have access to an e-beam welding facility. If you are TIG welding one part to another attached to a structure which cannot withstand high temperatures, it is possible to immerse part of a structure in liquid nitrogen while simultaneously melting another part of it a few cm away.

Press Joints, Screw Contacts Welds are not easily demountable. Very low resistance thermal contacts can be made by bolting two clean, gold-plated copper or silver surfaces together. Some people feel that it is better to have a contact where a screw is turning into the piece and galling slightly on the final turn, tearing the metal slightly at the threads and under the head to establish direct fresh oxide-free metal contacts. Not knowing which school of thought to believe, most of our joints are made with gold-plated surfaces bolted together with copper screws just in case.

Oxygen Annealing of Copper or Silver Rods To enhance the thermal conductivity of the metal connecting, for example, the sample to a nuclear refrigerant, it is possible to anneal the material in a partial pressure of about .01 mbar of oxygen at a temperature of around 700° C for about a day. This allows the impurity atoms (particularly magnetic impurities) to diffuse at a significant rate to grain boundaries, where they eventually react with oxygen and become immobilized. At too high a concentration of oxygen, the impurities react too soon and don't congregate on grain boundaries, too low they don't ever get reacted. At too low temperatures, diffusion takes forever, at too high temperatures, the oxide breaks up. With this procedure, it is possible to get RRR values of several thousand, often enhancing the thermal conductivity by as much as a factor of ten! The pressure of oxygen may be achieved by a slightly poor (in a carefully controlled sense) vacuum, or by using helium with a 10 ppm concentration of oxygen carefully introduced. Etch surface before annealing with acid (or on silver 30% hydrogen peroxide and concentrated acid, mixed half and half). A good reference on oxygen annealing is Ehrlich (1974).

Brazed Joints Many of the fill lines for a cryostat need a significant surface area at several heat sinks. Although this may be achieved by small sinter-filled volumes, it is often adequate to wrap a meter or so of .004 or .010" id cupronickel tube around a copper post and braze it together. It is advisable to add another wrap or two of the tube at each end around the post and tack it down with soft solder after the brazing is done, as the higher temperature of the braze has a tendency to weaken the adjacent bits of tubing. The soft solder is a superconductor with a rather high critical temperature, and is probably not good for making the primary thermal contact, however.

GE 7031 Varnish, Grease, Epoxy Wiring must be insulated, so one does not have the option of cooling by electron thermal conductivity. If very thin (awg 40 or smaller) copper wires or resistance wires are used from room temperature, they may be satisfactorily anchored at 4.2 K by wrapping around a metal post for a length of 5-10 cm and gluing them down with GE varnish or 1266 epoxy. If GE varnish is used, which softens formvar insulation with its solvent, or if there are any sharp edges or burrs on the metal post which might nick the insulation, it is wise to insulate the wires from the post with a thin layer of something such as cigarette paper. The wires may be clamped thermally at the 1 K plate, still, heat exchangers and mixing chamber of a dilution refrigerator. Usually, though, a couple of these points will suffice. To make a less permanent attachment, it is possible to clamp a group of wires between two smooth plates using grease to establish contact. If twisted pairs are used to reduce pickup, it is either necessary to untwist a short section, or to use a heat-sinking technique that doesn't involve clamping.

Coaxes are more difficult to heat sink. Cooner Wire Company, 9186 Independence Ave., Chatsworth, CA 91311, makes a coaxial cable with a braided stainless steel sheath about .5 mm in diameter, with a teflon dielectric and either stainless or superconducting inner conductor. This cable is easily heat sunk by clamping gently between two copper plates about an inch in length, as the thermal contact through the dielectric is comparable to the conductivity along the center conductor. Unfortunately, the shielding, rf characteristics, and susceptibility to microphonics are not nearly so nice as for UT20-SS or UT85-SS semi-rigid coax made by Uniform Tubes, with a stainless steel tube outer conductor, teflon dielectric, copperweld inner conductor (unfortunately magnetic). They also sell by special order stainless inner conductor (more lossy), and with large minimum orders beryllium copper conductors (very nice if you can lay your hands on a supply) or various superconducting combinations. Unfortunately the special order combinations tend to be more expensive than a typical individual low temperature group can afford because of the size of a minimum order. The copperweld inner probably requires breaking the coax open, soldering in a short section of metal foil insulated from ground by a thin dielectric, and losing the nice 50 ohm characteristic of the line. (It is probably possible to make stripline heatsinks which retain decent propagation characteristics with much more effort.) A neat trick which works at high, fixed frequencies is to tee into the line with a quarter wavelength of the same cable which has the inner and outer conductors shorted at the far end, permitting direct electronic cooling of the inner conductor. It is also possible to make your own coax by running a wire through a thin-walled tube and using teflon spaghetti as a dielectric, or injecting Stycast 1269 as a dielectric and thermal contact. (1269 is less lossy than 1266.) Twisted pairs can make good constant-impedance transmission lines as described in the sections on high frequency techniques, and are easier to heat sink.

A technique which seems promising, but which has not been developed, is the use of a series of a dozen or so striplines in parallel on a Kapton ribbon, with aluminum backplane and strips. This would be a good conductor at the room temperature end, and a low thermal conductivity superconductor at the low temperature end. Heat sinking could be achieved by simply clamping the backplane at various points, and electrical contact by silver paint at appropriate points (perhaps aided by sections of evaporated gold over the aluminum at intervals of 10 cm or so).

For convenience in wiring of the cryostat, it is useful to have some sort of terminal board at 4.2 K and perhaps at the mixing chamber, so that new changes can be made without redoing all the heat sinking, but merely unsoldering or unplugging a couple of pins in a convenient array. The cryostat should always be assembled initially with more leads than you can possibly imagine using, as there will eventually never be enough for all the things you want to hook up.

Sintering of Fine Metal Powders for Heat Exchange This is a subject of considerable importance for almost any apparatus at low temperatures. Appropriate recipes for successful sintering are given in Section 3.5 (cooling the sample).

3.1.3   Joining to Superconducting Wires

Small (typically .003-.005") Nb or NbTi wires which must be joined together with a superconducting joint which requires rigorously zero resistance, but which need carry only small currents, are typically needed for applications such as the flux transformer for a SQUID magnetometer. Spot weld these wires together, after first removing the Formvar insulation from the wire with STRIPEX or formic acid, or removing more tenacious insulations with a razor blade, then removing the copper or cupronickel cladding (if any) with concentrated HNO_3. The principal disadvantages of the technique are that the resulting joints tend to be mechanically quite fragile (they may be reinforced with Mylar or Kapton tape after welding, to provide both insulation and mechanical support), and that the exact amount of charge required to get a good weld is much more critical than with some materials. Too little energy results in a weak or non-existent weld, too much may vaporize several inches of wire! The technique is really only satisfactory for monofilamentary wires.

For the same small wires, particularly if rather higher current capacity of up to a few amps is necessary, superconducting crimp joints work quite nicely. The wire is stripped to the bare superconductor as in the preceding paragraph, then *momentarily* etched in 50% conc. HNO_3, 50% HF, as is a short (3-5 mm) length of Nb tube of 1 mm od, 0.5 mm id (obtainable from Biomedical Technology, Inc.). Then both are rinsed in distilled water. The tube is slid over the ends of the wires to be joined and squeezed tightly with a pair of pliers. This type of joint may also be used on multifilamentary superconductors. It will carry currents of up to about 1 ampere in a vacuum, and appears to be stable with currents of several amps if the joint is immersed in the helium bath.

Superconducting leads to a sample chamber It is sometimes convenient to use a superconducting lead to minimize thermal contact to a sample chamber for experiments such as heat capacity measurements. By using a copper or cupronickel clad multifilamentary superconductor and etching away all but a short section at each end of the wire with conc. HNO_3, one will have a bundle of small superconducting fibers on the order of 1 micron in diameter. All but one of these may be cut away, leaving a single superconducting filament which may be connected to electrically by simply soldering to the cladding on each end.

Demountable contacts may be made for low current uses by tapping holes for small screws in a Nb block and wrapping a wire under the head of a screw which crushes it against the block. The screw need not be a superconductor–brass or stainless works fine. It is best to use a separate screw for each wire rather than trying to pinch multiple wires under a single screw.

In the helium bath it is often necessary to connect leads to a superconducting magnet. One technique that works well is to overlap a length of several inches of the copper-clad leads and join with soft solder. This is even adequate for attaching a persistent current switch to a typical superconducting magnet, as the resistance of such a joint may be 10^{-6} to 10^{-8} ohms and the typical inductance of a magnet 10 Henries, giving an L/R decay time of 10^7 to 10^9 seconds. The much smaller inductances of typical heatswitch magnets or some experimental magnets may make such a technique inappropriate for use in these cases.

3.1.4   Heat Switches

It is often necessary to change the strength of thermal coupling between a sample chamber and the cooling system by large amounts. Usually one would like to maximize the ratio of on/off contact. There are several methods which are applicable in different temperature regimes. A technique which is very simple and useful in some ^{4}He experiments uses the superfluid properties of ^{4}He. It works well only between 1 K and 2.17 K. A small (.010 in id) cupronickel capillary is connected between the sample chamber and a 1 K refrigerator, with possibly a small cavity containing sintered copper at the ends. At low pressures the tube is an excellent conductor, but at 30 atmospheres the helium in the capillary solidifies, and as the solid helium is a rather poor thermal conductor the sample chamber is thermally isolated. Mechanical heat switches have been used with variable success at very low temperatures. A typical technique is to pressurize one gold-plated contact against another one, using a bellows or a metallic diaphragm actuated by liquid ^{4}He (essentially a perfect insulator at sufficiently low temperatures). Vibrations caused on opening or closing can cause heating, and the quality of contact on closure is not always very good, and different groups have had variable success. At higher temperatures, an automatic version has been used for pre-cooling apparatus. A rather stiff bellows is expanded by putting in a substantial pressure (up to several hundred psi) of ^{4}He gas. In this configuration the bellows maintains a good thermal contact with the outer walls of the vacuum can during the initial transfer of helium into the cryostat. When the apparatus eventually reaches 4.2 K, the gas liquefies, the pressure drops, and contact is removed. It is only useful for rather simple cryostats, not for dilution refrigerators or demagnetization cryostats where many thermally remote sections all need to be precooled. The same goes for an ingenious gas contact heat switch described by Torre and Chanin (1984).

Superconducting heat switches are the most generally used heat switch in the region well below 1 K. These take advantage of the dramatic difference in conductivity of a superconductor in the super and normal states. Several materials have been used successfully by different research groups. One that works particularly well has been described by the Jülich group, and uses aluminum as the material (Mueller *et al.*, 1978). It is operated with a low field (Tc=100 gauss for Al) and has a very good switching ratio, but the construction process is rather complicated because of the need to remove the aluminum oxide film on the surface and replace it with gold. Cornell modifications with fewer (2-6), shorter (6 mm), narrower (6 mm), and thicker (.3 mm) foils, using a simpler clamp arrangement, and with the copper end pieces in their turn bolted on rather than welded have worked quite nicely on a number of our cryostats. Design, performance, and plating recipe for the Jülich design are shown below:

Electroplating of Aluminum Thoroughly rinse the aluminum in water following each step. Use care in handling the very toxic cyanides; do not allow them to contact acids.

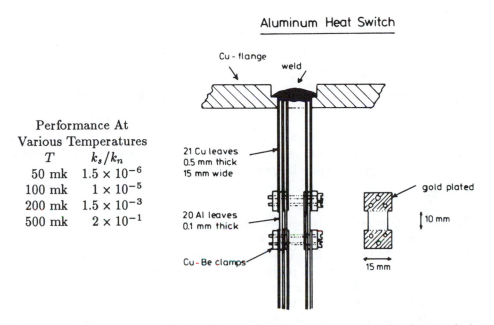

Aluminum Heat Switch

Performance At Various Temperatures

T	k_s/k_n
50 mk	1.5×10^{-6}
100 mk	1×10^{-5}
200 mk	1.5×10^{-3}
500 mk	2×10^{-1}

Figure 3.11. *A typical design for an aluminum superconducting heat switch, showing various construction details.*

1. Wash in alkaline cleaner at 75° C for 60 s (22 g/l of $Na_3PO_4 \cdot 12\ H_2O$ and 22 g/l of Na_2CO_3).
2. Acid bath for 15 s (equal volumes conc. HNO_3 and water).
3. Zincate solution at $(22 \pm 2)°$ C for 60 s (1 g/l $FeCl_3 \cdot 6\ H_2O$, 100 g/l ZnO, 525 g/l NaOH, 10 g/l $C_4H_4KNaO_6 \cdot 4\ H_2O$).
4. Acid bath for 30 s.
5. Repeat steps (3) and (4) until the aluminum is slightly but uniformly etched. The aluminum may be briefly dried to apply a protective lacquer on parts not to be plated, but do not touch the parts to be plated. The foils must be kept wet at all times following this step until completion of the process.
6. Zincate solution at $(22 \pm 2)°$ C for 10 s.
7. Copper strike; have electrodes connected and power on before immersion, so plating begins immediately; use copper anode and plate at room temperature for the first 2 min at 26 mA/cm^2, and at 13 mA/cm^2 for 2 min more; (41.3 g/l CuCN, 50.8 g/l NaCN, 30 g/l Na_2CO_3, 60 g/l $C_4H_4KNaO_6 \cdot 4\ H_2O$).
8. Gold plate; deposit 1 μm according to the manufacturer's instructions (we used Technigold 25 from M. Schlotter, Galvano-technik, Postfach 92, D-734 Geislingen, W. Germany).

Many groups have used high purity tin for heat switches. This material may be attached by soldering a thin foil between two more massive copper fingers.

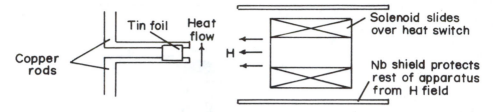

Figure 3.12. *View of a tin heat switch and the magnet assembly which slides over it to provide the switching field.*

The soldering operation is somewhat delicate because of a possibility of melting the foil away rather than soldering it in place if the timing is not precise. It is also necessary to use flux for the soldering operation, despite the possibility of contaminating the tin. If it is possible, it is desirable to arrange the heat flow direction perpendicular to the magnetic field from the heat switch magnet. Thus, if trapped flux lines are left in the metal (although type I, impurities, grains, etc. may permit trapping of fluxoids), the normal "core" will not act as a short circuit for thermal conduction through the switch.

This may also have the advantage of making the heat switch solenoid capable of being installed after the heat switch is assembled. A possible configuration is shown in Figure 3.12. A potential problem with tin heat switches is the formation of grey tin, a form of tin stable at low temperatures which will potentially turn the heat switch to powder if the region from 77 K to 4.2 K is traversed too slowly. If a superconducting shield is put around the heat switch solenoid, it must be located sufficiently far from the windings so that the screening currents set up in the shield do not too thoroughly cancel out the field at the center of the solenoid. Another neat design for heat switches which is particularly compact has been described by Schuberth (1984). Although not all of the bugs have been worked out, it seems worth pursuing.

Other popular materials are zinc and lead, both of which may also be attached by soldering. An ingenious use of lead heat switches has been made in several demagnetization cryostats by careful placement of the lead foil in the fringing field of the main magnet, such that it experiences about 1 Kgauss field when the magnet is fully energized. By the time the main field has been reduced by about 20%, the heat switch will stop conducting. In this fashion there is no need for a separate heat switch magnet. However, such a system is not applicable in systems where the field gradient is sharply localized spatially, as is commonly the case in modern systems.

For the small solenoids to actuate the heat switches, it is often useful to bypass the magnet with a shunt resistor, in the bath, having a few milliohms resistance. This effectively filters any ripple from the power supply used to operate the heat switch. This can be an important consideration as rather high conductivity materials run through the solenoid and can experience eddy current heating with microscopic levels of 6 Hz field variation. It also permits the ramping up and down of the field with much less care. The same thing applies to the main bath solenoid, but this is likely to be using a higher quality power supply, and itself has a much higher inductance, so that ripple is less likely to be a problem.

3.1.5   Manufacture of Flow Impedances for Refrigerators

Continuous cycle refrigerators (^{4}He, ^{3}He, or dilution) all have a returning flow of liquid, whose pressure must be reduced from the level required to condense the liquid from the gas, usually 0.3-1 bar. to the fraction of a torr at which gas is being extracted by the pumping system. For a continuous fill ^{4}He pot running at 1.4 K, with a heat load of 10-20 mW which is typical for a modest circulation dilution refrigerator system, one needs an impedance of about 4×10^{11}/cm^3. This is conveniently achieved by using a 1 m length of .1 mm id cupronickel capillary, which may be spiralled around a pencil or Q-tip to make a compact, tidy coil. If this appears to be a little too high an impedance, simply cut down the length.

For a ^{3}He or dilution refrigerator, the typical impedance required is 2-5 $\times$ 10^{12}/cm^3. This is easily fabricated by inserting a length of .010" stainless wire into a 30 cm length of .010" cupronickel tube. The stainless wire may be stretched slightly between pairs of pliers to reduce slightly the diameter if it will not fit, or if the impedance achieved is too high. Usually several tries may be needed to get the desired size of impedance. The impedance is easily measured by collecting bubbles of helium gas blown through at room temperature and collected in an inverted graduated cylinder filled with water.

3.2   Special Materials

by Jeffrey E. VanCleve

3.2.1   Thermal Conductivity

The thermal conductivities K of solids can vary by six orders of magnitude at liquid helium temperatures. Figure 3.13 illustrates typical K for six kinds of solids:

1. Large pure crystal (Sapphire, from Slack, 1962)
2. Glass (Vitreous silica, from Raychaudhuri, 1975)
3. Ceramic (Sintered alumina, from Tait, 1975)
4. Polymer (Teflon, from Lounasmaa, 1974)
5. Elemental metal (Oxygen Free High Conductivity Cu, a purified form with very high electrical and thermal conductivities. In general, the conductivities of metals vary greatly with purity and annealing. An intermediate value is shown, from Touloukhian, 1970.)
6. Alloy (Brass, from Denner, 1969) In the limit of low T ($< 2K$) there are some explanations for these differences:

Large Pure Crystals Phonons are scattered only by the boundaries of the crystal, so the phonon mean free path λ is a constant roughly equal to the diameter. Kinetic theory says $K = (1/3)Cv\lambda$, where v is the thermal carrier velocity and C the contribution to the heat capacity of these carriers. Since $C \propto T^3$, and $v =$ speed of sound (constant), $K \propto T^3$ and is quite large in the 1-10 K range.

Glass There are localized low-energy excitations in glasses which scatter phonons and increase the heat capacity at low T. The mfp due to this goes as $1/T$. Only the phonon T^3 term in the specific heat contributes to thermal transport, so $K \propto T^2$ and is small over the whole temperature range.

Ceramics Same as for the large pure crystals except λ is tens of microns, not millimeters. $K \propto T^3$ and is small over the whole temperature range.

Polymers, including Epoxies Although more complicated than glasses composed of atoms or small molecules, they exhibit similar behavior, and K is small and $\propto T^n$, where $1.5 < n < 2$.

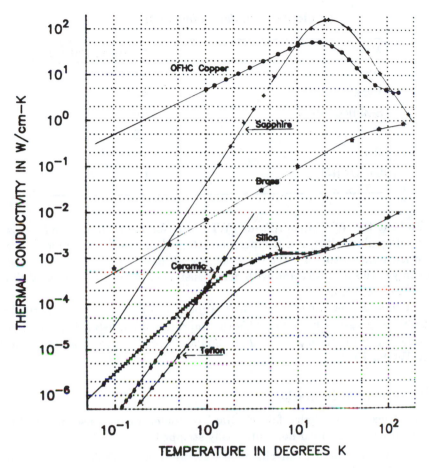

Figure 3.13. *Representative thermal conductivities of six kinds of materials.*

Elemental Metals Heat is carried by both phonons and electrons. Electrons scattering off impurities dominates at low T. λ is constant and the electronic specific heat $\propto T$. v is the fermi velocity. Hence $K \propto T$ and depends strongly on impurity and defect concentrations, but can be quite large in pure, annealed metals. It is sometimes worthwhile to anneal metal yourself to enhance the low-temperature conductance by a factor of 10 or more. Regular magnet wire will have a residual resistance of 1 percent of its room-temperature value (a residual resistance ratio or RRR of 100), while carefully annealed Ag and Cu can have RRR's of 1000 or more.

Alloys Electrons have a very short mfp, so $K \propto T$ and is smaller than that of pure metals.

Table 3.2. $\bar{K}$ *for Common Materials.*

Material	K-bar(W/cm-K)
Nylon	.0027
Pyrex	.0068
Stainless Steel	.10
Constantan	.18
Brass	.67
Cu(OFHC)	1.6

To beat the Wiedemann-Franz (WF) law (for metals), $K \propto \sigma T$ (σ is the electrical conductivity), use a superconductor. Well below the superconducting transition temperature T_c, the electrons are all paired and can't carry heat–so you can get thermal isolation with zero resistance, or switch a thermal link on and off with a magnet.

For cooling down the cryostat and keeping it cool, the average thermal conductivity, $\bar{K} = \frac{1}{T_2-T_1} \int_{T_1}^{T_2} K(T)dT$ is the important parameter. The values of $\bar{K}$ for some typical materials, with $T_2 = 300\,K$, $T_1 = 4\,K$ are listed in Table 3.2. (from White, 1979)

3.2.2 Practical Remarks

Al—Inexpensive, light, fair K. Hard to solder–weld or use metal O-rings. Easy to machine. Tapped holes are easily stripped.

Cu—Use OFHC, hard to machine (like trying to cut bubble gum accurately). Soft after annealing. Low susceptibility.

Brass—Easy to machine and solder. Can have cracks, especially in extruded bar stock. Brass is not a well-specified alloy, and compositions vary, often including trace elements that give unwelcome magnetic effects. Sometimes brass is recycled metal, giving unexpected treasures in the middle of a bar.

Stainless Steel—Hard to machine, reliable. Solder flux corrosive. Weld if possible.

Fiberglass—Light and fairly strong. Leak tight to liquid helium, porous to He gas at room temperature. Used in modern storage dewars.

Graphite—cheap but brittle. Pitchbonded graphite has the lowest K. For best results, reactor-grade graphite is used.

Alumina—brittle, hard, and impossible to machine.

Vespel—easily machinable, low K. SP-22 is loaded with graphite and is an even better thermal insulator. These resins, available from duPont, are very expensive

Table 3.3. *Thermal Contraction of Selected Materials*

Material	Contraction (per 10^4)
Teflon	210
Nylon	139
Stycast 1266	115
Alkali Halides	60-80
SP22 Vespel	63.3
Stycast 2850FT	50.8
Stycast 2850GT	45
Al	41.4
Brass	38.4
Cu	32.6
Stainless Steel	30
Quartz a-axis	25
c-axis	10
mean, for typical transducer	15
Titanium	15.1
Ge	9.3
Pyrex	5.6
Si	2.2
Pyroceram 45 (Corning ceramic)	2.0
Vitreous Silica	−.5

(about \$10/oz.), but are great for thermally insulating support rods.

3.2.3 Thermal Expansivity

Materials shrink as you cool them, and it's important to match thermal contractions in your construction materials wherever tight tolerances are important (seals, optical things). The induced stress can also cause sorrow. Samples break, transducers pop off, coax dielectric shrinks and splits. However, most shrinkage occurs by liquid nitrogen temperatures, so you don't have to cool all the way down to look at potential thermal contraction problems. For details, see White or Touloukhian. Table 3.3 presents the total contraction, in parts per 10^4, of some common materials between 293 K and 4 K. (From Touloukhian, 1970, and NBS 29).

Epoxy seals should fit around the outside of metal tubes, so they squeeze the metal and maintain the seal as the cryostat is cooled (see Recipes section). Polymers should not be used if their dimensions (within 2 percent) are critical.

3.2.4   Electrical Conductivity and Wires

The object is to reduce heat flow into the cryostat while keeping lead resistances low, or measuring things so that lead resistances don't matter (3, 4 terminal). For DC or low frequency leads, wire will do, and alloys like constantan, Advance (easy to solder), or Evanohm (ornery to solder) give you the thermal isolation you need. A typical lead resistance (proportional to thermal resistance below 10 K and at room temperature–where the WF law is valid) would be 50 to 100 Ω from the top of the cryostat to the sample area. For thermal anchoring, use GE varnish 7031 diluted with methanol. Most of these alloys have Ni in them–a problem if you're measuring very small magnetic fields, or use the wire in a low temperature calorimeter (Schottky anomalies).

If your leads must carry large amounts of current and Joule heating is a problem, use high-resistance wire down to the 4.2 K heat sink. From there, use superconducting wire big enough to carry the required currents. You can avoid the difficulty of soldering Nb and its alloys by covering a fine Cu wire with 50-50 Sn/Pb solder, or using Cu-clad Nb magnet wire and removing the Cu in the middle with acid. Else crimp wires together with CuNi tubing or Nb tubing (no heat capacity but about $60/ft).

Pure metals are not useful, since their large low-temperature conductivities mean you have to use unacceptably fine wire to get good thermal resistance. In addition, the low-temperature K of pure metals varies considerably, and you won't have a good handle on what your heat leaks will be.

For high frequencies and especially sensitive measurements, coaxial cable is required (See High Frequency chapter). SS coax, with the inner conductor clad in Cu, offers low attenuation and moderate heat leak if properly thermally grounded (most of the heat flows through the cladding). All-steel coax may be required if heat leaks are unacceptable, or superconducting coax can be had at great cost (usually a special order from Cooner or UTI).

Wire, while humble, is important. Here are suppliers of various kinds:

- Very fine (down to #52 = .8 mils) Cu wire–Hudson Wire Co., Ossining, NY.

- Fine superconducting wire–Supercon Inc., 9 Erie Dr., Natick, MA 01760.

- Big composite superconducting wire for magnets–Oxford Superconductors, 600 Milik St., Carteret, NJ 07008 (201) 541-1300.

- Resistance alloy wire–WB Driver, Newark, NJ. Driver-Harris Co., Harrison, NJ. Bergquist.

- General specialty wire–California Fine Wire. Sigmund Cohn, Mt. Vernon, NY.

Table 3.4. *Resistivities of Common Alloys (in $\mu\Omega$-cm)*

Material	Resistivity (295 K)	(4.2 K)
Brass	7	4
Constantan	52.5	44
CuNi	26	23
Advance	45	about the same
Evanohm	134	133
Manganin	48	43
Stainless Steel	71 to 74	49 to 51

- Indium wire (for O-rings, solder)–Indium Corp. of America, 1676 Lincoln Ave., Utica, NY 13503.

- Aluminum wire–SGA Scientific, Bloomfield, NJ.

- Precious or weird metals in wire, sheet, or bulk–A. D. Mackay Metals, 10 Center St., Darien, CT 06820, 203-655-7401. If you need ytterbium honeycombs (or equally exotic items), A. D. Mackay is a good place to start inquiring. Like Sigmund Cohn, they're a job shop, which means they don't actually make everything they sell, but rather find out where your material can be purchased and processed. You get what you want, but it might be cheaper to do your own shopping if possible.

- White (1979), lists resistivities of elements and alloys. Some data for common alloys are shown in Table 3.4.

3.2.5 Nuclear Magnetic Moments

If you're cooling a sample cell in the mK region (by demag or Pomeranchuk cooling), you don't want the entropy of the cell to drop while you're cooling it—this means heat must be removed and you won't be able to cool too well. If the cell (*not* the demag salt) is in a fixed field H while cooling, the entropy of the nuclear spins changes rapidly when $\mu H < kT$, where μ is the nuclear magnetic dipole moment for a particular isotope. A typical field is 10 Tesla. For $\mu < 1$ nuclear magneton, $\mu H/k$ is in the mK range. Hence the sample cell should be made of isotopes of nuclear spin $I = 0$ (which means μ is 0). Elements with stable $I \neq 0$ isotopes should have a small natural abundance of these isotopes.

The nuclear force is attractive, which favors spatially symmetric (hence spin antisymmetric) states. Thus nucleons pair to give zero angular momentum, and nuclei with even numbers of both protons and neutrons have no moments. These are also the most stable nuclei. See Segre (1964) for details.

If $I \neq 0$, the details of nuclear structure determine the magnitude of μ. For example, $I = 1/2$ for both H and ^{109}Ag, while μ is 2.8 for the former and $-.13$ for the latter. Hence hydrogen, which constitutes most of an epoxy or a plastic, must be avoided, while silver is an acceptable material above 1 mK, even though both have $I = 1/2$. Table 3.5., taken from Sagan (1985), lists elements with significant $I \neq 0$ isotopes. Moments may be found in the *CRC Handbook*. S_{max} is the spin entropy when $\mu H/kT << 1$, averaging over natural isotopic abundances. $T_{.5}$ is the temperature at which 50 percent of the entropy remains in a 10 Tesla field ($T_{.5}$ is proportional to H).

Titanium is strong, with low entropy, but can't be soldered. Sagan recommends drilling oversize holes in Ti and melting Ag in them so it alloys around the circumference of the hole but not in its center. One then drills out the Ag in the center and solders capillaries and posts to that. All polymers have H or F in them and must be shunned. Moisture adsorbed on surfaces gives unwanted H. The best filler for sample cells is $CaCO_3$, which is easy to obtain and grind into the proper shape. Si is also a good material for this.

3.2.6   Other Materials and Remarks

Epoxies Epoxies used in low-temperature work include Stycast, Scotchcast (3M), Epibond, and Araldite. Some epoxies are loaded with solids to make their expansivities match those of metals, like Stycast 2850 GT and FT. Stycast 1266 is particularly useful when you need a low-viscosity epoxy to fill in tight places. The Stycast epoxies are made by Emerson and Cumming of Canton MA 02021, but one usually orders epoxies from a local distributor. Instructions and some technical data come with the order. Our Ithaca supplier is Deanco Inc. (607) 257-4444.

Thermal Relaxation A general feature of amorphous materials (glue, glass, epoxy, plastics . . .) is the logarithmic time dependence of the linear (in T) term in the specific heat (see section 3.2.1). The low-energy excitations relax very slowly, and the heat flow from a cooling glass is governed by the slow (hours and up) relaxations of these internal modes rather than the link to thermal ground (see Zimmerman and Weber, 1981). This looks like a heat leak at very low ($<$ 100 mK) temperatures—phonons have equilibrated but the internal modes are oh-so-slowly giving up their heat. Thus one should try to minimize the quantity of glue, grease, varnish, and plastic (try a ceramic if you need thermal isolation) in the coldest part of your cryostat. Other explanations for the virtual heat leaks are relaxation of mechanical stress (caused, say, by tightening screws) or cosmic radiation (not much you can do about that . . .) In a dilution refrigerator you can probably remove the heat as it evolves and come back the next day after everything has settled down. At lower T, with one-shot systems, it's more serious.

Table 3.5. *Nuclear Entropy of Selected Elements*

Element	Isotope	Spin	Abundance (percent)	S_{max} (J/molK)	$T_{.5}$ (mK)
Aluminum	^{27}Al	1/2	100	14.9	4.6
Beryllium	^{9}Be	3/2	100	11.5	2.0
Boron	^{10}B	3	20		
	^{11}B	3/2	80	12.4	4.0
Calcium	^{43}Ca	7/2	.14	.024	1.4
Carbon	^{13}C	1/2	1.2	.063	2.4
Chlorine	^{35}Cl	3/2	75		
	^{37}Cl	3/2	25	11.5	1.3
Copper	^{63}Cu	3/2	69		
	^{65}Cu	3/2	31	11.5	1.58
Fluorine	^{19}F	1/2	100	5.76	9.1
Germanium	^{73}Ge	9/2	8	1.49	.81
Silver	^{107}Ag	1/2	52		
	^{109}Ag	1/2	48	5.76	.42
Helium	^{3}He	1/2	a)	5.76	7.4
Hydrogen b)	^{1}H	1/2	100	5.76	9.7
Indium	^{113}In	9/2	4		
	^{115}In	9/2	96	19.1	5.0
Lead	^{207}Pb	1/2	23	1.30	2.0
Magnesium	^{25}Mg	5/2	10	1.49	1.1
Niobium	^{93}Nb	9/2	100	19.1	5.6
Nitrogen	^{14}N	1	100	9.13	.88

Gases— ^{3}He The best way to obtain the magic gas is through the DOE. If you have a DOE contract, you can purchase it for substantially less than the commercial price, through Monsanto Mound Labs, Miamisburg, OH (513-865-3501). Otherwise, Isotec Inc., 7542 McEwen Road, Centerville, OH 45459 will sell or loan (at half price for a year) it to you. The cost is roughly $200/STP liter, depending on the size of the order. You also have to buy the bottle it comes in, which is $200 regardless of size. They also have deuterium at about $2/liter, less for large orders, and a wide variety of lesser-known but presumably useful (for laser gas mixes, etc.) isotopically enriched or pure gases. Write for a catalog or a quote.

Table 3.5.   *Nuclear Entropy of Selected Elements (continued)*

Oxygen	^{17}O	5/2	.04	.006	2.4
Potassium	^{39}K	3/2	93		
	^{41}K	3/2	7	11.5	.64
Silicon	^{29}Si	1/2	8	.271	1.9
Gold	^{197}Au	3/2	100	11.5	.24
Sodium	^{23}Na	3/2	11	11.5	3.8
Sulfur	^{33}S	3/2	.8	.088	1.1
Tin	^{115}Sn	1/2	.4		
	^{117}Sn	1/2	8		
	^{119}Sn	1/2	9	.96	3.5
Titanium	^{47}Ti	5/2	7		
	^{49}Ti	7/2	6	2.04	1.06
Tungsten	^{183}W	1/2	14	.829	.40
Zinc	^{67}Zn	5/2	4	.611	1.1
Zirconium	^{91}Zr	5/2	11	1.67	1.6

a) Entropy for pure ^{3}He
b) Epoxy is roughly .14 mole/cm^3 H. Elements that magnetically order have been omitted.

Tubes Metal capillary is useful for shielding wires, making your own coax, controlling gas flow, and for crimping together wires that can't be soldered. Cu, brass, CuNi, and SS capillary are common. SS and CuNi are the most frequently used. Both have low K. SS is welded or brazed, but you can soft solder small diameters with an acid flux. CuNi is easily coiled, solderable with a non-corrosive resin. Both SS and CuNi are slightly magnetic. CuNi should not be used to crimp if its Schottky anomaly is annoying (like in a calorimeter). Austenitic stainless is non-magnetic and thus less offensive than the martensitic and ferritic varieties. Unfortunately, working or heating austenitic stainless can alter its properties enough to give magnetic problems in very sensitive applications. Nb capillary, great for crimps and superconducting shielding, is usually a special-order item, which means lots of it at $60/ft. Uniform Tubes, the coax maker, sells Cu, SS, and Al capillary—and will probably make anything for a price. Another source is the Superior Tube Co., 1938 Germantown Ave., Norristown, PA 13404.

Cans and Shields Cans can be SS, brass, or Cu, but SS is the strongest and most reliable, and is suitable for metal O-ring flanges. Cu is used in low-melting solder-sealed cans on some older cryostats. If magnetism is a problem, use Al. For shields, we want to avoid eddy-current heating, which means a thin layer of metal on an insulating back. Cu on fiberglass or Ag on Mylar are recommended. In less critical conditions, copper sheet rolled into a cylinder and capped at one end does fine.

Adsorbers Materials with a large interior surface area due to pores and cavities can suck up an amazing amount of gas as you cool them, hundreds of times their volume of STP gas. If your can is sealed well and pumped out at room temperature, a layer of charcoal on the bottom can keep the pressure down in the micron range for days at 4 K. Charcoal starts desorbing He near 30 K so it's not useful then. While ill-characterized, charcoal is cheap. Vycor, a Corning product, is made by leaching the alkali out of an alkali-rich glass. It comes in a variety of shapes and sizes, but is brittle and hard to work. It's useful for studying the properties of gas layers themselves, or can be used to release and store gas in the cryostat by heating and cooling. Since it has a low K, it must be used in thin (few mm) sheets or tubes. The best way to get it is to call up Corning and see if they have some odd bits lying around—it's an intermediate step in a process leading to a commercially more important product. Be sure you get *porous* Vycor—the final product has the same name. Zeolite and Grafoil (exfoliated graphite) are similarly useful.

PrNi$_5$ (coolant metal) Has a low ($<$ 1mK) ordering temperature. Contact the Ames Research Lab, Material Preparation Center, Ames, Iowa. It is often soldered with cadmium because of cadmium's low critical field and good thermal conductivity (other solders superconduct and don't carry heat well), but it is not essential to use this expensive and toxic material. Cadmium can be obtained from Koch-Light Industries, Haverhill, Suffolk, England, or from A.D. Mackay. PrNi$_5$ often develops fractures after repeated thermal cycling, which impedes heat flow and reduces its effectiveness.

3.2.7   Recommended Reading

- The series of books on *Thermophysical Properties of Matter* by Touloukhian, Powell, Ho, and Klemens (Plenum, 1970) contains all the relevant thermal properties—thermal conductivity, specific heat, radiative properties, diffusivity, and thermal expansion—for hundreds of metals, insulators, and semiconductors, along with references to the original experiments and best estimate curves.
- Taylor, B.N., Parker, W.H., and Langenberg, D.N., 1972, *AIP Handbook of Physics* (McGraw-Hill, New York). Topical summaries of various fields of physics (mechanics, acoustics, etc.) with relevant materials data in each section.

- Smithells, Colin J.(ed.), 1976, *Metals Reference Book*, 5th ed., (Butterworths, Boston). Vast amounts of practical information on metals and how to work them, including soldering, welding, and electroplating.
- Corruccini, R.J., and Gniewek, J.J., 1960, *Specific Heat and Enthalpy of Technical Solids at Low Temperatures NBS Monograph 21.*

3.3   Vibration Isolation

by Roman Movshovich

There exist many types of experiments that are very sensitive to the unwanted vibrations of the apparatus. To name a few:

- Optical experiments, such as holography, interferometry, etc.
- Magnetization experiments of all sorts (flux noise).
- Nuclear magnetic refrigerators, where the heat is produced by eddy current heating.
- Mechanical resonators, where stray vibrations produce frequency and amplitude noise.

There are many sources of vibration, such as:

- Building vibrations that result from the wind, traffic and machinery. These couple to the apparatus through the support frame.
- Pumps that couple to the apparatus through the pumping lines.
- Everything else (see below).

The following is the result of an effort to gather knowledge about vibration isolation from several sources but mostly the experience of Cornell people.

3.3.1   Uncoupling Vibrations from the Building

Conventional isolation systems are passive and achieve their performance by behaving like soft springs. This is accomplished by using soft materials such as rubber, composite pads, springs, or using pneumatic systems. In contrast, active systems sense forces acting on the system or system motion and try to null disturbances by applying counteracting forces (Saulson, 1984). Today only a limited number of such systems have been built, and only for special applications at considerable expense.

It is desirable from the isolation standpoint to make the resonant frequency as low as possible, which has less effect on the apparatus, and greatly improves performance for all frequencies above the resonance as well. Experience has shown that considerable building motion occurs at frequencies as low as 2 to 4 Hz, with much of the energy between 5 and 15 Hz, the typical pendulum frequency of cryostat. This means that the resonance frequency should be of the order of 1 Hz and definitely below 2 Hz. This effectively rules out rubber or composite pads since their resonances typically fall right in the spectral region where the floor motion is the greatest, and as a result they act more like amplifiers than like isolators of floor vibrations. Steel springs offer another possible solution, but this system might suffer from difficulties associated with lateral

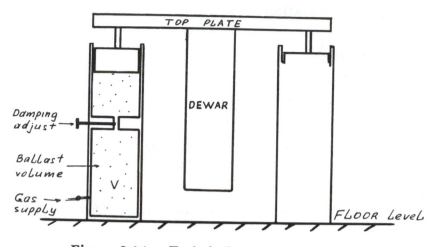

Figure 3.14.   *Typical vibration isolation system.*

instabilities and damping, that will inevitably degrade overall performance. This leaves the pneumatic system as the sole remaining alternative for good vibration isolation.

An air spring system offers several satisfactory solutions to the problem of vibration isolation. It effectively synthesizes the low frequency characteristics of a very long spring, while offering the distinct advantage of relatively small physical size. A pneumatic isolation mount can be thought of as a pressurized cylinder with an internal piston. The pressure acting against the piston provides the static upward force required to support the load. The compressible nature of the gas in the cylinder provides the low spring constant required for vibration isolation. In practice, the rolling diaphragm is used to provide a leakproof and almost frictionless piston seal. Air springs first and foremost achieve the goal of very low resonance frequency—often below 2-3 Hz and occasionally below 1 Hz. The schematic picture of the apparatus using pneumatic isolation is given in Figure 3.14.

The spring constant k is a function of the area of the piston and volume of the air in the chamber. A small piston in a large chamber will produce small changes in gas pressure with its excursions, small changes in force, and will yield a small k value—a soft, or very low frequency spring. Very soft air springs that serve up to several thousand kilograms can be made in very compact sizes. Generally the air volume required is several liters. A given air spring can work well over varying load conditions. This is accomplished by adjusting the gas pressure in the chamber to suit the applied load, which overall does not change the characteristics of the spring over a wide load range.

Air springs can be made as single or multiple chamber devices (Ried, 1977). A single chamber device will use the flexure resistance of its rubber seal or an external damper to damp the motion of the air spring around its resonance frequency. A multiple chamber device will use a choke-flow technique to cause the air in the spring chamber itself to contribute a damping force. In the former system, shown in Figure 3.14, as the air spring piston moves and causes a pressure change in the chamber to which it is directly attached, air is forced through a small restrictive orifice to the next chamber in the air spring. This type of system has the potential for the best performance, since the air-flow damping has the least spurious resonance. The design considerations in the second scheme are the ratios of the volumes of the two chambers and the restrictiveness (impedance) of the orifice. Adjustable impedance (needle valve) can be used to fine-tune the damping of the isolation system. Horizontal displacements are a different matter in that they involve the lateral distortion of the diaphragm rather than the simple rolling action. Hence the horizontal behavior for the rolling diaphragm-type pneumatic mounts is still 'spring-like' but typically much stiffer than in the vertical direction.

The Newport Research Corporation's horizontal isolation system (NRC catalog, 1984) is one way to solve the problem that gives a resonant frequency of less than 2 Hz. Typical characteristics of the NRC's systems are given in Figure 3.15. The XL-systems have horizontal isolation and the XK do not. The Micro-g vibration isolation systems from the Technical Manufacturing Corporation have similar design features. We tested the TMC air springs and found that they gave excellent vibration isolation.

Several home-built vibration isolation systems are shown. Figure 3.16 shows a diagram of the cryostat with a pneumatic vibration isolation system of a two-chamber type that was built by Steve Gregory at Cornell. An air spring itself is also shown in detail (Sagan, 1985). A soft metal bellows (#0126-100-400-080 from Flexonics Inc.) was used to isolate the pressurized gas chamber and at the same time avoid the not-air-like contribution that the stiff bellows would have given. Due to the small diameter of the bellows (2 in.=5 cm) the pressure of about 1.1 MPa (160 psi) was needed to float the apparatus. At the same time a very low resonance frequency of less than 0.1 Hz was achieved. A very important advantage of this scheme is that it provides a restoring force in the horizontal plane and therefore effectively isolates the system from horizontal vibrations.

Another, and somewhat less expensive, vibration isolation system also of a double-chamber type is shown in Figure 3.17. It was built at Cornell by Dave McQueeney. In his system the air springs used were manufactured by the Firestone Co., and go by the name "double convolution style Airstroke actuators #25." The volume of air in each of the springs is 83 in^3 = $1.4 \times 10^{-3} m^3$, and its cross-section area is 15 in^2 = $1 \times 10^{-2} m^2$. The current price is ~$90 each. The resonant frequency of the vertical motion of the system was calculated to be ~ 1 Hz. The measured resonance frequency of the system turned out to be ~ 3 Hz. One of the possible drawbacks of the scheme is that the isolation from the horizontal disturbances might prove to have a substantially higher resonant frequency. The analysis of the double-chamber design is given below.

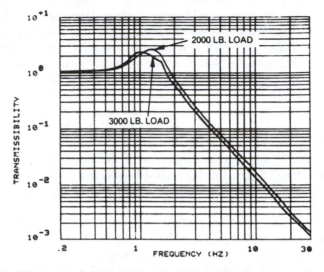

Figure 3.15a *Transmissibility for vertical direction for RS512-12
top supported by NRC XL4A-28 isolation support system.*

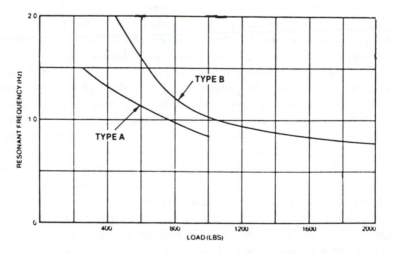

Figure 3.15b   *Variation of resonant frequency in vertical direction with load.*

3.3.2   Uncoupling Vibrations from the Pumps.

Vibrating mechanical devices, such as pumps and compressors, often operate in conjunction with apparatus sensitive to vibration, such as low temperature cryostats, instruments on optical tables, etc. Hence there exists a problem of minimizing the transmission of vibrations due to the pumps. For small pumping

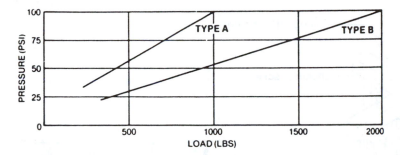

Figure 3.15c *Load carrying capacity of NRC's single type A and B Mount as a function of pressure.*

XL SERIES LEG SYSTEMS	
Support System	System Part Number
Two Type A Mounts and One Type B Mount	XLB2A-H
Four Type A Mounts	XL4A-H
Four Type B Mounts	XL4B-H
Six Type A Mounts	XL6A-H
Six Type B Mounts	XL6B-H
Eight Type A Mounts	XL8A-H
Eight Type B Mounts	XL8B-H

XK SERIES LEG SYSTEMS	
Support System	System Part Number
Two Type A Mounts and One Type B Mount	XKB2A
Four Type A Mounts	XK4A

Figure 3.15d *Table of NRC Mounts.*

lines up to 4 cm (1.5 in.) in diameter the usual approach for providing isolation is to approximate the-soft-spring-attached-to-a-large-mass by using a long piece of metal bellows tubing or rubber hose that is firmly held at some point(s) to a massive object such as a sandbox, a wall, or the floor. The spring and large mass system is quite effective for damping most of the vibrations. For the sandbox arrangement it is better to have both a piece of rubber hose and a metal bellows, with the rubber hose damping the high frequency vibrations and the bellows actually in the sand box, since it provides better coupling to the sand and damps longitudinal vibrations. See Figure 3.18.

However this approach when applied to the larger diameter flexible tubing rapidly becomes ineffective. The metal bellows must be much stiffer since thick

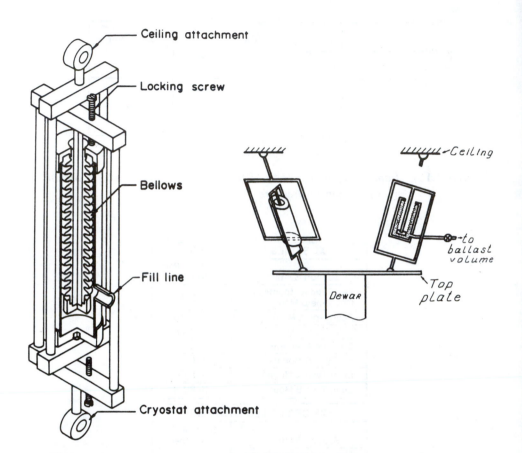

Figure 3.16.   *Vibration isolation system with steel-bellows air springs.*

walls are needed to support the large force arising from moderate pressure differences. To compensate for the stiffness, the bellows length must be increased
and soon becomes prohibitively large in order to approximate a soft spring. Two
simple designs shown in Figure 3.19 do a rather good job. The upper design in
Figure 3.19, for example, uses the ceiling as the large mass, and the low frequency
ceiling vibrations are then coupled to the cryostat through the metal bellows. In
the scheme in the lower part of Figure 3.19 the bellows act as the soft spring,
i.e., their natural vibrational frequency will be much lower than the frequency
of vibration of the pump.

The difficulties with using large-diameter metal bellows have led to designs
that use short sections of very flexible thin-walled metal bellows (Kirk and Twerdochlib, 1978). One scheme frequently used is the crossed-bellows design shown
in Figure 3.20(a,b). Bellows A and B offset the collapsing forces on the bellows

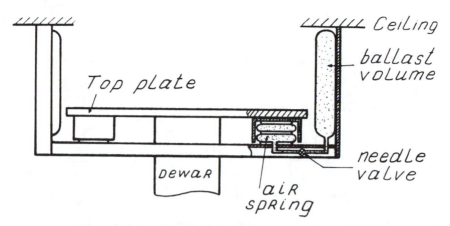

Figure 3.17.   *System with Firestone air springs.*

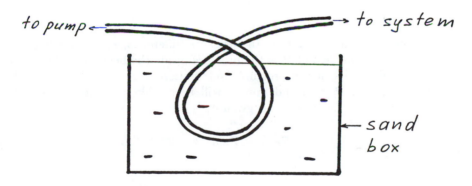

Figure 3.18.   *Sandbox isolation.*

b, and yet allow the output to move rather freely in three orthogonal directions. The rods, c, and the collars, a, prevent the whole assembly from collapse under the pressure difference. A supported double-gimbal metal bellows design, shown in Figure 3.21(a,b), is much more effective. Each bellows, b, with its associated double-gimbal, d, cable, a, and u-support, c, is free to move independently.

To demonstrate that the double-gimbal design is the most effective we have to make an analysis of the above arrangements. The equivalent spring diagrams of both designs are shown in Figure 3.20(b), 21(b) where each bellows is replaced by the spring of force constant k.

Consider the crossed-bellows setup first, depicted in Figure 3.20. Vibrations,

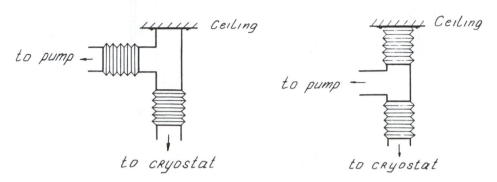

Figure 3.19.   *Pump vibration isolation for large tubing.*

perpendicular to the axis of the bellows, will cause the point where they are joined to undergo an angular displacement θ. The restoring force on this point comes from three contributions:

1. The contribution arising from the pressure difference, $P_0 - P_i$, where P_0 is the pressure outside and P_i is the pressure inside the bellows. If the bellows has a radius r, then the pressure contribution is $T = \pi r^2 (P_i - P_0)$. For vacuum inside, $P_i \approx 0$ and thus T will be sizable and the source of the largest restoring force, which is given by

$$F_T = -2T \sin \theta \approx -2T\theta. \tag{3.1}$$

2. A second contribution comes from the small stretching of each bellows, which is equal to $\sigma \approx (R/2)\theta^2$ (to second order). The restoring force due to the stretching of the springs is

$$F_S = -2k\sigma \sin \theta \approx -kR\theta^3. \tag{3.2}$$

3. The final contribution comes from the bending motion of each bellows since the base is held rigidly. The restoring force due to bending is equal to (Kirk and Twerdochlib, 1978)

$$|F_B| = (8kr/\pi)\theta^2. \tag{3.3}$$

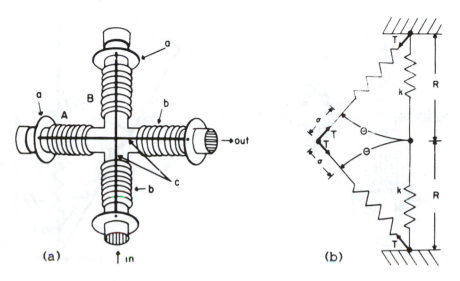

Figure 3.20.   *(a) cross-bellows design, (b) its force diagram.*

Now consider the equivalent spring diagram of the double-gimbal arrangement shown in Figure 3.21. This diagram is different from that in the original paper (Kirk and Twerdochlib, 1978) that contained an error. That error led to an extra factor of $1/2$ in the expression (7) for the restoring force. In this case the U-support rotates around the gimbal pivots by small angle θ. At a distance ϵ away the bellows pivots with a small angle θ'. (This distance should be made as small as practical.) The restoring force comes from the difference of these angles which is equal to $\theta' - \theta = \epsilon\theta/R$ from $\theta'R = \theta(R + \epsilon)$. Here R is the distance between the base of the bellows and the point where the metal string is attached to the elbow. Just as before there are three contributions to the restoring force:

1. The tension T' in the string is equal to the radial component of the force T which is due to the pressure. The restoring force is a tangential component of $T = \pi r^2(P_0 - P_i)$, and is equal to

$$F_T = -T\sin(\theta' - \theta) \approx -(\theta' - \theta)T \approx (\epsilon/R)T\theta. \qquad (3.4)$$

2. The stretching of the bellows $\sigma \approx \epsilon\theta^2/2$ results in the restoring force

$$F_S = -k\sigma\sin(\theta' - \theta) \approx -(\epsilon^2/2R)k\theta^3. \qquad (3.5)$$

3. And finally the restoring force contribution from the bending of the bellows is the same as in the crossed-bellows case.

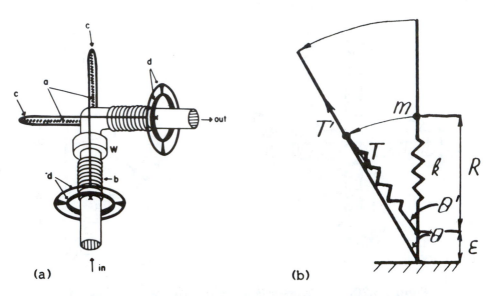

Figure 3.21. *(a) double-gimbal design, (b) its force diagram.*

Comparing the magnitudes of the restoring forces in the two designs above makes it clear why the double-gimbal arrangement is superior. The restoring force due to the pressure difference is a factor of $\epsilon/2R$ smaller in the double-gimbal arrangement. The factor is even smaller in the restoring force due to the stretching of the bellows: $(\epsilon/R)^2/2$. From this analysis we can readily see the ways of improvement: you want to have as small an ϵ as possible, and the larger the gimbal's U-support, the better. Typically, one can arrange $\epsilon \approx 1$ cm and $R \approx 50$ cm, hence the improvement in lowering the resonance frequency is $(1/100)^{1/2} = 0.1$ over the crossed-bellows design.

If we neglect all the restoring forces but the one due to the difference in pressure, then for the crossed-bellows arrangement with 15 cm (6 in.) in diameter, 20 cm bellows we get the resonance frequency of the isolating system to be

$$\omega = \left(\frac{2T}{Rm}\right)^{1/2} = \left(\frac{2\pi r^2 P_0}{Rm}\right)^{1/2} = \frac{13}{m^{1/2}}(\text{Hz} - \text{kg}^{1/2}). \qquad (3.6)$$

The same type of calculation gives for the double-gimbal design with the bellows 15 cm (6 in.) in diameter, the length of the U-support of $2R \approx 1$ m and $\epsilon = 1$ cm the resonance frequency of

$$\omega = \left(\frac{\epsilon T}{4R^2 m}\right)^{1/2} = \left(\frac{\epsilon \pi r^2 P_0}{4R^2 m}\right)^{1/2} = \frac{.84}{m^{1/2}}(\text{Hz} - \text{kg}^{1/2}). \qquad (3.7)$$

The difference between the two resonance frequencies is obvious: we would need an elbow of a prohibitively large mass to bring the resonance of the crossed-bellows arrangement below the low frequency vibrations of the building. At the same time it is very easy to satisfy that condition for the double-gimbal arrangement: an elbow of 16 kg will bring the resonance down to $\approx .2$ Hz.

3.3.3   Other Sources of Vibration

Sound is probably the next most important source of vibration noise. It turns out that just by gluing a piezoelectric crystal to the top plate of the not carefully sound-isolated cryostat and connecting headphones to its electrical output it is possible to hear people talking in the room. Also sound can couple not only to the vibrational modes of the solid construction itself but also to the sound resonance of the sample cell inside the cryostat. The resulting resonance can be easily observed by sweeping the frequency of a speaker positioned next to the cryostat and following the behavior of a vibration-sensitive parameter, for example the SQUID output in the non-uniform magnetic field. These types of resonances usually fall above .5 Khz. One of the ways to decouple the apparatus from sound is to put it inside a box with the walls made of a sheet of lead sandwiched between foam layers. This can reduce the sound coupling to the apparatus by up to 50 dB.

It is also necessary to take care that electrical leads to the apparatus do not transmit vibrations generated by the equipment outside of the setup. For example the leads must be firmly attached to large masses in several places. It is advisable to send electrical leads that go directly to the sample region, such as current leads of the superconducting magnet, through the sandbox. Otherwise the vibrations generated by the electrical fans will be transmitted to the apparatus.

On the whole, **beware**.

3.3.4   Structure Considerations

The apparatus should provide the required stability and rigidity for the applications of interest within the envelope of anticipated disturbances. Rigidity of the apparatus is determined by measuring the response of the system while subjecting it to dynamic force. The quantity measured is compliance, which is the ratio of displacement to force. Low compliance values are achieved by optimizing three factors: damping, stiffness and mass. Scaling up the basic stiffness of the structure obviously helps, as does increasing its mass, as long as it does not depress the resonant frequencies of the structure. Still more important is the damping behavior at the resonant frequencies of the apparatus. Besides being the primary factor in determining the compliance value, damping determines how quickly an induced disturbance will subside. Vibrational energy that is coupled into the cryostat acoustically, or produced by the equipment on top of the cryostat (vacuum lines) remains trapped there. Damping within the top plate is the only effective means of dissipation. Metal plates have very low damping, and it

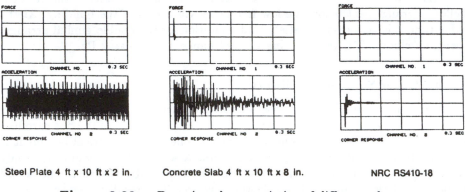

Steel Plate 4 ft x 10 ft x 2 in.   Concrete Slab 4 ft x 10 ft x 8 in.   NRC RS410-18

Figure 3.22.   *Damping characteristics of different plates.*

is always a good idea to sandwich them between two plates of "dead" material, such as wood, etc. For comparison the responses to δ-function-like disturbances of different plates are shown in Figure 3.22, with the best performance given by the NRC's honeycomb top (NRC catalog, 1984).

3.3.5   Evaluation of the Performance of the Cryostat

To evaluate the performance of the cryostat you can measure the vibration spectrum of the apparatus with an accelerometer hooked up to a spectrum analyzer, as shown in Figure 3.23. Then the spectrum is taken either directly, to see the actual state of the cryostat vibrations, or the response of the cryostat to a disturbance of a particular frequency is measured by using a variable vibration frequency source that can be constructed from a DC driven servomotor and a variable voltage source. In the former case the performance of the vibration isolation system can be tested by comparing the vibration spectra with and without the pumps turned on, and the whole system floating and not floating.

3.3.6   Analysis of the Double-chamber Air Isolation System

For a spring supporting a mass m the equation of motion is

$$m\frac{d^2x}{dt^2} = -k(X - Y) \tag{3.8}$$

where Y is the displacement of the floor, and X the displacement of the mass m. Solving this equation in frequency domain gives us

$$X(\omega) = Y(\omega)\frac{1}{1 - m\omega^2/k} \tag{3.9}$$

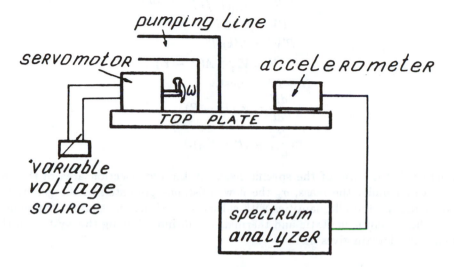

Figure 3.23.   *Setup for testing the vibration isolation.*

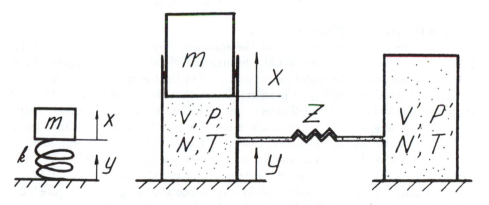

Figure 3.24.   *The diagrams of simple spring and double-chamber air support systems.*

which has a pole at the resonant frequency $\omega_{\text{res}} = (k/m)^{1/2}$

For an air support system connected to the reservoir through a flow impedance Z, and assuming no heat flow (isentropic system), we get a system of coupled equations:

$$T = T_{\text{eq}}(P/P_{\text{eq}})^{(\gamma-1)/\gamma} \tag{3.10}$$

$$T' = T'_{\text{eq}}(P'/P'_{\text{eq}})^{(\gamma-1)/\gamma}$$
$$PV = Nk_BT$$
$$P'V' = N'k_BT'$$
$$V = V_{\text{eq}} + A(X - Y)$$
$$N + N' = \text{const.}$$
$$\frac{dN}{dt} = \sigma_z(P' - P) \tag{3.11}$$
$$m\frac{d^2x}{dt^2} = (P - P_{\text{eq}})A$$

where γ is the ratio of the specific heats, A the cross-section area of the column of air under the mass, σ_z the flow resistance presented to the gas by the impedance Z, and all quantities with subscript "eq" are the equilibrium quantities: the system is in static and thermal equilibrium. Solving this system in the frequency domain gives us

$$X(\omega) = Y(\omega)\frac{1}{1 - m\omega^2/k} \tag{3.12}$$

where $k = PA^2/V_{\text{eff}}$, with

$$V_{\text{eff}} = (1/\gamma) \times \left(V + V'\frac{1}{1 + i\omega V'Z\eta/P}\right) \tag{3.13}$$

and η is the viscosity of the gas.

From this result we can see that the resonant frequency $\omega = (k/m)^{1/2}$ depends on the mass of the cryostat only through the effective volume V_{eff}, since the pressure of the gas in the system is proportional to the the mass m. On the other hand, the dependence of V_{eff} on pressure is such that the optimum impedance of the path connecting the two volumes scales with the pressure and, consequently, the mass of the system. We can easily derive the results for infinite impedance and no impedance at all. In the first case we get $V_{\text{eff}} = (1/\gamma)V$, corresponding to the single air chamber, and in the second case we get $V_{\text{eff}} = (1/\gamma)(V + V')$, the total volume of the system. Several curves of the transfer function $X(\omega)/Y(\omega)$ were calculated for different values of the impedance and are given in Figure 3.25. The two peaks corresponding to $\sigma_z = 0$ and to $\sigma_z = \infty$ are well pronounced.

It is very important to notice that the system shown in Figure 3.16 can be made to provide an effective horizontal isolation. The primary consideration must be to have the axes of rotation of the air springs go through the center of mass of the cryostat. In Figure 3.26, X is the displacement of the cryostat in the horizontal direction, and Y is the displacement of the ceiling. In that case we have: a, the constant distance between cryostat top plate and the ceiling, l, the length of the air spring, and b, the length of projection of the spring on the ceiling. Now, if the axes of air springs go directly through the center of mass of the cryostat, the forces created by compression and expansion of air springs

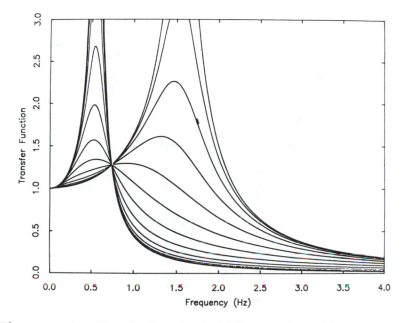

Figure 3.25.   *Transfer function for different values of impedance Z.*

will not generate the rotational motion of the cryostat in the vertical plane. In fact they will cause translational motion only. To calculate the restoring force on the cryostat, we have: $l = (a^2 + b^2)^{1/2}$, $\implies$ the change in the length of the air spring is $\Delta l = (b/l) \times (Y - X)$, and the restoring force for air spring 1 is $F = (b/l) \times (P_{eq} - P) \times A = (b/l)(PA/V_{eff}) \times \Delta V_{eff} = (b/l) \times (PA^2/V_{eff})\Delta l = (PA^2/V_{eff}) \times (b/l)^2 \times (Y - X)$. Here the meaning of all quantities is the same as in the analysis for the vertical vibration isolation. We are even better off than in the previous case due to the additional factor $(b/l)^2$, that pushes the resonant frequency lower yet.

3.3.7   Some Companies

- Technical Manufacturing Corporation, 15 Centennial Drive, Peabody, MA 01960, (617) 532-6330, Telex 951408. Manufactures excellent vibration isolation systems. TMC's engineering staff is very helpful in designing systems for special applications.
- Newport Research Corporation, 18235 Mt. Baldy Circle, Fountain Valley, CA 92708, (714)-963-9811; Telex 685535. Sells vibration isolation systems, especially the ubiquituous optical tables found in laser labs.

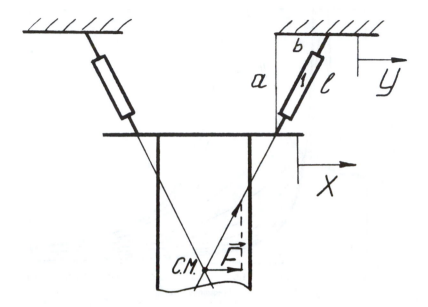

Figure 3.26. *Diagram for horizontal vibration isolation.*

- Ealing Corporation, Pleasant St., South Natick, MA 01760. (617)-655-7000. Sells the same stuff.
- Kinetic systems, Inc., 20 Aboretum Road, P.O. Box K, Roslindale, MA, 02131, (617)-522-8700. or 75 Massasoit St., Waltham, MA 02154, (617)-893-7747. Sells the same stuff.
- Norman Equipment Co., 9850 South Industrial Dr., Bridgeview, IL 60454, (312)-430-4000. Stocks and distributes Firestone isolators.
- U.S.Flex, 5025 Hampton St., Vernon, CA 90058. Sells very nice bellows tubing.
- Hyspan Precision products, Inc., 1685 Brandywine Ave., Chula Vista, CA 92011, (619)-421-1355. Sells bellows tubing.

3.4   Electric and Magnetic Isolation

by Robert S. Germain

The need to shield experiments from external interference is of growing importance in low temperature physics. This need is driven by two main concerns:

- Many of the experiments being performed or considered necessitate the measurement of very small signals.
- The quest for lower temperatures makes the problem of rf heating more acute as heat capacities decrease.

Because of space limitations, I will not spend a lot of time reproducing discussions about capacitive and inductive coupling which may be found in the references at the end of this section and in the section by John Denker. Rather, I will relate recipes actually used for wiring a cryostat and shielding low temperature experiments against external interference.

3.4.1   Electric and Magnetic Fields

At low frequencies, interference coupled in via electric and magnetic fields can be treated separately. Electric fields introduce noise via the mutual capacitance between noise sources and your signal lines. At low frequencies, a capacitive noise injector can be modelled as a current source. Measurements made around the lab by John Denker indicate that each meter of coax picks up several microamps of noise current. Thus, a circuit with a large load impedance is most susceptible. The standard method for suppressing this coupling is to surround the signal lines with a shield held at ground. This can be accomplished using coaxial cable, but one must be careful about where the outer is tied to ground. It may be preferable to make a balanced system, that is, one for which both signal lines have the same impedance to ground. The balancing aids in eliminating common mode interference. If the same noise voltage appears on each signal line, then an ideal differential amplifier reading out the circuit eliminates the common mode signal. Balancing is most easily accomplished by floating the source and carrying the signal over twisted pair with an external shield at ground for electric field shielding. Large currents flowing through the shield should be avoided because its effectiveness depends on being at a constant potential. Twisted pair is generally useful up to about 1 MHz. At higher frequencies, triaxial cable, coax with a separate outer ground shield, is preferable because of excessive losses in twisted pair.

Magnetically coupled interference arises from time varying magnetic fluxes passing through circuit loops (inductive coupling). The noise appears to be generated by an ideal voltage source. Hence large noise currents may flow in circuits with small loop impedances. Low frequency magnetic isolation is much more difficult to achieve than electrostatic shielding. A minimal precaution against this

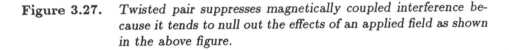

Figure 3.27.   *Twisted pair suppresses magnetically coupled interference because it tends to null out the effects of an applied field as shown in the above figure.*

Table 3.5.   *Some Typical Values of Magnetic Shielding Parameters*

	permeability	H_{sat}
co-netic:	25,000	7500 G
netic:	200	21,400 G

is the use of twisted pair to null out the total flux through the loop, the more twists per inch the better (Figure 3.27).

An ordinary metallic shield must be very thick to have any effectiveness against inductively coupled low frequency noise. High permeability materials such as mu-metal are useful for magnetic shielding, but you should make sure that the shielding material has been annealed and degaussed before use. If the shielding is magnetized, the dc field inside the shielded area will cause interference when leads vibrate. This can be an extremely serious problem if low signal levels are being measured. In the earth's field, a centimeter of line vibrating at 1 KHz with an amplitude of 0.1 mm will induce a signal of order a microvolt. For comparison, room temperature amplifiers commonly in use around the lab have input voltage noise of a few nanovolts/$Hz^{1/2}$.

In addition, be warned that advice from "magnetic shield engineers" should taken with a grain of salt. One experiment whose magnet had to be shielded ended up with an iron shield spot welded to the mu-metal outer. When we turned on the magnet, the iron was magnetized and its fringing field saturated the mu-metal, making the shield useless. The proper configuration for a magnetic shield set-up is a relatively low susceptibility, high saturation shield closest to the magnetic field and a high susceptibility, lower saturation shield some distance away from the first.

3.4.2   Superconducting Shields

One advantage that low temperature experiments have is that superconducting shields may be used. These provide nearly ideal shielding and are generally employed for the most sensitive experiments. We use niobium capillary for shielding twisted pairs from 1 K on down. These are injected with epoxy to prevent the leads from moving around inside and possibly inducing emfs due to trapped fields.

We use niobium tubing to shield inductors and SQUIDS in a convenient way. Lead foil is used for more irregularly shaped objects, but we are careful to solder up the seams in these packages. It is also difficult to make electrical contact from the lead foil to the niobium capillary which brings in the leads to the shielded component.

The possibility of trapping magnetic fields in a superconducting shield makes it advisable to surround the entire dewar with a mu-metal shield as the cryostat cools. This cuts down on the field which is present when the superconducting shields go through T_c.

This does not help, however, when large magnetic fields are applied in the experimental region inside the cryostat when, for example, a field is trapped for a platinum nmr thermometer.

The most sensitive cells or devices should be provided with their own mu-metal shields *outside* the inner superconducting shield. The temperature dependence of the mu-metal permeability is not known very well, although it decreases at low temperatures. Another material, cryo-perm, is rumored to retain large permeability even at liquid helium temperatures. In the past, Nb-Ti has been used as a shielding material, but above H_{c1} it allows flux creep. Based on the values for the pure materials Nb is a better bet for a superconducting shield (H_c(Nb) = 1980 G).

3.4.3   Wiring a Cryostat

At the top of the cryostat is the feedthrough from the top plate into the bath or vacuum space (Figure 3.28). We have found that the MS hermetic connectors available from Cannon or Detoronics are a nice way to bring shielded twisted pairs into a cryostat. These connectors can either be soldered into a flange or sealed into place with an indium O-ring. Avoid thermally shocking the connectors when soldering. Use a hot plate, heating the connector along with the bulkhead in which it will be mounted.

If possible, the cryostat should float with respect to ground when it is constructed. This makes it easier to locate and eliminate ground loops. When the one connection to ground is made, you should check the wiring of any 60 Hz equipment attached to the cryostat carefully. It is possible to encounter the following situation.

A diffusion pump attached to the support structure of the cryostat has a miswired plug so that the wires connecting the safety ground and the current return are interchanged. This leads to a 12 amp current flowing through the

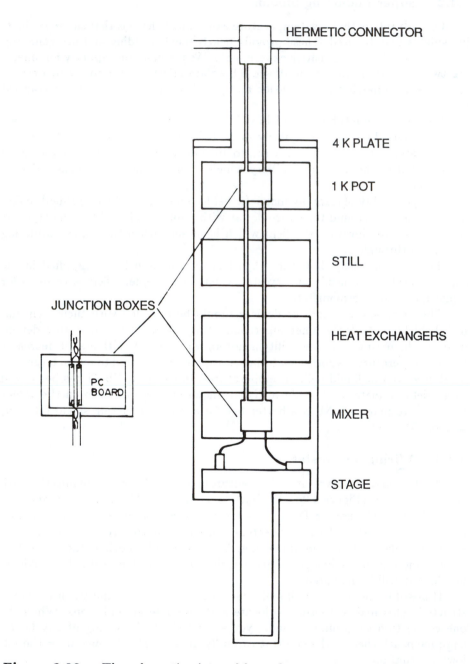

HERMETIC CONNECTOR

4 K PLATE

1 K POT

STILL

JUNCTION BOXES

HEAT EXCHANGERS

PC BOARD

MIXER

STAGE

Figure 3.28.   *The schematic pictured here shows some general features common to the wiring of many cryostats.*

cryostat structure causing 7 volt, 60 Hz oscillations of the cryostat with respect to ground.

It is very difficult to fix shorts in cryostat wiring. Therefore extra leads should be provided to serve as spares when the inevitable shorts occur. One helpful technique used to avoid shorts is to thread twisted pairs through #30 teflon spaghetti (alpha tubing) before inserting into Cu-Ni capillary. The formvar insulation on most of the wire we use is fairly fragile and the process of drawing the pair through metal tubing will almost certainly rub off some formvar if this precaution is not taken. We may switch over to using superpolythermalese or duPont ML insulation some day because it is tougher stuff, but formvar insulated wire is more commonly available and formvar is much easier to remove.

When starting from scratch, I prefer to bring leads down from room temperature in vacuum to avoid bath level dependent effects such as changes in cable capacitance. In keeping with the general principle of modularity, there should be a junction box for low frequency lines at low temperatures. This may be a metal enclosure containing a circuit board connecting leads going up with leads continuing down. Depending on the degree of isolation between channels required, the junction box may or may not be subdivided into separate shielded compartments. There is usually another junction box on the mixing chamber which also serves as a convenient point for heat-sinking leads. Junction boxes made out of niobium are used for the most critical lines.

It is difficult to reconcile the two goals of reducing the heat leak down leads and maintaining shielding integrity. Above mixer temperature, it may be sufficient to clamp the capillary securely to heat sink it and rely on conduction through epoxy injected into the capillary to cool the wires within. Below 50 mK, you can take advantage of the very poor thermal conductivity of superconducting niobium to reduce your heat leak. From mixing chamber temperature on down, only Cu-Ni clad superconducting wire should be used. The Cu-clad wire has much too large a thermal conductivity. Thus the use of niobium capillary serves two purposes: first, to shield, and second, to preserve the thermal isolation of different segments of the cryostat.

3.4.4   RF Shielding and Filtering

Another aspect of the emi problem for low temperature experiments is the injection of rf (megahertz and up) energy into the cryostat. A common rf induced problem is a heat leak caused by a local radio station which has the property of decreasing when the radio station turns down its power at night. FM stations often ride in over resistance bridge wiring whose signals are very low frequency. Since gaps in shielding of order $\lambda/20$ are significant, making an enclosure rf tight means making it water-tight also. The use of conductive gasket material is required when a totally rf tight system must be constructed. Since it is difficult to achieve effective rf shielding, the principle of shielding as few different areas as possible should be followed. A standard procedure is to place the entire cryostat inside a shielded room and running filtered lines to the outside world via a feedthrough panel. Such rooms commonly provide 120 dB of attenuation from 50 KHz to 1 GHz. The existence of this large rf-free area means that individual lines inside the cryostat do not have to be perfectly rf-tight. A related precaution which may help even if there is no shielded room is to put shields around known rfi sources. It is easier to shield one source than many vulnerable circuits.

Whenever filtering does not induce unwanted changes in the signal you are looking at, it is desirable to filter lines as heavily as you can. When choosing components for use in filters, keep in mind that the parasitic reactances present in all devices will cause them to self resonate sometime. A 100 mH inductor in combination with a 10 ohm resistor may seem like just the thing to roll off the response at 160 Hz, but if it self-resonates at 50 KHz due to the interwinding capacitance of the coil it won't do anything above that frequency. Therefore, a stacked approach using some components to give the low frequency rolloff and others to keep that filter attenuation up at higher frequencies should be used. Inside the cryostat, series resistors at 4 K in signal lines in combination with the capacitance present in the twisted pair give a low pass characteristic to the system. This added loop resistance has the additional advantage of reducing the current noise induced by magnetically coupled interference. Also, the epoxy injected into shielded twisted pair is quite lossy in the rf region.

A very important component in the elimination of unwanted high frequency radiation inside a cryostat is the use of emi filters on the lines going into the top plate. These are pi section filters which remain effective up through GHz frequencies. We generally use emi filters from Erie (pt. no. 1233-000) which start having significant attenuation at 1 MHz and reach 60 dB insertion loss at 100 MHz. These filters are mounted in a Pomona box separated into two compartments by a piece of double sided copper clad circuit board in which the emi filters are soldered or silver epoxied. The use of the ground plane is important because of possible capacitive coupling between input and output. Some thought should be given to the circuit you have when both emi filters and large series resistances are used. One can get significant voltage division of ac signals going down such lines.

The dewar itself can serve as a shielded enclosure if all the lines going into the top plate are filtered properly. An emi tight connection can be made from the top plate to the dewar by replacing the usual rubber O-ring with a flexible conductive composite available from Cho-merics Inc.

High frequency lines needed for experiments cannot use emi filters, but the principle of bandwidth limitation still applies. Band pass filters on such lines reduce the unwanted rf in the cryostat. Filters can be installed at low temperatures, but the use of capacitors at millikelvin temperatures should be approached with caution because of possible temperature dependence in the dielectric constant. Even for use at one degree and above, the capacitors and inductors used should be tested in a storage dewar.

Returning to the environment outside the cryostat, there are frequently preamplifiers between the top-plate connectors and the interface to the outside world with its lock-ins and other local oscillators. They also serve to isolate filters at the

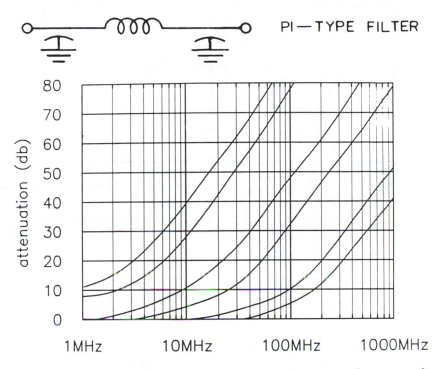

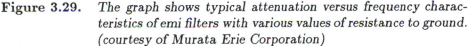

Figure 3.29. *The graph shows typical attenuation versus frequency characteristics of emi filters with various values of resistance to ground.
(courtesy of Murata Erie Corporation)*

cryostat head from those at the wall. At the shielded room feedthrough panel,
the emi filters on all the low frequency lines keep computer clocks and FM radio
stations out of the presumed rf-free region.

3.4.5 Quick Ground Loop Fixes

If a circuit has two ground references, there is the possibility of current flow
along the ground line which is unrelated to the signal. When this is unavoidable,
there are several ways to reduce the problems caused by the resulting ground
loop.

Isolation transformers break the circuit at dc and couple the signal inductively. When working at low frequencies, you must analyze the circuit to make
sure that the impedances in your system allow this scheme to work.

Baluns are related to isolation transformers, but allow dc continuity while
suppressing common mode signals. Even better, they are simple to make: Just
wind the coax carrying the signal around a ferrite bagel as many times as possible.

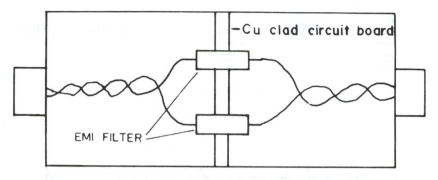

Figure 3.30.   *The feedthrough panel into a shielded enclosure should include emi filters mounted in a ground place as shown above. We generally use Pomona boxes for modularity of construction.*

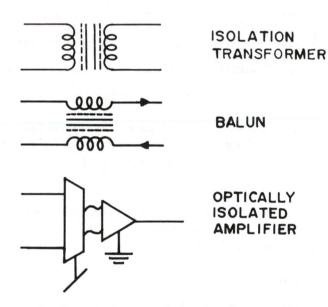

Figure 3.31.   *Some of the options available for breaking ground loops may be more or less attractive depending on whether dc response is required.*

From the circuit diagram you can see that a current flowing in the ground side only will see a large inductance, but a signal will not.

Opto-isolators break the ground completely like isolation transformers, but also can respond at dc without any modulation/demodulation scheme. Extremely linear opto-isolators may be purchased these days from Burr-Brown.

The fact that these options are available does not mean that you can be careless about grounding. None of these methods may be enough in some cases. When this is true, a combination of careful grounding and other means of isolation may be required to achieve a sufficiently low noise level.

3.4.6   Sources of Materials and Equipment

- Ad-Vance Magnetics, 625 Monroe St., Rochester, IN 46975, (219) 223-3158. Magnetic shielding materials.
- Biomagnetic Technologies (formerly S.H.E.)
- Cho-Merics, 77 Dragon Court, Woburn, MA 01888, (617) 935-4850. Conductive caulks and composite elastomers.
- Detoronics Corp.,10660 E. Rush St., P. O. Box 3805, So. El Monte, CA 91733 (213) 579-7130, (800) 423-4314. Hermetic MS connectors.
- Erie Technological Products, Ltd., Trenton, Ontario, Canada, (613) 392-2581. EMI filters.
- Ferronics, Inc., 52 N. Main St., Fairport, NY 14450. Ferrite cores, bagels.
- ITT Cannon, 10550 Talbert Avenue, P.O. Box 8040, Fountain Valley, CA 92728, (714) 964-7400. All kinds of MS connectors.
- Oxford Instruments, NA, 3A Alfred Circle, Bedford, MA 01730, (617) 275-4350. Until recently Oxford carried small quantities of small capillary which they shipped to customers very quickly upon request. For some reason, they have abandoned this valuable service. Let us hope that they can be encouraged to resume it.
- Perfection Mica–Magnetic Shield Division, 740 N. Thomas Drive, Bensenville, IL 60106, (312) 766-7800. Magnetic shielding.
- Ray-Proof, 50 Keeler Avenue, Norwalk, CT 06856, (203) 838-4555. Shielded rooms.
- Tek-nit. More conducting composites and rf gasket material.
- Uniform Tubes, Inc., 7th Ave., Collegeville, PA 19426. Hardline, all kinds of capillary.

3.4.7   Sources of Further Information

- *RF Design*, Cardiff Publishing Company, P.O. Box 6275, Duluth, MN 55806: Articles about rf and emi as well as many ads for shielding materials.

- *IEEE Transactions on Electromagnetic Compatibility*: Best if you want to know how to ground a Saturn V rocket or how to avoid frying the electronics on an F-15 when it's struck by lightning, but has some interesting articles.

3.5   Cooling the Sample

by Eric T. Swartz

3.5.1   Internal Time Constant of the Sample

Suppose you have a sample connected on one side to a constant temperature bath, but the sample hasn't equilibrated yet. To calculate how long that will take, assume that the sample is a cylinder attached at one end. Specifically, Figure 3.32 shows the geometry. We will use the following notation.

$$c = \text{specific heat in Joules/cm}^3\text{K}$$

$$A = \text{cross-sectional area of the sample in cm}^2$$

$$l = \text{length of the sample in cm}$$

$$\Lambda = \text{thermal conductivity in Watts/cm K}$$

$$D = \Lambda/c$$

To eliminate unnecessary grunge, we will model the above geometry by lumped elements as is shown in Figure 3.33.

Here, $C = (1/2)cAl$, and $1/R = \Lambda A/l$. The 1/2 is a guess; the heat capacity is distributed, and not all of it contributes. I really don't like to turn this into an electrical model, because I can never figure out what to do with the capacitors. It is much easier to use the definitions of thermal conductance and heat capacity. One Watt across a conductance of one Watt/Kelvin produces one Kelvin temperature drop, and one Joule into one Joule/Kelvin raises the temperature one Kelvin.

In our model, we define the temperature of the sample to be $T_1(T)$ and the temperature of the bath to be T_0. Then the heat flux across the thermal conductance is $(T_1 - T_0)/R$, and $d(T_1 - T_0)/dt = (T_1 - T_0)/RC$. The result is the thermal time constant which is (suggestively) RC.

$$RC = 4cl^2/\pi^2\Lambda. \tag{3.14}$$

The $4/\pi^2$ was put in to make the result "exact." Note that the guess of 1/2 was quite good. The exact value of RC is obtained by solving the heat equation.

$$dT/dt - D\,d^2T/dx^2 = Q. \tag{3.15}$$
$$(D = \Lambda/c, \text{and is the Diffusivity})$$

The forcing function Q (here, the heat input per unit volume) is zero, and I will take the initial condition to be $T(x,0) = T_0 + T_1 \sin(\pi x/2l)$. This automatically

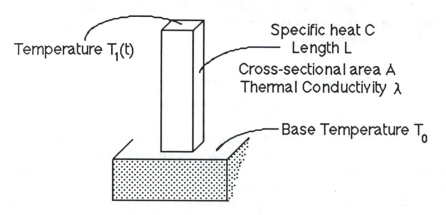

Figure 3.32.   *Thermal model of sample.*

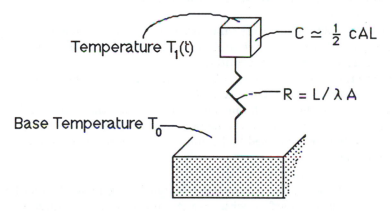

Figure 3.33.   *Effective thermal model of sample.*

satisfies the boundary conditions:

$$T(0,t) = T_0, \text{ and}$$
$$dT(l,t)/dx = 0. \tag{3.16}$$

The solution is precisely

$$T(x,t) = \left(T(x,0) - T_0 \right) e^{-t/RC} + T_0, \tag{3.17}$$

using the value of RC shown above. Other methods of solution are separation of variables, and looking in Carslaw and Jaeger(1959).

In the rest of this section the analyses will be much less explicit. The example shown above demonstrates the type of analysis which can be done, and which must be done if quantitative results are sought.

3.5.2 Thermal Boundary Resistance

The next problem in getting your sample cold is getting the heat through the interface between the sample and the sample holder. This is a critical problem in the millikelvin temperature regime, or when the interface cannot be made to be an intimate contact. The following are useful references on thermal boundary resistance: Little (1959), Pollack (1969), Cheeke, Ettinger and Hebral (1976), Katerberg, Reynolds and Anderson (1977), Kaplan (1979), Simons (1974), Schmidt and Umlauf (1976), Wolfmeyer, Fox and Dillinger (1970), Matsumoto, Reynolds and Anderson (1977), Schmidt (1977), Peterson and Anderson (1972 and 1973), O'Hara and Anderson (1974), Schumann, Nitsche and Paasch (1980), Reynolds and Anderson (1976). There are four different geometries into which most experimental situations can be mapped.

1. The sample is in good electrical contact to the sample mount; consequently quite good thermal contact usually results. **Such contacts should not, however, be taken for granted.**
2. The interface between the sample is a nice interface, like that between a vapor-deposited metal and a crystal.
3. The sample is glued to the mount, in which case there are two interfaces, each between a crystal (maybe metallic) and a glassy material.
4. The sample is being cooled by helium, or the sample is helium being cooled by some heat exchanger.

First, I will discuss the parameters relevant to a general thermal boundary resistance. Given an interface with area A and a heat flux $\delta Q/A$ across that interface there will be a temperature drop at the interface. The magnitude of the temperature drop is:

$$\delta T = \delta Q \, R_{bd}/A, \tag{3.18}$$

where R_{bd} is a boundary resistance; R_{bd} has the units K/(W/cm^2). A boundary resistance of one K cm^2/W means one Watt across an interface of one cm^2 will cause a one K temperature drop at the interface.

3.5.3 Heat Transfer Across an Interface between Two Metals

It is a common misconception that metals that are in electrical contact are automatically in good thermal contact. The thermal conductance of a metal contact, or any series of contacts and bulk thermal resistances is easily and not too poorly calculated using the Wiedemann Franz law. That is, $\Lambda = LT\sigma$, where σ is the electrical conductivity and L is the Lorentz number, usually taken to

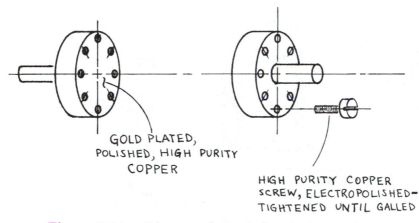

GOLD PLATED,
POLISHED, HIGH PURITY
COPPER

HIGH PURITY COPPER
SCREW, ELECTROPOLISHED-
TIGHTENED UNTIL GALLED

Figure 3.34.   *Diagram of a typical contact geometry.*

be about $2.5 \times 10^{-8} \, \Omega W/K^2$. This means if a contact has a resistance of $10^{-9} \, \Omega$ at 1 K, then the thermal resistance at low temperatures will be $(1/25T)K^2/W$. At 1 mK, 1 $\mu\Omega$ means one nano-Watt will cause a temperature rise of .4 mK.

I will describe some of the ways to decrease the electrical resistance and therefore the thermal resistance. The best contact is made by welding two metals together and then annealing the weld. Soldering is no good below the superconducting critical temperature because the paired electrons do not contribute to thermal transport. Above the superconducting critical temperature soldering or brazing provides very adequate contact. Soft solders superconduct below a few K, and silver solders typically superconduct below $\approx$ 50 mK. BT braze, which is a copper-silver alloy, does not superconduct; you may be able to find some additional brazes that don't superconduct. Contact resistance between two pressed contacts is a function of the area of the contacts, pressure applied to the contacts, and cleanliness of the contact. The area that is relevant is the microscopic area of contact. The standard procedure for making a nice contact is to polish both contact faces and then gold plate them to prevent oxidation. Additionally, the gold is soft and won't superconduct. Then the contact faces are bolted together using copper screws and tightened so that the screw will gall in its threads. The effect of galling the screw is to remove the oxide layer and to get good intimate contact. Electropolishing the screw and the tapped hole may also help. A typical contact is diagrammed in Figure 3.34.

Another clever way to get very good contact is to make use of differential thermal contraction. I will describe a technique which I will attribute to Doug Osheroff at AT&T Bell Labs, but there may have been others involved and similar designs by others. I would like to acknowledge Bob Richardson and Glen Agnolet for describing this to me. The idea is to make a matched, tapered pair of dissimilar metals and position the metal with the lower expansion in the center. When the pair is cooled the taper fit will tighten and the very large pressures

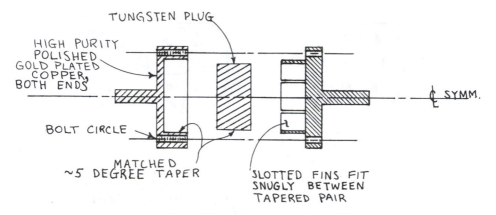

Figure 3.35.   *Diagram of improved contact geometry.*

that result will force the metals together. This joint will separate easily at room temperatures. Alternatively, polished and gold plated copper leaves can be put between the metals and squeezed together by the taper. This is diagrammed in Figure 3.35.

3.5.4   Nice Thermal Boundary Resistances

The thermal boundary resistance at an ideal solid-solid interface at temperatures below a few K is quite well described by acoustic mismatch theory, due originally to Khalatnikov. The concept is quite simple for an interface between two solids. We assume that phonons are the major contributor to the thermal transport (i.e., at least one of the materials is an insulator). One calculates the heat flux across the boundary between two different materials at different temperatures and defines, as above, $R_{bd} = \delta T/(\delta Q/A)$. To calculate δQ, one integrates the contribution of each individual phonon to the heat flux, assuming that the probability of any phonon being transmitted can be calculated by using the acoustic analog to Snell's law and the Fresnel equation. (It's just like reflection and refraction of light at an interface between two materials with different indices of refraction.) The result is that the boundary resistance will be a minimum between materials with matched speeds of sound and densities. Matching speeds of sound is about the same as matching Debye temperatures. A good engineering number for order of magnitude calculations is that the thermal boundary resistance between a metal and a dielectric is 30 $T^{-3}(K^4\text{cm}^2/W)$. It is worth noting that it is hard to make an interface that is perfect enough to have a thermal boundary resistance as low as the acoustic mismatch value. A factor of two increase above the calculated value is usually an adequate rough estimate for vapor-deposited metals on dielectrics. This has already been incorporated into the engineering estimate above. There has been some success in using ultrasonic soldering techniques to improve the contact at the interface by driving the metal into microscopic cracks in the dielectric. This technique can improve the contact in some cases by as much as a factor of three. Vapor deposited metal films can be used to thermally anchor leads and parts of the cryostat when very good thermal contact is needed. Properly deposited films can be soldered to, and even soldered together. For general information about solid-solid interfaces, see Little (1959), Pollack (1969), and Cheeke, Ettinger and Hebral (1976), as well as many of the papers by Anderson, *et al.* listed earlier.

3.5.5 Heat Flow Across a Metal-epoxy-metal Sandwich

The thermal boundary resistance between crystals and glasses, for example between metals and epoxies, is slightly anomalous at temperatures above about .5 K. This has been quite well characterized by A. C. Anderson and others, and is due to a frequency dependent mean-free-path in the glass. If you need good estimates for the total thermal resistance of a carefully prepared metal-epoxy-metal sandwich, see: Schmidt (1977), and the many papers by Anderson *et al.* listed earlier.

3.5.6 Kapitza Resistance

Acoustic mismatch theory can be applied only with moderate success to the interface between ^{4}He and metals, and with some added complication. At very low temperatures the Kapitza resistance gets to be so high that a cube of copper immersed in helium at 1 mK would have a time constant for equilibration on the order of years. Because that time scale is completely unacceptable, great efforts are made to increase the surface area of contact between the helium and the copper. This increase in surface area is most easily achieved by sintering fine powders of metal under pressure; diffusion of the metal across the surfaces of the granules produces a single porous sponge. This diffusion will occur at temperatures far below the bulk melting temperature. Because of the large surface area to volume ratio, and the good electrical and therefore thermal contact between the sinter and the copper, the resulting contact between the helium and the bulk copper is greatly improved. Typically, successful cooling in the temperature range near 1 mK is achieved with surface areas of around 10 square meters per cm^3 of helium to be cooled. There will be an optimal size sphere depending on the temperature and geometry, because the electrical and therefore thermal conductivity of the sinter decreases with decreasing sphere radius. The surface area to volume ratio for spheres of radius r with about a 33% packing fraction is $(.33)4\pi r^2/(4/3)\pi r^3 = 1/r$. That means one cm^3 of 1000 Å spheres will have a surface area of 10 m^2. Packing fractions around 40-50% are more typical, but much of the surface area is lost in the sintering process. For accurate values of the thermal conductivity, surface area, and packing fraction, see Busch, Cheston and Greywall (1984) and Keith and Ward (1984). The most often used powder in recent years is 700 Å silver powder from Vacuum Metallurgical Company, in

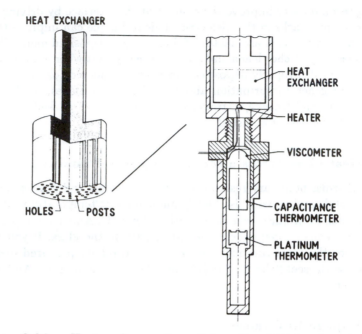

Figure 3.36.   *Heat exchanger from Osheroff and Richardson (1985).*

Japan, with North American distributor Ulvac North America Corp., 105 York St, P.O. Box 799, Kennebunk, ME 04043. The actual size of the particles seems to be larger than the 700 Å nominally quoted by the manufacturer.

I will describe a typical geometry of a metal to helium heat exchanger to bring out the relevant parameters. Figure 3.36 is a sketch of a nice heat exchanger, from Osheroff and Richardson (1984).

The posts of very pure copper or silver are to minimize the thermal path from any point in the sinter to the base. Similarly the holes in the sinter minimize the thermal path from the helium bath to the helium in the sinter. The more holes and posts the better is the contact between the helium and the helium near the sinter or between the sinter and the base. On the other hand, increasing the number of holes and posts, decreases the surface area, and therefore the contact between the sinter and the helium next to the sinter is degraded. Similarly, the larger the sphere size in the sinter the greater is the thermal conductivity of the sinter, again at the expense of surface area. Earlier literature has often indicated that because of the long wavelengths of the thermal phonons at millikelvin temperatures, using powder sizes of less than 10 μm is counter productive. This statement is purely theoretical, and is in fact wrong. All of the best heat exchangers serve as counter examples. Another parameter to consider is the sintering process. Sintering at higher temperatures, say at 700 K will increase the conductivity of the sinter at the expense of a large decrease in the

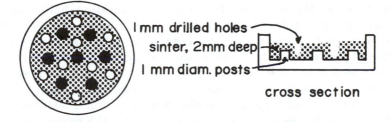

Figure 3.37.   *Heat exchanger due to Eric Smith.*

surface area. No sintering will lead to a very poor thermal conductivity in the sinter. Typically, a 370 K to 500 K sinter temperature is used, depending on the metal used for the sinter.

The Kapitza resistance between ^{3}He and metals at low temperatures does not behave according to acoustic mismatch theory. R_{bd} has been observed to have only a T^{-1} temperature dependence as opposed to a T^{-3} dependence. In some cases R_{bd} has even been observed to decrease with decreasing temperature. In terms of making nice heat exchangers, this is good. The anomalous Kapitza resistance is observed only for ^{3}He on metals, and not for ^{4}He on metals, and even a few percent ^{4}He in the ^{3}He will ruin the enhanced contact. Many have studied the effects of impurities on and in the sinter, and the consensus is the contact is enhanced by impurities both on and in the sinter. Thus, other relevant parameters may be the amount of junk on the surface of the sinter and the amount of impurities in the sinter. Impurities in the metal will cause a lower sinter thermal conductivity, and junk on the surface of the sinter may block the pores in the sinter and consequently reduce the area.

The above concepts are well demonstrated by the following two explicit examples of heat exchanger design. The first heat exchanger I will describe is due mainly to Eric Smith at Cornell. Some of this section on heat exchangers is taken from Eric's unpublished description of this heat exchanger and the principles behind heat exchanger design. This heat exchanger has on the order of 6 m^2 surface area; it can be made as shown in Figure 3.37.

We machine a disk of coin silver as in the figure. (Coin silver is 90/10 silver/copper.) Although the thermal conductivity of coin silver is not as high as pure silver or pure copper, its mechanical and machining properties are much better. Since the thermal conductivity of coin silver is so much better than the sinter, and since the geometry of the coin silver is not unfavorable, it would not be a significant improvement to use pure copper or pure silver. The posts are made by using an end mill that has its center ground out, so that drilling into the coin silver leaves a post. Alternatively, milling crossed rows of slots will work. The posts should be separated by no more than a couple of mm. We then further roughen the surface by sandblasting or repeatedly electropolishing and electroplating. We then pack $700\,\overset{\circ}{A}$ silver powder at a pressure on the order of a hundred bars into the coin-silver disk at room temperature. Subsequent heating in a hydrogen atmosphere to 50° C or 100° C may improve the performance. Some variations on this design have been used successfully. The surface area can be increased by increasing the depth of the disk, as long as the length of the posts is increased to the full depth of the sinter. It is also a good idea to press the powder only 2 mm at a time to minimize the variations in density. After the pressing operation, about 1 mm diameter holes must be drilled interdigitated between the posts to decrease the mean distance between the helium in the sinter and the helium in the bath.

I will describe two more modifications of this design that might be useful. By machining and filling both sides of the disk with sinter, a counterflow step heat exchanger can be made; such heat exchangers have been used at Cornell in many of the dilution refrigerators designed by Eric Smith. Use of both sides will alternatively increase the surface area by a factor of two, so by making both sides about 6 mm deep, on the order of 30 m^2 are produced. Further increases in surface area can be achieved by using multiple disks heat sunk to a common base. Thus the simplicity of design can be combined with very high surface areas and very short mean distance from sinter to bulk metal.

The second heat exchanger design I will describe is the heat exchanger built by Doug Osheroff and others at AT&T Bell Labs in 1984 for his sample chamber. The heat exchanger in his mixing chamber is also quite good, although the sinter is coarser. Neither of these two heat exchangers is described in print in any great detail. I will do my best to describe his sample chamber heat exchanger. Figure 3.36 is a sketch of that heat exchanger, from Osheroff and Richardson (1984).

The goal is to transfer heat between a silver platform and liquid ^{3}He. The space in which the helium sits has a 1.000" inner diameter; the platform was machined to a diameter of .970", leaving room for the helium to flow around it. The thickness of the platform was about .25". Forty .040" diameter high purity silver posts were welded into the platform in an regular pattern, and the result was sintered in a hydrogen oven for an hour at 400° C. 36 g of Ulvac silver powder type II was compressed under 200 bars pressure to a packing fraction of 45%, and it was not sintered. Then 20 .040" diameter holes were drilled into the result equally spaced in order to improve the contact between the helium in the bath and the helium near the sinter. The length of the posts, holes and sinter was about 5/8". The resistivity of the result was $18\,\mu\Omega$ cm at 4 K. It was determined that heat treating at 50° C for about an hour didn't change the surface area (about 2 m^2/gram), but the resistivity at 4 K was lowered to about $12\,\mu\Omega$ cm. From a private communication with Doug Osheroff, I learned that he thought a heat treatment at about 100° C might be better. Another improvement Doug suggested was to drill out the center .25" for better helium conduction to the bath. One comment Doug made was that if holes were drilled into the sinter after sintering, the porosity of the sinter near the holes would be drastically reduced.

By drilling before sintering, the porosity was not affected.

Some of the relevant parameters in this heat exchanger are the following:

- The total surface area of the sinter is about 50 m^2.
- The mean distance in the sinter to a silver post is about .08".
- The mean distance from the sinter to a drilled hole is about .11".
- The thermal conductivity of the sinter is about $2 \times 10^{-3}T$ (W/cm K^2).
- The thermal conductivity of the helium in the pores is discussed by Rutherford, Harrison and Stott (1984).
- The quasiparticle mean free path is measured (see Greywall (1984)) to be $(1000\,T^{-2})$ K^{2}cm. (So the mean free path in the pores is the pore size for $T < 10$ mK.)
- The thermal conductivity in the helium below 10 mK is about $30/T$ (ergs/sec cm).
- The silver posts had a residual resistance ratio of 4000, so they conduct heat very well.

The measured value of the Kapitza resistance in zero magnetic field at saturated vapor pressure is on the order of $R_{bd} \approx (250/T)K^2m^2/\mathrm{W}$, for $T <\approx 1$mK.

The same company that sells the 700 Å silver powder also sells a nominally 300 Å copper powder. However, this powder oxidizes rapidly at room temperature. This powder is usable for sinters if it is first heated in a hydrogen atmosphere at temperatures of around 140° C for an hour or so. It will turn from black to pink, with the evolution of water vapor on the walls of the container. Induction heating in a graphite crucible is a convenient mechanism for carrying out the reaction. The copper powder, as the silver powder, appears to have a larger than advertised particle size. In the past few years, several groups have experimented with catalytic platinum powder, available from a number of chemical or metallurgical supply houses. This platinum powder typically has a surface area in excess of 20 m^2 per gram. It does not adhere well to bulk platinum with room temperature compaction, but does stick moderately well if compacted in a vacuum oven at 135° C, with little or no loss in surface area. At somewhat higher temperatures there is a rapid loss in surface area (private communication, Y. Takano). Takano warns that annealing in a hydrogen furnace will cause a disastrous decrease in surface area because the platinum catalyzes the reaction between the hydrogen and the adsorbed oxygen, which causes a very large pulse of heat—enough to locally melt the platinum. One trick is to coat the walls of the exchanger with platinum powder and anneal in hydrogen. This makes this surface-sinter adhere very well to the walls. Then the rest of the powder can be pressed into the exchanger; it will stick well to the annealed sinter with the vacuum heat treatment described earlier. This yields the best combination of surface area and adhesion. The surface area lost in the first anneal is only a small fraction of the total surface area of the exchanger. The other trick is to anneal the exchanger in argon and let in hydrogen very slowly so the hydrogen-oxygen reaction occurs at a controlled rate. It is not yet clear whether heat transfer is improved over the use of similar volumes of silver powders.

At higher temperatures (1 K pot, still, ^{3}He Pot), one often needs only a few hundred square cm of surface area, as the Kapitza resistance is low at high temperatures. Copper powder of 1-5micron size (from Aremco Products, Inc., 23 Snowdon Ave., Ossining, N.Y. 10562, for example) works well if sintered at 700-800° C for an hour or two in hydrogen or vacuum. (Use OFHC copper for the container if you do it in hydrogen, or it will probably be porous after the sintering procedure.) This gives a mechanically stronger sinter than the low temperature, small particle sinters (which lose much of their area if heated this high), and works best if the sintering is carried out with an applied pressure during heating, as there is some shrinkage during heating. 50-100 psi is sufficient, if the powder has previously been packed to a 40-50% packing fraction at room temperature. Another popular material for making sinters is small copper flakes made for printing ink, available from Alcan Ingot and Powders, Div. of Alcan Aluminum Corp., PO Box 290, Elizabeth, NJ 07207, as Druid Copper MD-60 or MD-90. For this material, sinter in hydrogen at 400° C for several hours, increasing the temperature slowly, as there is a stearic acid binder which must be allowed to vaporize slowly. Some silver paints also work well. Many investigators have found that mixing a slurry of acetone or chloroform and the metal powders can aid in the packing of the sinter; such solvents come out of the structure easily because of their high vapor pressure. As these solvents are less compressible than air, they don't tend to make the structure elastically spring apart with the removal of pressure during the original packing. If you don't use these fluids, leave the packing pressure on for extended periods (an hour or two) before removing the sinter from the press.

For more information on Kapitza resistance see for example: A. C. Anderson and W. L. Johnson (1972), H. Schubert, P. Leiderer, and H. Kinder (1982), K. C. Rawlings, and J. C. A. van der Sluijs (1979), J. P. Harrison (1974), and Rurterford, Harrison and Stott (1984).

3.5.7 Thermal Conductivity of Links and Supports

Data for calculating the thermal conductance of links and supports in the cryostat exist in many places. I will not reproduce the data here—to do so would require many volumes, but I will include a list of references for finding these data. See Johnson and Stewart (1961), Crooker (1981), White (1979), and Rose-Innes (1964).

3.5.8 Other Considerations

In addition to the thermal conductances and heat capacities of the sample and parts of the cryostat, there are other factors which influence the thermal time constant and the ultimate temperature of the sample. Many of these are discussed elsewhere in this book. I will try to collect them here. The most obvious consideration is the heat leak directly to the sample through the leads. Other heat leaks include:

1. radiation
2. conduction through residual gasses
3. r.f. heating
4. eddy current heating if anything is vibrating in a magnetic field
5. direct vibrational coupling to vibrations outside the cryostat
6. direct high energy radiation from the sun and from cosmic rays
7. internal heat generation due to radioactivity
8. and internal heat generation due to structural relaxations with large time constants.

Heat leaks through the leads and all of categories 1. through 5. have been described elsewhere in this book. Direct vibrational coupling is easy to understand—vibrations couple to phonons very well since phonons are also vibrations. This is most important in situations where the sample can resonate with outside vibrations and where the sample is only weakly anchored to the bath. This can happen, for example, in a quasi-adiabatic specific heat measurement, in which a small sample is suspended on wires and can have a thermal time constant of several minutes. Methods of vibration isolation are described in Section 3.3.

There is little one can do about the heat load due to cosmic radiation. Presumably the heat load is less at sea level and in the basements of tall buildings.

The heat load due to the internal radioactivity of certain samples can prove to be quite a nuisance. When these samples must be used, some consideration should be given to isotope enrichment when a more stable isotope is available. See the paper by G. Steward (1983) for more information on how to deal with radioactive samples.

It is well documented that amorphous solids exhibit structural relaxations that have very long time constants; this has the effect of adding a virtual heat leak into cryostats. The time dependence of the heat leak is thought to be logarithmic, so it can last many weeks before it becomes negligible. Nevertheless, many cryostats that perform very well have large amounts of glassy epoxies, and other cryostats that have little epoxy still show large long-time-structural-relaxations. Due to differential thermal contraction, there can be large stresses in welds and solder joints and even in bulk metals; all of these can add to the heat leak. It has also been suggested that hydrogen impurities in metals can lead to a long term heat leak due to an ortho to para transition. (See Schward *et al.*, 1983.) Thus, care to minimize hydrogen impurity is suggested. Since the exact source of the dominant long-time-relaxational-heat-leak is not well understood, my advice is to carefully consider the stresses that will be created upon cooling, and minimize these whenever possible. The advantages of using glasses and epoxies may well outweigh the disadvantages of the possible added heat leak. A good way to see if the long-time-relaxational-heat load is a problem is to cool down the cryostat and see what happens. Here is a good example of where modularity is essential.

Chapter 4

Experimental Techniques and Special Devices

Peter L. Gammel, Gane Ka-Shu Wong, Mark R. Freeman,
Thomas J. Gramila, David Thompson, Bryan G. Statt,
Jeffrey S. Souris, Timo T. Tommila, and John S. Denker

4.1   Bridges

by Peter L. Gammel

A brief survey around the basement of Clark Hall shows that there are about four bridges associated with each cryostat. Bridges for resistance thermometers, capacitive level detectors, strain gauges (pressure and temperature measurement), and susceptibility bridges for paramagnetic salt thermometers are typical examples. While commercial bridges are now becoming more common, there is still about a 60-40 mixture of homemade and commercial bridges.

Bridges allow accurate measurement of component values with short time constants and very low power dissipation. In addition, problems associated with long cables connecting the electronics to the device being measured (which is down inside the dewar) can largely be overcome with bridges. Bridges also simplify the measurement since, to first order, a bridge on balance is independent of the amplitude, of frequency, and of excitation.

Capacitance measurement with tunnel diode oscillators, Colpitts oscillators, or microwave resonators can be very successful but will not be treated here. Nor will bridges of the Wheatstone variety, which are covered in detail in Hague and Ford, *Alternating Current Bridge Methods.*

4.1.1   Cryostat Schematic

Figure 4.1 is a rough schematic of how one might find a 4-terminal resistor wired in a cryostat if the advice of Bob Germain (Section 4.3) were followed. Even with these attempts to reduced stray couplings, many problems remain. Working from the top, we may find rather long leads leading from the top of the cryostat down into the vacuum can. The temperature profile in this part of

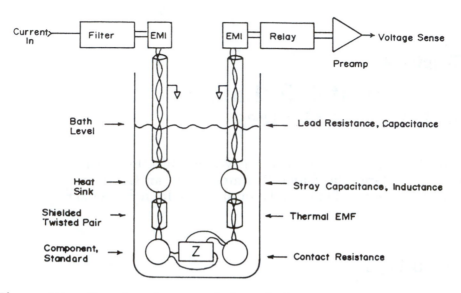

Figure 4.1.   *Overview of a cryogenic 4-terminal measurement showing sources of trouble.*

the dewar will change with bath level whether these actually pass through the helium bath or not. The stainless steel coax we use for capacitance bridges has a lead resistance of about 300 Ω and a capacitance to ground of 200 pf for a typical run of four feet to the mixing chamber. For twisted pairs in a resistance bridge the lead resistance is 10–100 Ω.

As the temperature profile changes, thermal contraction may well change the capacitance to ground by several percent (teflon, a common dielectric, shrinks almost an inch over 3 feet from room temperature to 4 K). For example, a three foot length of Cooner coax changes from 143 pf to 146 pf when dunked in liquid nitrogen. As we will show later, this directly affects the signal for bridges run off balance. The resistance of the leads may also change. Resistance wire changes resistance by a few percent on cooling. A length of 30.7 Ω/ft advance wire changes from 198.5 Ω to 195.2 Ω when dunked in liquid nitrogen. In general, however, lead resistances are less of a problem.

Having been strung from room temperature to the cryogenic environment, the leads continue through a series of junction boxes and heat sinks to lower temperatures. Often the leads are unshielded in the junction boxes, leading to extra capacitance between drive and detection lines. We often use a strip line heat sink without ground planes. For a 1″ run, the stray capacitance is about 10^{-3} pf. A ground plane will substantially reduce this. In addition, there may be additional inductance and capacitance if they are wound around a post for heat sinking (as is common with carbon thermometers).

Stray capacitances are often a major problem. We might expect a capacitive coupling of about 100 pf for unshielded wires running down a pumping line together. This corresponds to an impedance of 10^9 Ω/f where f is the measurement frequency. For resistance bridges, f is typically 10–100 Hz in order to keep this impedance large compared to the resistance of interest. Capacitive coupling becomes significant for resistances over 1 MΩ. Although this coupling should not appear in the same phase as the resistance being measured, 1 MΩ is a practical upper bound for accurate resistance measurement. Consideration of excitation levels and amplifier performance further justifies this estimate. It might be tempting to try to make a DC resistance measurement to eliminate the problems of stray capacitance, but then one must remember that the thermal emf's are going to be on the order of 10 mV. DC schemes have been used where the polarity is changed occasionally to compensate for thermal emf's (much like an AC measurement at extremely low frequency). We do this when measuring the RRR of bulk metal samples where the resistance may be between 10^{-6} and 10^{-9} Ω.

There may be contact resistances of up to a few hundredths of an ohm at the sample itself. A low temperature standard is often used in capacitance bridges eliminating thermal regulation of a room temperature standard. Also, for some capacitors (we use polystyrene) the temperature dependence of the capacitance seems to be much smaller. However, we still need to be careful of dielectrics. Capacitance measurements are often carried out to the 10^{-7} or even 10^{-8} level. At this level, the characteristic $\ln(T/\omega)$ dependence of the dielectric constant of polystyrene (which comes from its glassy behavior), at a level of 10^{-3}/decade is important—as is its amplitude dependence. So as a general rule, we try to mount standards at a point in the cryostat which remains at fixed temperature rather than as close to the sample as possible.

4.1.2   2, 3 and 4 Terminal Measurements

The principle of 2, 3 and 4 terminal measurements is shown in Figure 4.2. The small r's are lead resistances and the large R's are the resistances to be measured. In all cases we will assume that the output impedance of the current source and the input impedance of the voltage amplifier are infinite. These conditions will be examined later. The 2 and 4 terminal measurements are just $V = IR$. In the 2 terminal case, the input impedance of the preamp eliminates r'', so the measured $v/i = R_x + r' + r$. In other words, the lead resistances are important. In the 4-wire case, again the input impedance of the preamp eliminates r'' and the output impedance of the current source eliminates r and r' so we just get $v/i = R_x$. This is the best measurement. Many germanium resistors are even shaped like a Π with the top for current leads and the bottom for voltage leads.

To save wires, a 3 terminal measurement may be used, especially on less crucial thermometers. I have shown a 3-wire bridge. The voltage source V is coupled in via a transformer. Again, the preamp measures the voltage at the corner of the bridge. For a null, $R_s + r_2 = R + r_1$ or $R_s = R + r_1 - r_2$ (R_s generally a decade resistance box). If the lead resistances are roughly equal (as

2 WIRE

4 WIRE

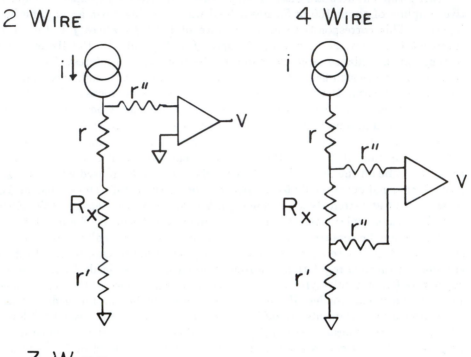

3 WIRE

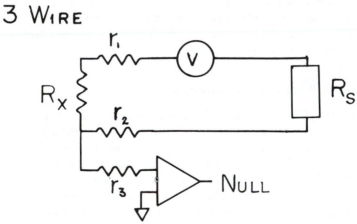

Figure 4.2.   *The principle of 2, 3 and 4 terminal measurements per text.*

would be expected for twisted pairs or even wires strung together), this works quite well.

In actual circuits, the lead resistance should be approximately equal, or at least roughly temperature independent even for a 4-terminal measurement. For example, if the preamp used has a finite input impedance Z, then there is a correction of order $2rR/Z$. This may be important if r has a large temperature dependence. Similar conditions hold for the current source.

It is often better to have the lead resistance less than the unknown resistance. This is more of a constrain due to the feedback schemes used in self balancing bridges. Basically, the feedback loop must be carefully designed if most of the voltage drop is across the leads and the feedback is based on the voltage drop across the sample.

In general it is convenient to use as few wires as possible in a cryostat. One needs to be especially careful, however, when ganging components together. Consequently, one should avoid using a single current line for many resistors in series. Otherwise, the other resistors in the line would play a role in the lead resistance. This could make the lead resistances unequal and very temperature dependent. For unimportant resistors, however, a common ground return is often used in the 3-terminal mode although this makes things worse if a short develops.

4.1.3   Excitation Levels

The excitation level which may be used in a resistance bridge is a trade-off between power dissipation and noise. For noise sources, assume the voltage noise of the preamp to be 10 nV/$\sqrt{\text{Hz}}$ (not especially small). The thermal noise from a resistor is

$$(\langle v_n^2 \rangle)^{1/2} = \sqrt{4kTR}/\sqrt{\text{Hz}}$$

Write the current source as $i = g_m V$ with g_m taken to be at room temperature. This gives a thermal noise

$$g_m^{-1} \approx 10 \text{ k}\Omega$$
$$\bar{v}_m \approx 13 \text{ nV}/\sqrt{\text{Hz}}.$$

The low temperature resistor R_x, assumed to be at 1 K, has a thermal noise

$$R_x \approx 10\text{k}\Omega$$
$$\bar{v}_x \approx 0.8\text{nV}/\sqrt{\text{Hz}}.$$

To avoid heating, calculate for a 1% temperature rise due to the measurement. The heat input is

$$\dot{Q} = i^2 R$$

and the temperature rise is

$$\frac{\Delta T}{T} = \frac{\dot{Q}}{T}\frac{R_k}{A}$$

where R_k is a Kapitza boundary resistance and A is the area. Assuming

$$R_k T^3 = 8 \times 10^{-4} \text{ K}^2\text{m}^2/\text{W}$$

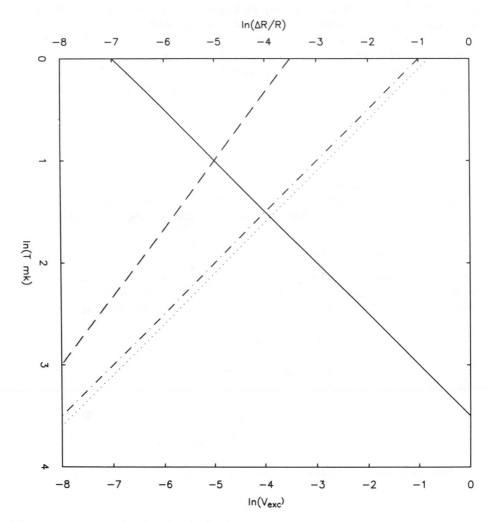

Figure 4.3. *Excitation levels for low temperature resistance measurements. To avoid self heating, the excitation level should be below the solid line. The noise from various sources is given by the dashed lines.*

which is a reasonable value for a copper-epoxy boundary and $A = 10^{-4}$ m^2 gives an acceptable current

$$i = T^2 \times 10^{-2} \text{ ma}$$

if we use

$$R_x = 10k\Omega$$

and

$$\frac{\Delta T}{T} = 1\%.$$

The voltage drop across the resistor is the solid line in Figure 4.3.

The dot-dash line is the voltage noise in the preamp divided by the voltage drop across the resistor which is an estimate of the precision of the measurement. The dotted line is the same for the room temperature current source. The dashed line is the limitation for ideal components which is just the thermal noise in the resistor being measured at low temperatures. One can do better than this but it requires noise thermometry. This is part of the reason resistance thermometers are not generally used for precision work below 100 mk.

The same consideration could well be applied to capacitance bridges where the dissipative part of the signal comes from the dielectrics. The commercial capacitors we use have an equivalent series resistance of about .1 Ω at 1 kHz (for a 1 μf capacitor). This gives a loss tangent of about 10^{-3}. There are dielectrics which are substantially better than this. The loss tangent of a single crystal sapphire is of order 10^{-7}.

4.1.4   Components

From the preceding discussion, it is clear that accurate components are essential for good bridge performance. This section is a review of ones we commonly use.

In Figure 4.4 a typical bipolar current source is shown. The current flowing through the load is

$$I_L = U_1/R_1 + V_L(R - R_3 - R_1)/R_1 R_3.$$

If

$$R_3 = R - R_1 \qquad \text{then} \qquad I_2 = U_1/R_1.$$

For

$$\Delta R = R - R_3 - R_1$$

the effective output impedance is of order

$$R\frac{R}{\Delta R}.$$

A fairly precise resistor has a temperature coefficient

$$\frac{\Delta R}{R} \approx 10^{-5}/^\circ C.$$

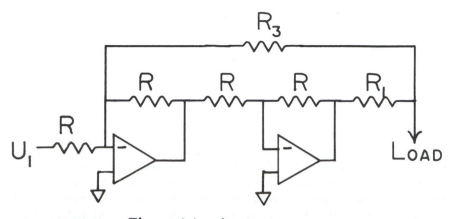

Figure 4.4.   *A current source.*

With $R = 10$ K we cannot expect an effective output impedance of much more than 10 MΩ due to fluctuations in room temperature. In addition, R must be stable since its temperature fluctuations feed directly into the measurement. A stability of 10^{-4} to 10^{-5} is typical for a precision resistor. Stability of U is easier to maintain since voltage sources stable to one part in 10^{-4} to 10^{-5} are easily obtained.

A good preamp is required on the measurement end and the OP-16E makes a good, cheap choice. Its specified input impedance is $Z_m \approx 10^{12}$ Ω. A blocking capacitor C and a shunt resistor R block the input bias current which may be an appreciable fraction of the measurement current at low temperatures. To be effective, the resistor must satisfy

$$R_s \leq V_{os}/I_{ob} \approx 10 \text{ MΩ}$$

which gives an approximation of the useful input impedance. Perhaps more to the point, we should take into account the temperature coefficients of the offset voltage and bias current. The usable input impedance is then

$$Z_m = \frac{TCV_{os}}{TCI_B} \approx 1\text{MΩ}.$$

Thus one cannot make the approximation of infinite input impedance for large R.

Figure 4.5 is a suggestion of how to get the best performance from the amplifier. The noise figure as a function of source resistance will have a minimum value

$$R_{\text{opt}} = \frac{e_n(\omega)}{i_n(\omega)} \approx 2 \text{ MΩ} \qquad (100 \text{ Hz} - -10 \text{ kHz})$$

where $e_n(\omega)$ and $i_n(\omega)$ are the voltage and current noise of the op-amp. Since a bridge will generally have $R_x \ll R_{\text{opt}}$, it may be necessary to insert a transformer to obtain the best noise performance.

There are a few other components which are useful in bridges. The ratio transformer is the most commonly used. Ratio transformers are voltage dividers with a ratio essentially independent of temperature or amplitude to better than 10^{-6}. There is an excellent discussion of their design and operation in Hague and Ford. At very low frequencies, ratio transformers have low input impedance and nonexact ratios, but above 60 Hz or so they are pretty good. (Hence we generally use them in capacitance bridges, but less frequently in resistance bridges). A brief summary of the specs for a typical large Singer-Gertsch ratio transformer follows:

$$R_{\text{out}} = 2\text{--}5 \ \Omega$$
$$\phi_{\text{out}} = .003^\circ$$
$$R_{\text{in}} = 50 \ \text{k}\Omega \qquad (60 \ \text{Hz})$$
$$130 \ \text{k}\Omega \qquad (100 \ \text{Hz})$$
$$500 \ \text{k}\Omega \qquad (400 \ \text{Hz}).$$

These ratio transformers are best used near 400 Hz where the transformer self-resonates. At this frequency, the input impedance is maximal and the output impedance is low.

In some cases, we have used op-amps to buffer the input and output of a ratio transformer to minimize the effects of loading. However, recall that the gain stability of a unity gain buffer is just the inverse of the open loop gain which is of order 10^{-6}. This may be a problem for very high resolution capacitance bridges, but is generally OK. In fact, it is typical to do this in susceptibility bridges which we generally only read to about four places and which tend to run at a rather low frequency.

One might try to use optical isolators in cases where grounding is a problem but they should be kept out of bridge circuits if at all possible. Even linearized optical isolators are only good to about 0.1%.

The last bridge component to be considered is the drive oscillator which must provide a constant frequency and amplitude of excitation. Although the bridge balance point should be independent of frequency and amplitude, bridges which are either run off balance or require high resolution need a stable drive (see later on capacitance bridges). We have often used a stabilized Wein Bridge oscillator as shown in Tietze and Schenk. This works from 10 Hz to 1 MHz with less than 0.1% distortion.

4.1.5   The Anderson Bridge

The schematic of an Anderson bridge with quadrature balancing is shown in Figure 4.6. Concentrating on the in phase component initially, the drive oscillator is attenuated and coupled into the toroidal transformer T. The op-amp is coupled to one of the matched secondaries and maintains a constant current $i = V/R_s$ through the resistor R.

The ratio transformer is coupled in through another secondary. The voltage drop across the ratio is

$$V' = \alpha V$$

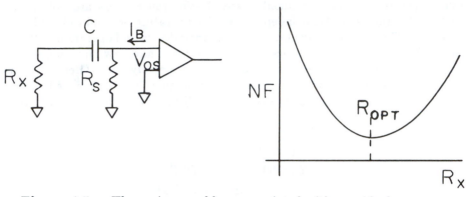

Figure 4.5.   *The various problems associated with non-ideal preamps.*

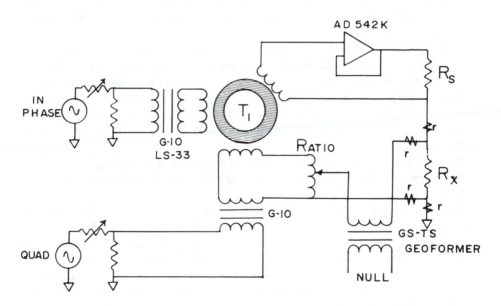

Figure 4.6.   *An Anderson bridge with quadrature balancing.*

and the current through R is

$$i = \alpha V / R_x.$$

The same current flows through R_s and R_x on balance. On balance therefore,

$$R_x = \alpha R_s.$$

Note that the balance condition is independent of the lead resistances and the excitation level.

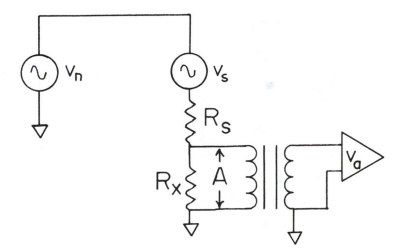

Figure 4.7.   *Equivalent noise circuit for the Anderson bridge.*

To implement quadrature balancing, a second transformer is added. With both phases balanced, the condition that no current flows in the detection arm on balance is exact. This sort of bridge is commonly called a double potentiometer. In fact, when Anderson first proposed this, he claimed it would reduce the noise due to the room temperature resistors R_s. Figure 4.7 shows an equivalent noise circuit.

The output voltage noise from the op-amp is v_n and v_s is the thermal noise from the resistor R. The total voltage noise across A is

$$V_n = \left(\langle v'^2 \rangle\right)^{1/2} \approx \left(v_s^2 + v_A^2\right)^{1/2} \frac{R_x}{R_s + R_x}.$$

Anderson claims $R_s \ll R_x$ but in fact $R_s \approx R_x$ so the noise reduction is small. The total output voltage noise is v_A combined with the transformer noise in the primary and secondary as well as the voltage noise of the final amplifier. As in our discussion of coupling to amplifiers, one wants

$$R_x n^2 \approx R_{\text{opt}}$$

where n is the turns ratio of the transformer.

4.1.6   Other Resistance Bridges

To help computer interfacing, a self-balancing feature is preferable since the Anderson bridge operates properly only when on balance. A slightly different bridge which can incorporate feedback and self-balancing is shown in Figure 4.8. The op-amp holds a constant potential αV at the bottom of R_x . When

$$iR_x = \alpha V.$$

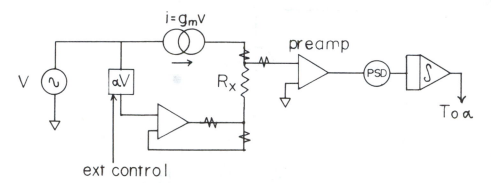

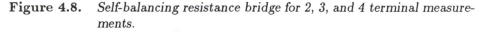

Figure 4.8.   *Self-balancing resistance bridge for 2, 3, and 4 terminal measurements.*

the top of R_x is at ground and the output is a null.

Therefore, the balance condition is

$$g_m R_x = \alpha$$

Off balance, the signal in the resistive phase is amplified and transformed into a dc level by the p.s.d and integrator. α can be controlled with this level. In the simplest version, α may be an analog multiplier (for example, Analog Devices now sells a rather inexpensive 14 bit multiplying DAC). The integrator output voltage is then proportional to the unknown resistance. Any other servo scheme may be used of course.

The particular bridge may also be wired for 2 or 3 terminal operation depending on the location of the preamp and the op-amp relative to various lead resistances. An inexpensive version has been constructed that uses as the current source described earlier, an Evans card lock-in (from Evans Associates) and a preamp using two OP-17's.

Figure 4.9 shows the block diagram of the LR-400 (from Linear Research), an excellent commercially available bridge. Its measurement frequency is 15.9 Hz and its feedback loop has an effective DC gain of 10^5. Both phases are automatically nulled and the input stage of the preamp changes to noise match to the source impedance. Balance occurs when

$$iR_x = V$$

where V is voltage across the transformer. The error due to lead resistance is $< 10^{-4}$. Resistance changes of one part in 10^5 at 15 mK can be measured with this bridge.

The feedback scheme does not need to involve analog circuiting, of course. In several capacitance bridges and a few resistance bridges, we have used a computer to control the feedback as shown in Figure 4.10. An Anderson bridge is used in

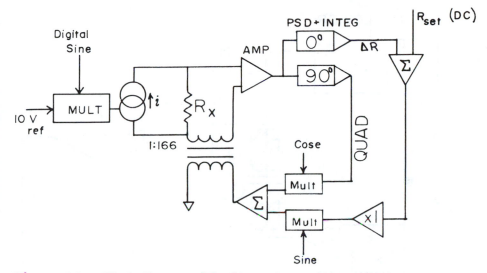

Figure 4.9. *Block diagram of the Linear Research LR-400 self-balancing bridge.*

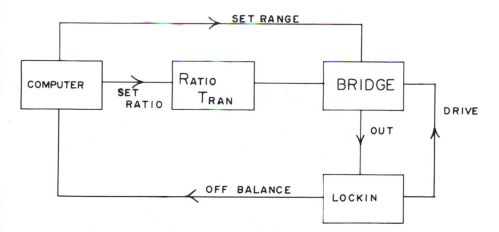

Figure 4.10. *Typical block diagram for computer control of a bridge.*

this diagram. The 'set range' line controls several mercury relays which select the value of R. The computer also controls the ratio transformer.

Due to the finite lifetime of the relays in a ratio transformer, such a bridge is generally allowed to run off balance. When the ratio is changed, the computer estimates the change in off balance signal due to the change in ratio. The computer uses this and the off balance signal to provide a continuous interpolation

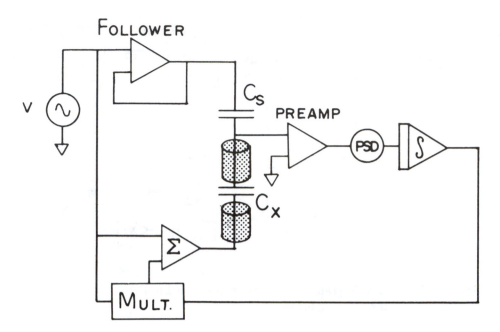

Figure 4.11.   *Simple auto balancing capacitance bridge of the type used for level detectors.*

of the bridge balance point.

A final example of a self-balancing bridge is shown in Figure 4.11. This is a particularly simple and inexpensive design for use with capacitive level detectors. In such detectors, changes in the level of some liquid (the dielectric constant of liquid helium is 1.05, and that of liquid nitrogen is 1.3) causes changes in C of order 5%. The shaded regions indicate coaxes leading to the low temperature environment.

For the empty detector, the capacitances are chosen so that

$$C_s \approx C_x.$$

When C_x increases, the integrator output is multiplied by the drive signal, attenuated, and summed with the drive signal to maintain a null condition. The integrator output voltage is proportional to the change in capacitance. In this design, the phase sensitive detector is particularly simple and inexpensive. The preamp supplies the signal and its inverse which are fed into an AD751201KN analog switch. The switch is controlled by the drive oscillator which operates as a half-wave rectifier at the drive frequency. The switch output is buffered and integrated to give the DC level.

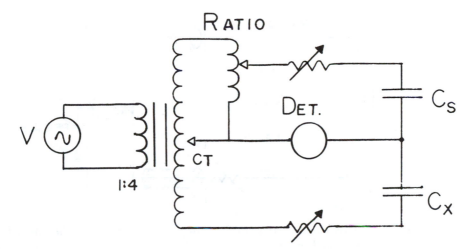

Figure 4.12. *The asymmetric capacitance bridge.*

4.1.7   Ratio Transformer Bridges

The preceding section was largely devoted to resistance bridges which were reasonably (10^{-4}) sensitive and could be incorporated into feedback loops. In this section we turn to high sensitivity $(10^{-6}$–$10^{-8})$ capacitance bridges. These may be used on strain gauges, glass capacitors or susceptibility thermometers. (i.e., in inductance mode). The low loss means that these bridges run at high excitation levels, typically of order 1 V. The excitation frequency is also higher, typically 1 kHz.

In Figure 4.12, an asymmetric capacitance bridge is shown. C_s is a standard and the balance condition is

$$RC_s = C_x$$

where R is the ratio. The variable resistors adjust for imbalance in the line resistance. The dependence of the balance on the amplitude and frequency of the excitation are roughly

$$\frac{\Delta R}{R}\frac{f}{\Delta f} = 9.6 \times 10^{-4}$$

and

$$\frac{\Delta R}{R}\frac{V}{\Delta V} = -4.6 \times 10^{-5}.$$

These rather severe dependencies on drive level and frequency can be reduced by using the balanced configuration of Figure 4.13. The balance condition for this bridge is

$$(1 - R)C_s = RC_x.$$

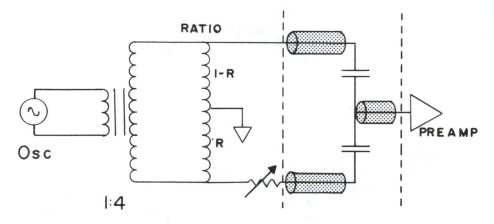

Figure 4.13.   *The symmetric capacitance bridge. The area within the dashed lines is at low temperatures.*

The standard is chosen so that $C_s \approx C_x$ and the bridge is very nearly symmetric (on balance $R = .5$). For this bridge

$$\frac{\Delta R}{R}\frac{f}{\Delta f} = 3.8 \times 10^{-6}$$

and

$$\frac{\Delta R}{R}\frac{V}{\Delta V} \approx 1.5 \times 10^{-6}.$$

These dependences on frequencies and amplitude, shown in Figures 4.14 and 4.15, are almost two orders of magnitude better than for the asymmetric configuration.

The Wein bridge oscillator for the drive has temperature coefficients

$$\frac{\Delta f}{f} \approx 1.6 \times 10^{-3}/^{\circ}\mathrm{C}$$

and

$$\frac{\Delta V}{V} \approx 7 \times 10^{-5}/^{\circ}\mathrm{C}$$

which give a bridge noise of 3×10^{-8}. This is comparable to the maximum stability we have had.

4.1.8   Dealing with Long Leads

Figure 4.16 shows the effects of coaxial cable on a bridge. For ideal voltage sources, the circled capacitances do not enter. However, if the bridge is off balance, the potential seen by the preamp is

$$e = V\frac{C_1 - RC_2}{C_1 + C_2 + C_L}.$$

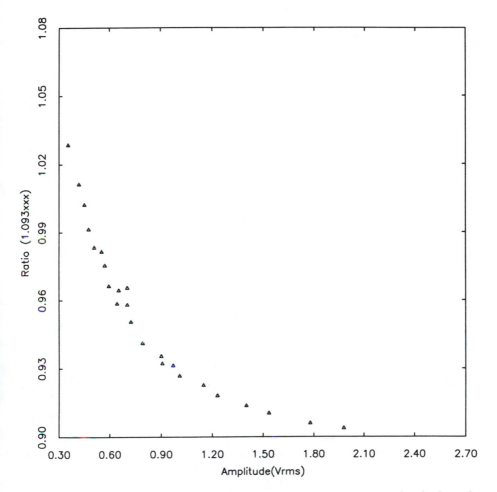

Figure 4.14. *Dependence of balance ratio on excitation level for the symmetric bridge.*

Note that if the bridge is allowed to run off balance,

$$e \sim V \frac{\Delta R C_1}{C_L}.$$

In this case, the excitation voltage (and frequency for a resistance measurement) as well as the stray capacitances all feed directly into the measurement. However, if the bridge is always near balance so ΔR is small, their stability is less important. Although the balance condition is unaffected, for many of our applications,

$$C_1 \approx C_2 < 10 \text{ pf}$$

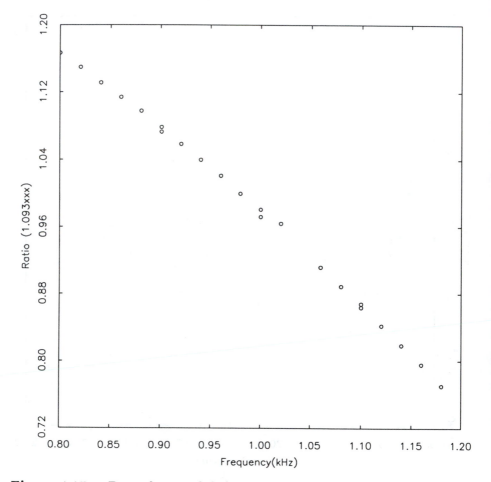

Figure 4.15.   *Dependence of balance ratio on drive frequency for the symmetric bridge.*

$$C_L \approx 200 \text{ pf}$$

and line attenuates the signal by a factor of 10–100. (For a resistance bridge, the low frequency makes this unimportant if $\omega RC \ll 1$.)

One way to get around this shunt capacitance is to use low temperature amplifiers to reduce the length of the final coax. Unfortunately, for millidegree and microdegree experiments, placing the amplifier at 1 K reduces the length of the cable only by a factor of 3 at best.

Another solution is to use an active guard as shown in Figure 4.17. If the op-amp has an open loop gain $A_\nu(0)$ and bandwidth $1/z_0$, this will effectively

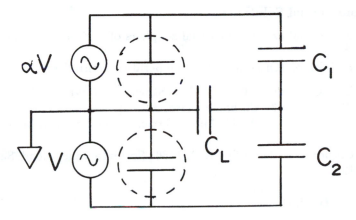

Figure 4.16.   *Shunt capacitances in the symmetric bridge.*

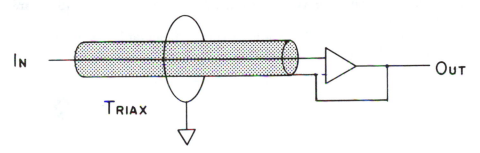

Figure 4.17.   *An active guard for cancelling the effect of shunt capacitances.*

reduce

$$C_L \rightarrow C_L/(1 + A_\nu(0)).$$

Unfortunately, this only works for frequencies

$$\omega_0 \ll \frac{1}{A_\nu(0)z_0}.$$

For a 741 this rolloff occurs at 10 Hz and for an OP-16 it occurs at 200 Hz.

There are some very fast line drivers which rolloff at 5–10 kHz. These still have problems: First, unless you are very careful, they oscillate. Second, they tend to have a large voltage noise, typically of order 100 nV/$\sqrt{\text{Hz}}$. If the line driver isn't increasing the S/N by more than a factor of 10 (for a 10 nV/$\sqrt{\text{Hz}}$ preamp), it's doing more harm than good since this noise is fed into the detector. In general, you can't beat shunt capacities.

4.1.9 Commercial Solutions

Below is a partial list of commercial suppliers of bridges and components.

Resistance Bridges (Often including temperature controllers)

- Lake Shore Cryotronics, 64 E. Walnut Street, Westerville, OH 43081.
- Linear Research, 5231 Cushman Place, Suite 21, San Diego, CA 92110.
- Oxford Instruments, NA, 3A Alfred Circle, Bedford, MA 01730 (617) 275-4350.
- B.T.I. Corporation (formerly SHE), 11661 Sorrento Valley Road, San Diego, CA 92121.

Capacitance Bridges

- General Radio Company, West Concord, MA

Components

- GenRad Inc., 300 Baker Ave., Concord, MA 01742 (617) 369-4400. (Standard Resistors and Capacitors)
- Singer-Gertsch Instr., 3211 South La Gienga Boulevard, Los Angeles, CA 90016. (Ratio Trans, Ratio Bridges)

4.2   Torsional Oscillators

by Gane Ka-Shu Wong

Recent disciples of a crafty low temperature physics professor at Cornell have been imbued with the philosophy "if you've got it, oscillate it." The physical manifestation of this madness is the **torsional oscillator**—a mechanical resonator which is used as a very sensitive micro-balance. Its main low temperature property is its high quality factor $Q \approx 10^6$. This leads to a low noise bandwidth and, consequently, a very high period and amplitude stability ($\delta P/P \approx 10^{-9}$ and $\delta A/A \approx 10^{-4}$).

Under ideal circumstances, torsional oscillators are modeled as simple harmonic oscillators in which the resonant period is proportional to the square root of the moment of inertia and the resonant amplitude is proportional to quality factor times the drive. Changes in P and A which result from the addition of a sample to the cell are then measures of the mass and dissipation of the added sample.

The basic experimental configuration is a torsion rod with some sample cell attached to one end of it. The oscillator is driven by a capacitive transducer and its response is detected by another capacitive transducer. Torsional oscillators differ from one another primarily in the design of the sample cell. It is not possible to discuss every variation in detail and so I will restrict this write up to the design, construction, installation, and operation of one simple representative torsional oscillator.

4.2.1   Past Applications

It is important that you do not form a bias toward the one type of oscillator that I will describe. Thus, here are some of the other designs which have been used in the past, as well as the experiments to which they were applied. The sample cells are illustrated in Figure 4.18.

Low temperature torsional oscillators were first used by Andronikashvilli for his classic experiments on the density of the normal component of helium II. A stack of aluminum plates, each separated from its neighbor by 0.21mm, was suspended from the bottom of a phosphor-bronze wire (not a torsion rod). The oscillator Q was not reported but it must have been very low because the period was resolved to only 0.08%. At the oscillator period of 25 seconds, the viscous penetration depth of the normal component was 0.38mm—larger than the plate separation. The normal component was thus viscously locked to the plates, while the superfluid component was not, and the normal component density was deduced from the measured period shifts.

Section 24.5 in Landau and Lifshitz's volume on Fluid Mechanics contains a succinct derivation of the viscous penetration depth $\delta = (2\eta/\rho\omega)^{1/2}$. Cells with more complicated internal geometries, say ones packed with a superleak, can

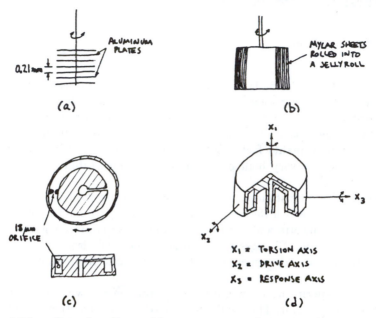

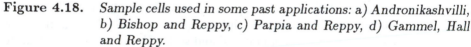

Figure 4.18. *Sample cells used in some past applications: a) Andronikashvilli, b) Bishop and Reppy, c) Parpia and Reppy, d) Gammel, Hall and Reppy.*

also drag some fraction χ of the superfluid along with it ($\chi = 0$ ideally). This is analogous to carrying the superfluid around in a bucket. For example, $\chi \approx 0.76$ is found for a Vycor substrate. It is not a trivial matter to predict χ factors in general and, in practice, they must always be measured.

Reppy and coworkers made a big improvement in the Q's of these oscillators by using (BeCu) torsion rods rather than torsion fibres. This made possible a number of very sensitive experiments. Bishop and Reppy used a cylindrical sample cell, filled with a jelly-rolled Mylar sheet, to measure the superfluid density of ^{4}He films. The idea was much like Andronikashvilli's and the superfluid mass resolution was 1 part in 10^4, even though the films were only a few atomic layers thick.

Parpia and Reppy observed critical velocities in superfluid ^{3}He using an aperature cell—a cell with an annular flow geometry in which the region of high velocities is localized by blocking the flow channel with a thin partition containing a small 18μm orifice. A break in the slope of the observed drive versus oscillator amplitude curve was then attributed to the presence of a critical velocity.

Gammel, Hall and Reppy used an AC gyroscopic technique to observe persistent currents in the B phase of ^{3}He. The sample cell had a toroidal flow channel, filled with a fibrous air filter material (to lock the normal component). The an-

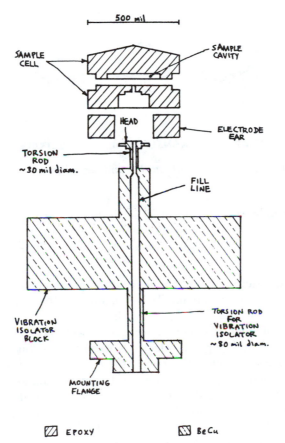

Figure 4.19. *Exploded cross-section of a torsional oscillator.*

gular momentum associated with the persistent current was detected by driving one of the floppy modes and observing the resultant precessional motion.

Torsional oscillators need not be restricted to helium studies and, in fact, one need not even study the torsional mode. Resonant bars have been popular with some of the gravity wave people and, to the best of my knowledge, these people hold the record for the highest Q (of order 10^9).

4.2.2   Design

Torsional oscillators are home-brewed devices. Figure 4.19 is an exploded cross-section of the hypothetical torsional oscillator which I will discuss. For

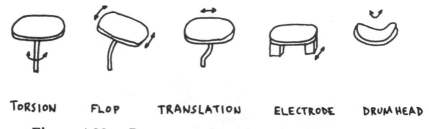

TORSION   FLOP   TRANSLATION   ELECTRODE   DRUMHEAD

Figure 4.20. *Resonant modes and associated deformations.*

clarity, the electrode ears are shown as separate exploded pieces—but they can also be machined as part of the sample cell. The mountings for the movable half of the electrode structure, usually attached to the vibrational isolator block, are not shown.

Resonant Frequencies One of the first design parameters to be chosen is the torsional mode resonant frequency. It is generally in the range 300Hz–3000Hz. Below this range, vibrational noise from the building becomes a problem. We have not tried really high frequencies because we feel that resonances from some of the smaller structures (e.g., the electrode ears) may pollute the response.

Even within this so-called safe range, the oscillator itself may have a multitude of resonant mode frequencies. Some of the more common modes, and their associated deformations, are depicted in Figure 4.20. In order to prevent these modes from being excited by low frequency vibrational noises, which would generate both unwanted signals and unwanted cell velocities, the torsional mode is designed to have the lowest resonant frequency.

For a simple pancake cell of radius r and height h, with a torsion rod of diameter a and length L, the resonant frequencies $\omega = (K/I)^{1/2}$ are given by these moments of inertia I and spring constants K:

$$\text{torsional mode } I = (\rho\pi r^2 h)r^2/2 \tag{4.1}$$
$$K = \pi G a^4/32L$$
$$\text{floppy mode } I = (\rho\pi r^2 h)\left([L + h/2]^2 + r^2/4 + h^2/12\right) \tag{4.2}$$
$$K = 3\pi E a^4/64L.$$

For these equations, ρ is the density of the sample cell, G is the shear modulus of the rod, and E is the Young's modulus of the rod. These formulae may be modified to account for the fill line, the electrode ears, and so on. In order to make f_{floppy} as large as possible, use a short torsion rod and put all the little appendages (e.g., the electrode ears) as close to the torsion rod as you can.

Table 4.1.   *Temperature Dependence of P and Q*
for an Empty Cell BeCu Torsional Oscillator

Temperature	Resonant period	Quality factor
300 K	2534μsec	4000
77 K	2429μsec	78000
4 K	2416μsec	260000

Do not spend much time calculating moments of inertia and elastic constants to excruciating precision. E, G, ρ and all of the physical dimensions are usually known only at room temperatures and there are always machining errors. The truth of the matter is that the resonant frequencies are predictable to about 5 or 10% only. Both I and K must be measured if you need to know them precisely.

Materials Selection The quality factor is greatly influenced by the choice of materials. BeCu (used commercially to make springs) is the current standard for the torsion rod material. It has good thermal conductivity and a very large Q. Typical (perhaps even conservative) figures for a BeCu rod with an empty sample cell are given in Table 4.1.

The figures at 77 K and 4 K were taken with the vacuum can pumped out. As a general rule, low temperatures improve Q dramatically. However, since a low Q implies that the oscillator losses are very small, you now have to be concerned with those little loss mechanisms that you might normally have dismissed as esoteric. That $Q \approx 260000$ value at 4 K can go down to $Q \approx 100000$ with a helium leak of well under 1μm Hg—a cynic might thus use an oscillator as a leak detector.

BeCu is available from Kawecki Berylco Industries Inc., P.O. Box 1462, Reading, PA 19603 (215-921-5123). There are a number of variations, but we generally use BeCu 25. At room temperatures, the density of BeCu is 8.23 g/cm^3 and the shear modulus G is 5.3×10^{11} dynes/cm^2 while the Young's modulus E is 1.31×10^{12} dynes/cm^2.

The sample cell must be made of a low density material in order to see a measurable period shift when the cell is filled with helium. Stycast 1266 epoxy (density 1.18 g/cm^3) is often used but it is rather lossy. To avoid making matters any worse than they have to be, the joint between the epoxy cell and the BeCu head must be carefully made in order to get a large contact area. The idea is to avoid subjecting the cell to excessive stresses as that will lead to inelastic deformations and hence losses.

There is, however, nothing really magic about the epoxy. For example, we have recently made a resealable sample cell out of an aluminum can which is sealed to the BeCu torsion rod head by an indium O-ring.

4.2.3 Construction

Torsional oscillators are somewhat delicate devices and reasonable care must be excerised during their construction.

Torsion Rod The torsion rod, base, and head are all machined from the same piece of BeCu. The machining speed is the same as for OFHC (oxygen free high conductivity) copper and Rapid-Tap must be used as lubrication. Drill the fill line before you cut the torsion rod or else it will break when you try to drill it. Use a brand new drill. Take short (about one drill diameter) brisk cuts and clean off the chips between each cut. For turning the torsion rod down to its final radius, use a freshly sharpened thin ($\approx$40mils) parting tool. Secure the tool close to the cutting tip in order to avoid vibrations. Do NOT take axial cuts along the torsion rod. Starting at the end closest to the head, cut radially down to the final dimension, pull the tool bit back, slide it along axially, and then take the next radial cut.

After machining, the torsion rod should be ultrasonically cleaned with tricloroethylene (so what if it causes cancer in laboratory animals), acetone, and isopropyl alcohol. It should then be electropolished in a copper solution and ultrasonically cleaned again. Finally, it must be annealed at 316° C for about 8 hours in a vacuum furnace. When the time is up, turn the heat off and let the piece cool down overnight inside the furnace. Heat treated BeCu is very hard and thus difficult to machine, even more so than stainless steel! Do not even try it because you will also introduce surface defects which will affect the temperature dependence of the quality factor. BeCu will also shrink a few percent on heat treating so be sure to machine the matching epoxy sample cell only after heat treating the BeCu torsion rod.

Sample Cell For this design, the sample cell is made by joining together two separate pieces. To get a nice uniform gap, make an aluminum mandrel which is the negative of the desired gap. Ultrasonically clean the mandrel and then cast epoxy over it. Machine the cast epoxy with the mandrel in place and then dissolve the aluminum mandrel away in a NaOH bath. Since NaOH also reacts slowly with Stycast 1266, you should have machined off as much of the aluminum as possible in order to reduce the etching time required.

The pieces are glued together with Stycast 1266. The bottom piece of the sample cell is glued first to the BeCu head. An aluminum jig should be made to keep the pieces aligned while the epoxy is setting. To prevent the fill line from being accidentally epoxied, it can be blocked off with a drop of Duco cement. The Duco cement can be dissolved away with acetone but, since Stycast 1266 also dissolves slowly in acetone, be sure to use only a very small drop of Duco cement. Moreover, make an epoxy fillet around the joint to ensure that it will be leak tight. Do not get epoxy onto the torsion rod itself as that will degrade the Q.

For cells that must be operated under high pressures ($\approx$30 bars), it helps to have *machined* a radially deep notch around the circumference in order to accommodate an extra deep epoxy fillet. The epoxy fillet seems to adhere better

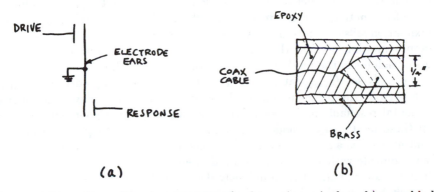

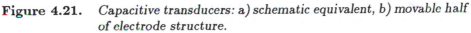

Figure 4.21.   *Capacitive transducers: a) schematic equivalent, b) movable half of electrode structure.*

to a machined epoxy surface than to the kind of smooth surface formed when the piece is cast against an aluminum mandrel.

Electrodes Two capacitive transducers are required, one for the driver and one for the detector. Schematically, they look like Figure 4.21a. The two electrodes ears form the grounded half of the capacitors. A conductive surface may be silver painted on the epoxy ears. Humidity and thermal cycling tend to degrade the silver surface so unused oscillators are kept in dessicators and the silver paint is refinished every year or so. The other half of the capacitors consists of movable electrodes mounted onto the vibrational isolator block. The typical structure is depicted in Figure 4.21b. The brass outer casing is grounded and acts as a shield. Machining is straightforward. Be sure to finish off the electrode surface so that the inner and outer electrodes are flush.

The capacitor gaps are typically 3–5 mils and the resultant transducers have capacitances of 1–3 pF. Smaller gaps tend to spark. The gaps are set by inserting mylar sheets of known thicknesses between the electrodes.

4.2.4   Installation

The golden rule for installing a torsional oscillator is that nothing should move unless it is meant to move.

Vibrational Isolation Vibrational noise can be coupled into the oscillator in many ways. All physical connections to the walls, floor, and ceiling are a problem. Pumping lines can be a real problem. Acoustic noise from your neighbor's loud stereo can be a problem. Even the boiling from the liquid nitrogen jacket in your dewar (if you have that type of dewar) can conceivably be a problem. Yes, it is possible to get a little bit paranoid about this.

A vibrational isolator is basically a mechanical oscillator, with a resonant frequency $f_{isolator}$ much less than f_{cell}, which is mounted between the cell and the outside world. The net transfer function is, approximately, the product of the transfer functions for a harmonic oscillator at $f_{isolator}$ and at f_{cell}. This approximation fails only near f_{cell}, where the net transfer function is smaller than this approximation because the cell steals energy from the isolator. In effect, a vibration isolator is the mechanical analog of a low pass filter.

We usually set $f_{isolator}$ at 100–200Hz. One might be tempted to set $f_{isolator}$ much lower but the problem is that the isolator will usually have a number of fundamental resonant modes, each with its own series of higher harmonics. If one of these fundamental modes has a very low resonant frequency, what you will end up with is a series of narrowly spaced resonances—one of which will be likely to coincide with your torsional mode frequency.

The torsion rod structure previously depicted already has a vibrational isolator block. Such blocks need not be machined as part of the torsion rod. In fact, it is often more convenient to make the block as a split ring and then clamp it in place.

Wires and Capillaries Be careful not to mechanically short out your vibrational isolator. We run the wires and capillaries along the various torsion rods axially, in order to mimimize their contribution to the corresponding moments of inertia. Then we tie everything down with dental floss. Some people will also lock everything in place by smearing everything with high vacuum silicon stopcock grease, which hardens into a glassy solid at low temperatures. In the old days, Apezion N grease was used instead of silicon stopcock grease but we eventually discovered that it degraded into a powder upon repeated thermal cycling.

4.2.5   Operation

Most drive circuits are some variant of a phase locked loop (PLL). In Reppy's group, they are called TOADS boxes (or Torsional Oscillator Automatic Drive Systems) so that they can be painted green and have little black frogs drawn on them.

A Simple Drive Circuit Figure 4.22 depicts a circuit which was used by Dave Bishop in his work on superfluid ^{4}He films. Technically, it is not a phase locked loop—it simply drives the torsional oscillator in a self resonant mode. The oscillator response is amplified, filtered, phase shifted, and then fed back to the oscillator drive via a zero-crossing detector and another amplifier.

The filter is a bandpass, centered near the resonant frequency of the torsional oscillator, but with a much smaller $Q \approx 100$ so that the frequency of the loop is determined by the torsional oscillator. The phase shifter is there to account for the fact that the oscillator response need not be in phase with the drive. It is adjusted for maximum oscillator response. The zero-crossing detector is nothing more than a comparator and is there to ensure that, no matter what the oscillator response is, we have a constant amplitude signal feeding the drive amplifier.

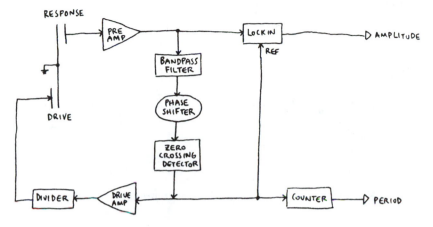

Figure 4.22. *A simple drive circuit.*

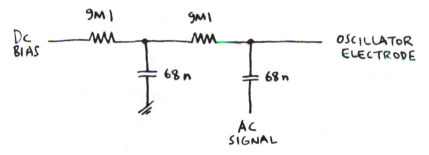

Figure 4.23. *Electrode bias circuit.*

Phase Stability By differentiating the phase with respect to frequency, at resonance, we can show that $\delta P/P = \delta\theta/2Q$. For a period stability $\delta P/P \approx 10^{-9}$ and a nominal $Q \approx 10^6$, we will then need $\delta\theta \approx 0.1$ degrees. Phase stability in the drive and detection electronics is thus a key ingredient in the proper operation of a torsional oscillator.

Electrode Biasing The rectilinear force exerted by the capacitor, $F = CV^2/2d$ (where d is the capacitive gap spacing), is not linear in the voltage. To make things linear, and to amplify the force, a high voltage (90–360V) DC bias is superimposed on the AC excitation so that $V = V_0 + \delta V$, $\delta V \ll V_0$. A typical biasing circuit is shown in Figure 4.23.

The capacitors actually contribute to the restoring force. By differentiating

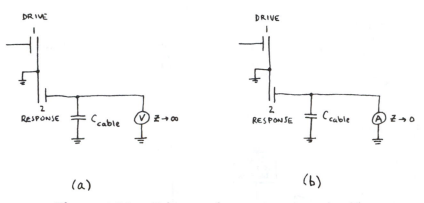

Figure 4.24.   *Voltage and current preamp circuits.*

the force expression by d, we can show that $K = C(V/d)^2 r$ (for the torsional mode, with the electrodes configured as above). Thus, if the bias voltage is not stable, period fluctuations will result. Batteries are the preferred bias voltage source. They can, however, drift with time and temperature and we have started putting regulators on our batteries for better long term stability.

Warning: you must disconnect the DC bias when pumping down your vacuum can because the capacitive gaps tend to spark near those pressures where the mean free path of the ambient gas equals the gap spacing ($\approx$2mm Hg at 77 K, $\approx$ 100μm Hg at 4 K).

Preamplifier Response The schematics in Figure 4.24 represent the circuits seen by the AC signal in the case of a voltage and a current preamp. There is always some cable capacitance $C_{\text{cable}} \approx 200$pF which is much larger than the transducer capacitance $C_2 \approx 1$–3pF. Assume that the capacitor gap spacing $d_2(t) = d_2 + (\delta d)e^{jwt}$. Then for the voltage amp, the condition $dQ_2/dt = 0$ gives:

$$\delta V/V_0 = (C_2/[C_2 + C_{\text{cable}}])\delta d/d_2.$$

And for the current amp, the condition $dV_2/dt = 0$ gives:

$$\delta I = -(j\omega V_0 C_2)\delta d/d_2.$$

The important difference is not the 90 degrees phase shift but the fact that the voltage amp's output is degraded by the ratio $(C_2/[C_2 + C_{\text{cable}}]) \approx 10^{-2}$, while the current amp's output is independent of the cable capacitance.

For the unloaded cell at ^{4}He temperatures, with $V_0 = 360$V DC and $\delta V \approx$ 10–100mV peak-to-peak, one typically requires voltage amp gain settings of $\approx 10^3$ and current amp gain settings of $\approx 10^{-9}$ A/V.

Quality Factors The easiest way to measure the quality factor Q is to keep the drive constant and then measure the resonant amplitude, which is proportional to Q. However, if what you are trying to measure is dependent on the oscillator velocity, it is better to rig up some sort of a feedback loop to keep the resonant amplitude constant and then measure the required drive amplitude instead. This method has the added advantage that, as you sweep through some dissipation feature, the response signal will not drop below the threshold required for the proper operation of your lockin or PLL.

In order to calibrate the system gains, we often do a ring-down experiment wherein the drive is disconnected and the exponential amplitude decay is monitored on a chart recorder. Using the definition of Q as the ratio of the energy stored per cycle over the power loss, it is easy to show that the decay time constant $\tau = PQ/\pi$. After any change in the system, the response amplitude will exponentially drift to its final value with this same time constant—which can be painfully slow. The period, fortunately, is usually stable after approximately one such time constant.

Temperature Backgrounds For some experiments, the response is dominated by the sample and the behavior of the unloaded oscillator is not crucial. As a matter of principle, however, one should always measure the temperature dependence of the period and quality factor for the unloaded oscillator. These so-called temperature backgrounds have been found to exhibit an anomaly near 30 mK, as shown in Figure 4.25. Experience with different sample cells has led us to conclude that the anomaly is a property of the BeCu torsion rods, and not the sample cells. In fact, all of the other materials that we had made torsion rods out of also show this same anomaly—but at different temperatures. For BeCu, above the anomaly, the temperature dependence of the period is roughly $\delta P/P \approx 5 \times 10^{-6}$ per degree K. Thus a period stability of $\delta P/P \approx 10^{-9}$ requires a temperature stability of ≈ 0.2 mK.

These period and amplitude backgrounds are not reproducible upon thermal cycling. Residual gases (e.g., remnants of your exchange gas) which condense onto or desorb from the oscillator during thermal cycling may affect the period by as much as 1 part in 10^3. Just as with resistive thermometers, a new calibration must be done for each run.

Nonlinearities We have observed nonlinear behavior in some of our unloaded oscillators. This was detected by stepping the drive amplitude and watching for variations in the resonant period and quality factor (or more correctly, the ratio of the response to drive amplitudes). In one particularly bad case, strains on the torsion rod which were as small as 10^{-6} caused period shifts as large as 1 part in 10^5. The quality factor was found to increase, actually diverge, as the drive amplitude was decreased. The actual shapes of the P and Q versus response amplitude curves were somewhat pathological but a further discussion is beyond the scope of these notes (and beyond the scope of our current understanding).

Lacking any explanation for the non-linearities, the usual solution has been to build another oscillator. To be absolutely safe, one can always run his or her oscillator in constant amplitude mode (as opposed to constant drive mode). The observed period shifts can then be correctly interpreted as due only to the

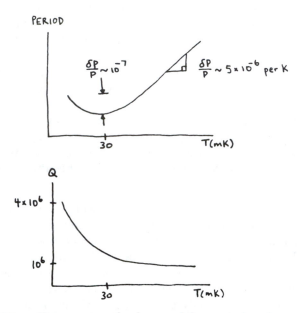

Figure 4.25. *Temperature background for period and quality factor.*

added sample. However, if the oscillator's intrinsic damping mechanism is not proportional to the velocity, then the resonant amplitude will not be simply proportional to the quality factor times the drive. Changes in the drive required to maintain the constant amplitude are still valid indications of the added dissipation due to the sample but more quantitative information may be difficult to extract.

Limitations The book by Braginsky, Mitrofanov, and Panov contains a very interesting discussion on the fundamental limitations of any simple harmonic oscillator. The first issue to consider is what determines the oscillator Q. Even in a perfect crystal, there will be loss mechanisms (dubbed thermoelastic dissipation, phonon-phonon interactions, and phonon-electron interactions in Braginsky's text). In practice, however, the oscillator Q is determined by defects both in the crystal and on the surface. The exact details of these processes are far from being well understood. Nevertheless, this is one of the reasons why the BeCu torsion rods that we use must be electropolished and annealed.

Given a particular oscillator Q, the resolution will then be fundamentally limited by the fluctuation-dissipation theorem of statistical mechanics. This same theorem, applied to electrical systems, results in Johnson noise for a resistor. In practice, only the gravity wave people have come close to achieving this fundamental limit. Our oscillator resolution is limited by external vibrations, temperature stability, bias stability, phase stability, and who knows what else. We feel that there is still room for improvement in our torsional oscillators.

4.3   Cryogenic Electronics

by Mark R. Freeman

The environment in which ultra-low temperature experiments are performed is such that the level of noise coming out of many types of transducers can be made smaller than is possible in any other situation. In addition to the extremely low level of stray electric and magnetic fields inside a properly designed and constructed cryostat, thermal noise levels at micro- and millikelvin temperatures are orders of magnitude lower than at room temperature. The advantage of lower noise is, of course, that it facilitates the measurement of smaller signals—provided that the signal (and hence noise) levels can be amplified cleanly *within* the cryostat before being exposed to the harsh room temperature environment. This requires an active device which functions at low temperatures and itself contributes little noise. Although it may seem obvious, it is worth mentioning that cooled electronics should be used only when the physics demands it, as the increase in complexity is considerable for most experiments.

This section is a condensation of the experience of members of the low temperature group at Cornell, and is intended as a starting point for experimentalists. We are familiar mostly with superconducting quantum interference devices (SQUIDs), in particular commercial ac biased units and dc biased devices which we obtain through a collaboration with IBM. We have done a limited amount of testing of Si and GaAs field effect transistors, which are able to operate at low temperatures because they do not rely on carriers generated by thermal activation. The physics behind all of these devices is extremely interesting, but knowledge of it, although useful, is not essential for applications. Keep in mind that a black box approach is usually more efficient when electronics is used as a tool and not pursued as an end in itself. The literature in this area is extensive and widely scattered. I have limited the references to a small number of informative sources.

Cryogenic electronics is at the moment in a phase of explosive growth, fueled by advances in materials research. Small bandgap, high mobility semiconductors and the new high T_c oxide superconductors are likely to spawn many devices designed for liquid nitrogen temperature operation.

4.3.1   Passive Components for Use at Low Temperatures

Knowledge of how things change upon cooling is important to all designers of objects intended for use at low temperatures. It is relatively easy to find passive electronic elements with predictable and reliable low temperature properties. It is a good idea to check all new or unfamiliar components. We make routine measurements with an RLC bridge and a "yardstick" with a few wires which inserts into a liquid helium storage dewar.

Metal film resistors are a good choice for low temperature circuits. These are made with alloys or cermets in order to obtain a small temperature coefficient of resistance, with the result that the values normally change by less than 10% upon cooling to 4 K. If a very constant (to better than .01%) resistance is required at low temperatures, it is necessary to avoid things like the Kondo effect (magnetic impurities) and magnetoresistance (if magnetic fields are used).

Among readily available capacitors, the mica and polyester film varieties have no cryogenic pathologies, apart from small ($<$ .1%) variations with temperature below 4 K (due to glassy states in the dielectric). Small capacitors can be made which are extremely insensitive to temperature and have very low loss, using a crystalline (quartz, sapphire) or vacuum dielectric.

The inductors and transformers used at low temperatures are often wound on nonmagnetic cores. Materials such as ferrite, with high permeability at room temperature, usually suffer from strong temperature and frequency dependences and eddy current losses. Powdered iron cores, such as the ones available from Micrometals, are a better choice if moderate permeability is required. Superconductors are very useful in making specialized transformers for use at low temperatures. More efficient coupling between primary and secondary is achieved when a layer of insulated superconducting foil is wrapped on the core to guide the magnetic flux lines. The insulation allows the ends of the foil to overlap without shorting and thereby screening out all of the field. DC (flux) transformers, and inductors with very high Q, can be wound with superconducting wire. In low noise applications these coils must be placed inside superconducting shields for protection against magnetic pickup. Screening then reduces the inductance, by a factor which in the case of concentric cylindrical elements is approximately $(1 - A_{coil}/A_{shield})^2$, where A denotes a cross-sectional area.

Variable passive elements can also be used inside cryostats. This is accomplished typically with a mechanical linkage such as a long shafted screw drive (more precise and stable than a push rod). Less direct mechanical methods are also useful, such as bellows or diaphragms activated by liquid helium, and piezoelectric devices. Variable tubular capacitors, available from Voltronics, can be cooled. Inductors can be made with sliding cores or, in some applications, a variable number of turns. When components are not required to be continuously variable, they can be switched discretely by reed relays or by superconducting/normal shunts.

4.3.2   Active Components at Low Temperatures

Semiconductors At present, most semiconductor devices used in cryogenic applications were designed for use at room temperature, but make use of physical principles which are sufficiently insensitive to temperature to allow them to operate cold. A semiconductor electronics culture unique to low temperatures is only beginning to emerge, based on III-V and II-VI compound semiconductors and epitaxial growth techniques.

Surveys of the use of cooled commercial devices have been published by Lengeler (1974), and by Kirschman (1984). Some of the devices considered by

Lengeler are no longer available (such as the Plessey GAT-1, which was quite popular in the low temperature community, but can be replaced by the P35-1101 of the same manufacturer), but the basic information is still very relevant. Some key points are as follows. Si and Ge bipolar transistors are to be avoided, because dropping minority carrier lifetimes degrade their performance as temperature is lowered. Even more seriously, the carriers freeze out so that these devices do not work at all below about 100 K. We are not aware of any commercial bipolars made with better low temperature materials, such as n-type GaAs, InP (III-V), or even more unlikely, HgCdTe (II-VI), or gray-Sn. The field effect is clearly a better physical basis for cryogenic devices because it works at any temperature. Si MOSFETs will work at 4 K provided that the source and drain contacts are degenerately doped. Driven by the radar and microwave communications markets, GaAs FETs have become quite common (Plessey, NEC, many others), and it is already possible to buy monolithic GaAs integrated circuits (Harris, NEC) and heterojunction FETs (NEC). The latter are of the type of high electron mobility structure, made by modulated doping of AlGaAs/GaAs during epitaxial growth, which claim the record for low noise performance at GHz frequencies.

How are device characteristics affected by cooling? The transconductance of most FETs increases by a factor of two or so. Changes in device noise are usually disappointing. The temperature in the channel of a FET will hang up at tens of Kelvins because there is little area of contact to the substrate through which to sink the dissipated power. A decrease in thermal noise of about a factor of three is about the best that can be obtained. In addition to this there is $1/f$ noise which is often significant into the megahertz region (for MOSFETs; the flicker noise in JFETs is much smaller). This noise can actually increase as the temperature is reduced, depending on the distribution of trap states which give rise to it. Empirically it is found that the corner frequency (where the $1/f$ noise spectral density equals the white noise contribution) scales inversely as the gate area of the FET. Thus many FETs (in particular all of the GaAs devices so far) are best suited to RF and microwave applications. Refer to Motchenbacher and Fitchen for details about device noise and noise measurement.

Many other semiconductor components are useful at low temperatures. Low noise mixers and switches can be made with cooled diodes. Varactors as tunable capacitors, FETs as variable resistors (or capacitors), and tunnel diode oscillators are other examples.

Superconductors The rf SQUID historically has been the most common cryogenic amplifier used by low temperature physicists. A complete system, consisting of the SQUID sensor, a probe to mount it in and a controller to read it out, can be purchased for about $10K (Biomagnetic Technologies, Cryogenic Consultants). Basically, what this combination does is convert an input current to an output voltage, with a forward transresistance of up to 2×10^7 volt/amp. The device noise referred to the input is of order $10\,pA/Hz^{\frac{1}{2}}$. The BTi SQUIDs are typically used between dc and 20 kHz, and have a dynamic range of order 140 dB in a 1 Hz bandwidth. This large range is achieved by a "flux-locked-loop"—the signal is nulled at the SQUID by feedback from the controller, enabling the system to remain linear for inputs of up to 100 μA.

The rf SQUID is a superconducting ring broken by a single Josephson junction. In very basic terms, an rf SQUID magnetometer measures the screening by this ring of an inductor to which it is coupled, part of a tuned tank circuit. Small amounts of magnetic flux are exactly cancelled by supercurrents induced in the SQUID ring, but when the induced currents approximate the critical current of the junction, the effective screening is a strong function of applied flux. By choosing an appropriate rf level, then, the flux applied to the SQUID by a second inductor, the input coil, can be sensitively measured. Flux passes into the center of the SQUID ring when the critical current is exceeded, in such a way that the response is periodic in the flux quantum, $\phi_0 \approx 2 \times 10^{-15}$ Weber. Details about rf SQUIDs can be found in Giffard, Webb and Wheatley (1972). Day-to-day use is straightforward and is adequately described in the BTi manual.

DC SQUID technology has improved dramatically in the past seven years or so, resulting in practical devices many orders of magnitude more sensitive to magnetic flux than typical rf SQUIDs. A major breakthrough was the thin film planar geometry of Ketchen and Jaycox (1982). This is illustrated in the upper left of Figure 4.26. The open area of the SQUID loop is defined by a superconducting washer. Thin film inductors deposited on top of this structure efficiently couple signals into the SQUID, as most of the flux is forced through the hole by the screening action of the washer. The operating characteristics of a dc SQUID, described briefly below, are also shown in Figure 4.26, with the analogous properties of a FET included for comparison. Each device has a current–voltage relationship which is modulated by an additional parameter (magnetic flux or electric field in these instances) to produce a family of characteristic curves. Under appropriate bias conditions, the amplitude of this modulation corresponds to a large power gain—hence the utility of these devices as amplifiers.

Commercially available dc SQUIDs miss the state-of-the-art in sensitivity by an order of magnitude or two. $15K will buy a sensor and control electronics which include flux locking circuitry with a bandwidth of 100kHz (BTi). The equivalent flux noise referred to the input is about $10^{-5}\phi_0/Hz^{\frac{1}{2}}$. Any kind of feedback scheme running to room temperature has a bandwidth of at most a few MHz, because phase shifts in the fed back signal kill it off. However, it is possible to use SQUIDs in an open loop circuit; that is, as small signal amplifiers, to frequencies above 100 MHz when the devices are appropriately designed. Some groups fabricate their own thin film dc SQUIDs, often achieving superior performance. To be fair, the lowest noise is found only in small sections of the flux-voltage curve, and therefore applies only to the small signal amplifier mode. Flux locking schemes give an average of the noise performance over half a period of the curve. The popularity of these devices is growing rapidly, and it is to be expected that state-of-the-art will come onto the market within the next few years.

A dc SQUID is a superconducting ring broken by two Josephson junctions. Like the rf SQUID, its operation can be understood crudely in terms of screening

Thin film dc SQUID (top view)   **JFET (cross section)**

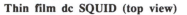

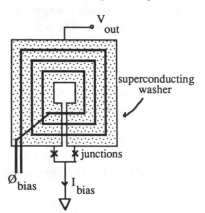

SQUID I-V characteristics   FET I-V characteristics

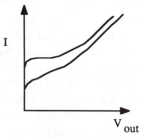

(typical scale: 10μV, 10μA)

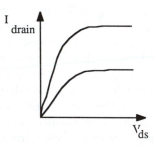

(typical scale: 10V, 100 mA)

Transfer characteristic for I_{bias} = constant   Transfer characteristic for V_{ds} = constant

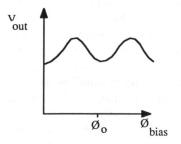

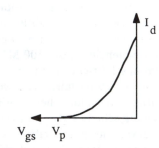

Figure 4.26.   *SQUID and FET Characteristics.*

currents set up in the ring in response to externally applied flux. Again the behavior is periodic in the flux quantum. Unlike the rf SQUID, however, the dc SQUID can be probed directly by hooking two leads across it. It is operated with a constant current (or voltage) bias—hence the name. Screening currents "add in" and modulate the measured voltage (or current). Currents in the ring actually circulate at the Josephson frequency, so what is measured is a time-averaged quantity. A nice introduction to dc SQUIDs is the article by Ketchen (1981).

Just as diodes are the building blocks of transistors, SQUIDs are built from Josephson junctions, which are of course useful in their own right as two-terminal devices (for example as mixers and switches).

4.3.3   Installation and Operation Notes

This subsection is divided arbitrarily between rf and dc SQUIDs. Apart from biasing and readout considerations, the comments apply equally to both. The use of SQUIDs is largely an exercise in careful shielding and wiring, which just reflects the susceptibility of small signals to interference problems. Refer to the Electric and Magnetic Isolation section (Section 3.4) for general information.

Typically rf SQUIDs sit in the helium bath. This is convenient because the probe package is quite large (5/8 inch diameter by 4 inch length) and is usually connected to a semi-rigid triaxial line which carries the rf and feedback signals. Whenever the length of this triax is changed, the tank must be re-tuned to resonate within the tuning range of the controller. This is usually done with a small inductor made by winding copper wire around a Q-tip, and/or a microwave chip capacitor, attached inside the Nb shield. The tuning procedure must account for the fact that the resonant frequency shifts on cooling. If it is out of range at 4 K, a quick fix can be effected by changing the tuning capacitance inside the rf head.

It is very important to take defensive measures against microphonics. All components and wiring must be rigidly mounted, even to the extent of anchoring the wires within shielded twisted pairs by injection with epoxy or grease. Superconducting shields should be used whenever possible, and a mu-metal shield used during cool-down to minimize the amount of trapped magnetic flux. Machined Nb junction boxes are good for shielding splices made between wires inside the cryostat. Boxes assembled from tinned circuit board (the solder superconducts) have been used with success, but carry more risk of noise due to flux line motion. The ideal shield would be a single-crystalline Type I superconductor. In practice, however, things like Nb capillary are used as drawn, not annealed, to avoid thermal shorts. Tinned Cu-Ni capillary is a very poor substitute for Nb—the solder can flake off with bending or just from thermal cycling. The Nb can which surrounds the SQUID is very effective. On the nuke (nuclear demagnetization) rig, the unit used for the LCMN thermometer sits right on top of the compensation coils for the main magnet, and is able to remain flux-locked during demags. The devices are reasonably insensitive to the changes in bath level and temperature which occur during normal cryostat operation. Wrap a copper wire around the Nb capillaries used to shield the signal lines running from the vacuum can to the SQUID, in order to heat sink them when the bath level is low (and if epoxy is used to seal the capillaries into the feedthrough, make sure it's Stycast 2850).

Commercial rf SQUID controllers can be run on battery power ($+/-$ 24 V), which can help to eliminate grounding problems. Symptoms of ground loop interference are drifts in the dc output of the controller and frequent unlocking of the feedback loop. A typical symptom of rf interference is a noisy, low amplitude triangle pattern. RFI often gets in through the signal input circuit when there are other, unfiltered lines running into a cryostat. Demanding applications may require an rf-tight shield to contain the controller and the rf head. This can be, for example, a welded aluminum box with emi filtered feedthroughs for signal and power, and a Type N bulkhead connector for the rf head. This is particularly useful when the cryostat is not in a shielded room, or when cross talk between multiple rf SQUID controllers must be eliminated.

A central element is the SQUID package itself, which usually incorporates Nb and mu-metal shields. The chip can be mounted compactly on a circuit board with wire-bonds made to solder pads for the attachment of leads. The tunnel junctions are sensitive to damage from electrostatic discharge, so remember to ground the soldering iron and yourself. Figure 4.27 shows a mounting scheme, due to Roukes, which has been used extensively by us in the past. Twelve pin ceramic 'headers' hold the chip. Wire-bonds are made between the pads on the chip and the header pins, which plug into sockets on the circuit board. It's a good idea to have a cover which fits over the chip to protect the SQUID and the wire-bonds from stray fingers and soldering tips. Future designs will have the chip mounted directly on a ground plane on the circuit board, for better heat sinking, and have stripline running as close as possible to the input coil for high frequency work. The interest in heat sinking arises from the possibility of achieving quantum limited noise performance from present-day SQUIDs at mixing chamber temperatures. The devices dissipate little power (nanowatt ballpark), but the heat goes into a very small area, perhaps 10 μm x 10 μm. The temperature rise due to Kapitza boundary resistance can therefore be substantial. Eventually we may see hermetic SQUID packagings, complete with fill line for ^{3}He cooling of the chip.

Mark Ketchen at IBM Yorktown has designed SQUIDs and packagings which isolate the input from the output, a particularly nice feature for rf work. He places the input coil pads on one side of the chip, away from everything else, and then mounts the SQUID on a flat milled into a Nb cylinder, which fits snugly into a Nb sleeve. The input leads are thereby isolated from most of the other wiring by a Nb wall.

Another essential element of a dc SQUID system is the readout scheme. The output noise level of a dc SQUID is usually too small to be measured directly by a room temperature amplifier, say 100 $pV/Hz^{\frac{1}{2}}$. For frequencies between about 10 kHz and 100 kHz, this can be matched to a low noise FET amplifier at room

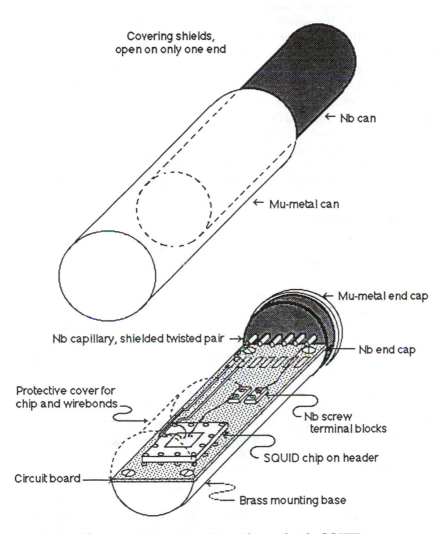

Figure 4.27. *Mounting scheme for dc SQUID.*

temperature by using a cooled transformer. The SQUID output impedance of a few ohms must be stepped up to a few tens of kΩs to hit the sweet spot of the FET, and the resulting voltage step-up increases the SQUID output noise voltage to the required level. At lower frequencies, the 1/f noise of the FET becomes dominant and an rf SQUID readout might be needed. At higher frequencies, input impedances are reduced by parasitic capacitance, so the FET must be cooled to decrease its noise. There is also the possibility of reading out one dc

SQUID with another. This is an option only when a loss of dynamic range can be tolerated, as the output noise of the second SQUID is still small and simply will be swamped by the amplified noise of the "front end."

Quite often the most desirable physical location for an active device inside a cryostat (say, right next to the experiment to minimize parasitic impedances) is disallowed by temperature considerations. If the power dissipation of the device is too large for the cryostat to cool properly, then clearly it must go in a higher temperature section. If the device has an optimum operating temperature which is warmer than the surrounding part of the cryostat, it can be mounted on an insulator with a weak thermal link to the cryostat. The optimum operating temperature can then be maintained with thermal regulation using a heater. We have encountered this situation with FETs which had minimum noise at about 40 K, and with high critical current SQUIDs which became hysteretic below 4 K.

4.3.4   Circuit Examples

In this subsection we present two examples, a FET amplifier for a capacitive transducer at audio frequencies, and a dc SQUID amplifier for an inductive transducer at rf.

Vacuum gap capacitors are often used to detect movement at low temperatures. The capacitor gap is changed by attaching one plate to, for example, a diaphragm or the 'bob' of a pendulum. The noise generated by such a transducer (thermal in the absence of extraneous vibrations) is extremely small, as the capacitor has a low dissipation factor in addition to its low temperature. The sensor noise as seen by a preamp is further reduced by voltage division from the cable capacitance. The system sensitivity in this case is determined by the noise floor of the preamp. The application of cooled JFETs to this situation is considered by Moster *et al.* (1987). (JFETs are chosen for low $1/f$ noise). They point out that the signal-to-noise ratio is independent of cable length in the usual case, which has the cable capacitance larger than that of the transducer, and current noise is dominant. The current noise is expected to drop with temperature as it is due to gate leakage. A cascode amplifier using a Teledyne Crystalonics ultra-low noise JFET is shown if Figure 4.28. The thermal noise of the big gate resistor (100 MΩ) dominates at room temperature. The current noise is so low, estimated 1 fA/Hz^{1}/2 at 1 kHz, that the devices are paralleled to reduce voltage noise, 1 nV/Hz^{1}/2 at 1 kHz. A quick factor of two in sensitivity is won by cooling the gate resistor to 77 K. Further gains become much harder to demonstrate. No evidence of decreased current noise was seen for these devices.

The thermal noise of an NMR pickup coil at millikelvin temperatures is sufficiently low that a dc SQUID must be used to match it. The SQUID input impedance is low, so it is best to use a series-tuned tank. The circuit is shown in Figure 4.29. Series Josephson junctions are included as limiters for pulsed NMR, to protect the SQUID and shorten the recovery time. Three sources of noise are important in determining the circuit elements for optimum sensitivity—the thermal noise of the tank, and the voltage and (circulating) current noises of the

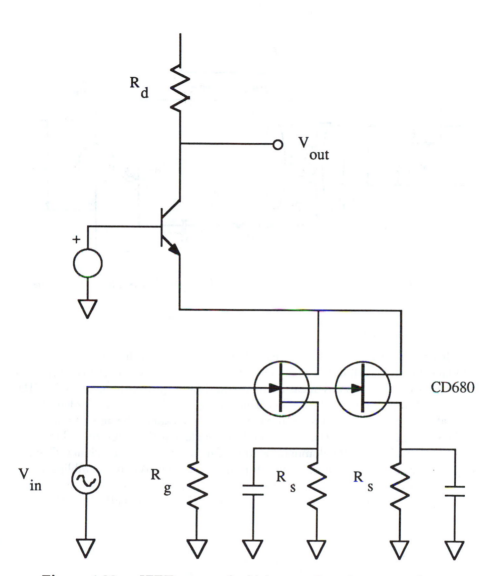

Figure 4.28. *JFET preamp for high source impedance transducers.*

SQUID (for more information on SQUID noise see Clarke, 1980). The matching condition turns out to be $\omega \times L_{input}/r \approx 1$. The noise temperature of this system is about $10 \times f$ mK, where f is in MHz. A cooled FET second stage amplifier must be used to achieve this in practice because of the bandwidth required for most pulse work.

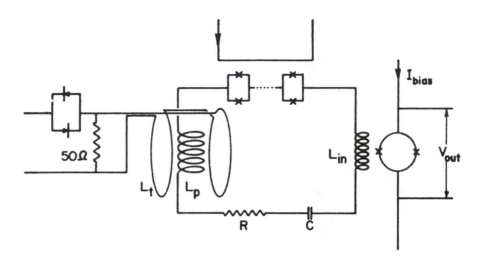

Figure 4.29. *dc SQUID used in NMR circuit with Josephson junctions*

4.3.5   Comments

Before using a dc SQUID or any other cryogenic amplifier, it is essential to convince yourself a) that it is really necessary, and b) that it will really work. It is tempting to think that any measurement could be improved by using a SQUID; but, in fact, nothing could be further from the truth. For example, cooled pickup coils attached to giant annular mixing chambers cannot be used to improve medical imaging, because the human body is an electrical conductor and thermal noise from the body is the dominant source of noise in existing machines. Clearly, it is essential to take all noise generators into account. When assembling input circuits, matching networks, and the like, it is important to check components carefully at low temperatures because you can't predict exactly how they will behave.

4.3.6   Suppliers

- Micrometals, Inc., 1190 N. Hank Circle, Anaheim, CA (714) 630-7420.
- Voltronics Corp., East Hanover, NJ (201) 887-1517.
- Plessey Microwave Materials, San Diego, CA 619/571-7715. (GaAs FETs)
- California Eastern Laboratories, Santa Clara, CA 408/988-3500. (NEC semi-conductors)
- Teledyne Crystalonics, 147 Sherman St., Cambridge, MA (JFETs)
- Biomagnetic Technologies, Inc., San Diego, CA 619/453-6300 (SQUIDs)
- Cryogenic Consultants, Ltd., London, England 01-743-6049 (SQUIDs)

4.4   Nuclear Magnetic Resonance

by Thomas J. Gramila

This chapter is a discussion of some of the techniques used in making NMR measurements on a sample in a cryostat. There will be no discussion of the physics of NMR; the reader is referred to the texts by Slichter (1978) and by Abragam (1961). Rather the emphasis will be on getting rf in and out of the cryostat as well as a short description of some room temperature electronics. Frequencies are generally assumed to be low enough that a simple coil can still be effective for creating and detecting magnetic fields; say below 100 MHz. For frequencies much above this readers are referred to the section on high frequency techniques. A valuable reference on general techniques of NMR, FFTs, etc. is the book by Fukushima and Roeder (1981) *Experimental Pulse NMR, a Nuts and Bolts Approach.*

4.4.1   Tank Circuit Fundamentals

The detector of the radio frequency magnetic fields generated by the sample is a coil, which has an inductance L approximately given by $L = (rN)^2/(22.9\ell + 25.4r)$ for a simple N turn solenoid of length ℓ and radius r, where L is in μH, and the length and radius are in cm. The losses in the coil can be represented by a series resistance r, which is generally a function of frequency. There are then two voltage sources, V_s and V_n; as illustrated in Figure 4.30 V_s, the signal, is the voltage induced by the spins; it is proportional to the number of turns in the coil, and thus proportional to the $\sqrt{L}$. V_n is the Johnson noise of the resistance and is given by the Nyquist formula

$$V_n = \sqrt{4kTRB} \qquad (4.3)$$

where B is the bandwidth. A 50 Ω resistor at room temperature has about 0.9 nanovoltz/$\sqrt{Hz}$.

The resonant frequency of the circuit containing the coil is tuned by adding a parallel capacitor whose value is chosen so that at the frequency of operation ω, we have $\omega L = 1/\omega C$. This maximizes the amount of power that is delivered by the spins, since the signal voltage source now drives the minimum possible impedance, which is just r.

Simple circuit analysis for the case when $\omega L >> r$ shows that the tuned tank can be equivalent to a voltage source with a source resistance R (Figure 4.30). The size of the voltage source is $Q \times (V_s + V_n)$, where Q is $\omega L/r$, and the value of the resistance R is $Q \times \omega L$. (This noise voltage is the same as we would get from a resistance R at the tank circuit temperature.) Our voltage signal to noise is just QV_s/QV_n, which at fixed frequency is proportional to the square root of Q (that is, proportional to $N/\sqrt{r}$). The Q you can use in the tank can be

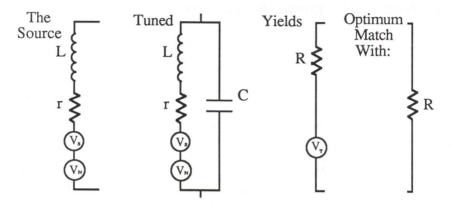

Figure 4.30.   *Schematic for the tuning of a simple tank.*

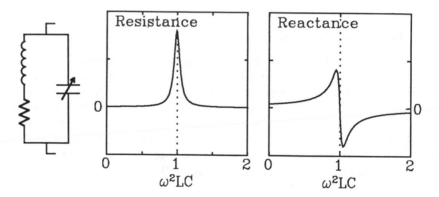

Figure 4.31a. *The output impedance of a simple tuned tank as the capacitor is tuned through resonance.*

limited by the width or Q of the NMR line that you are going to look at. Since the tank circuit itself has a ringdown time $\tau = Q/\omega$, too high a tank circuit Q will not allow you to accurately track the signal.

We have yet to deliver any of the signal power to the amplifier. Every amplifier has an optimum source impedance, R_{opt}, which is the source impedance that minimizes the effect of the amplifier noise in relation to signals from the source. For low impedance devices (i.e., 50 Ω) this is often the actual input impedance of the amplifier; while for FET input devices the actual input impedance can often be orders of magnitude larger than R_{opt}. Since the tank has a source impedance R (equal to $Q\omega L$), the best signal to noise is obtained by choosing an amplifier with as low noise as possible whose R_{opt} is equal to R.

Fortunately it is possible to design the tank circuit to have an arbitrary source impedance less than $Q\omega L$, without affecting its S/N or its ringdown characteristics. Figure 4.31a shows the real and imaginary parts of the tank circuit source impedance as the parallel capacitance is varied from the resonant condition (dashed line). As the parallel capacitor is "mistuned" the real part of the source impedance decreases and the imaginary part grows. By choosing the appropriate value for the parallel capacitor one can obtain an arbitrary value for the real part of the source impedance (as long as it is less than ωL); then, as illustrated in Figure 4.31b,

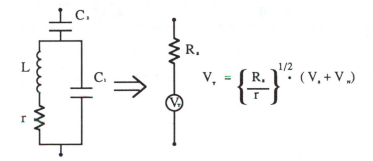

$$V_\tau = \left\{ \frac{R_s}{r} \right\}^{1/2} \cdot (V_s + V_N)$$

Figure 4.31b. *Tank tuning scheme for achieving arbitrary source impedance, $R_s < Q\omega L$.*

the imaginary part of the source impedance can be cancelled by adding a series inductor or capacitor. This circuit is now a voltage source with a source resistance R_s, where both the signal voltage and the Johnson noise have been transformed up by a factor $(R_s/r)^{1/2}$. The S/N is still proportional to the square root of the coil Q and the ringdown time still goes like Q/ω. We can, however, impedance match to any amplifier whose R_{opt} is less than $Q\omega L$. It is possible to purchase inexpensive 50 Ωs amplifiers which are already packaged in "connecterized" boxes and have been optimized for low noise, and even in some instances for pulse recovery and dynamic range. This makes it possible, without spending any time calculating bias currents or soldering resistors, to connect together a set of prepackaged amplifiers that typically outperforms commercially available NMR recievers.

4.4.2   Effects of Coax on Tanks

One significant complication when doing NMR in a low temperature environment is the necessity of using a length of coaxial cable between the coil and the electronics outside the cryostat. The requirement that it have low thermal conductivity generally means that it is very lossy at radio frequencies. One has two choices; either put the entire tank circuit in the bottom of the cryostat, or include the coax as a part of the tank circuit.

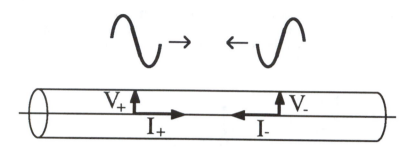

Figure 4.32. *Schematic of traveling wave representation of a coax.*

The primary advantage of having the entire tank cold is that one can obtain much higher Q's, which can substantially improve the signal to noise for narrow lines. Even good hardline coax such as UT-85-SS has a loss of 10 dB per 100 feet at 50 Mhz; five feet of this will limit the Q to about 10 at best and represents a significant heat load for the cryostat. Having the tank in the cryostat forces you to work at a fixed frequency, where both the frequency of operation of the tank and the quality of the match to 50 Ω are likely to change upon cooling down. Note here that the coax must be terminated with 50 Ω (or whatever the coax impedance is) otherwise the coax is effectively part of the tank circuit. Single frequency operation is avoidable if one is willing to design into the cryostat the necessary linkage to room temperature that would permit varying the tank tuning while cold. While this can be done it is by no means simple.

The choice of allowing the coax to be included in the tank circuit allows one to vary the frequency of operation; however, analysis of the tank circuit performance is complicated. In particular, the effect of the coax typically cannot be modelled by the addition of simple lumped circuit elements, such as its capacitance to ground. Transmission line properties of the coax must be included. Because of the loss in the coax, this is true even at low frequencies. The remainder of this section will discuss properties of coaxes and the optimization of tank circuits that include coax.

A typical coax can be thought of as a transmission line with waves traveling in both directions (Figure 4.32), where the ratio of the voltage in each wave to the current in that wave is the characteristic impedance of the coax Z_0 (typically 50 Ω). The impedance the coax presents at any point is simply the ratio of the sums of the voltages in the two waves divided by the sums of the currents. Since the currents are flowing in opposite directions this can be written

$$Z = (V_+ + V_-)/(I_+ + I_-) \tag{4.4}$$

where the subscripts + and − indicate the direction of travel of each wave. Since the voltage and current in each wave for a 50 Ω coax are simply related this can be rewritten as

$$Z/Z_0 = (V_+ + V_-)/(V_+ - V_-). \tag{4.5}$$

The peculiar properties of coaxes are perhaps best illustrated by a lossless section of coax which is a quarter of a wavelegnth long at some frequency ω. The right traveling wave (V_+) at the right end of the coax is 90 degrees out of phase with V_+ at the left end of the coax. The termination at the end determines the phase and amplitude relationship between the incoming and reflected wave. Suppose we terminate the coax with an open circuit. Then the two voltage waves must be in phase and of equal amplitude to obtain $Z = \infty$. The reflected wave then suffers an additional 90 degree phase shift upon traveling to the left end of the coax. At this point, V_+ and V_- are 180 degrees out of phase. Our equation for the impedance tells us that at the left end the coax has an impedance of zero, a short circuit. Thus a $\lambda/4$ cable transforms an open circuit into a short; similarly it transforms a short into an open circuit.

What happens for the coil termination? Similarly the coil in the cryostat will have its impedance transformed by the coax. The voltage across the coil will be the sum of the two voltage waves, where the amplitude and phase of the reflected wave will be determined by the impedance of the coil. The maximum voltage we could possibly develop across a coil is $2V_+$ which we would obtain for a coil of infinite inductance.

This voltage that develops across the coil induces the current that generates the magnetic field. This means that for a given coil one develops the largest field by maximizing the amplitude of the traveling wave in the coax. This does not necessarily correspond to developing the largest possible voltage or current at the top of the cryostat; it is achieved by impedance matching to your source at the top of the cryostat.

The effects of the lossy coax are most evident in choice of coil inductance and in determination of signal to noise. In general the losses in the cold sample coil are small compared to the loss in the coax. This distribution of loss determines the optimum coil impedance at a particular frequency.

If all the loss were in the coax, then for a given transmitter power, varying the size of the coil inductance does not change the size of the voltage wave in the cable, as long as we continue to impedance match at the top of the cryostat. We then want to choose the coil inductance which for a given voltage wave maximizes the size of the magnetic field. Since for a given coil volume the field is proportional to the current in the coil times the square root of L, we can calculate this optimum inductance. For a fixed incident voltage wave, the field in the coil can be shown to be proportional to

$$B \propto \left[\frac{L}{\left(1 + \left(\frac{\omega L}{Z_0}\right)^2\right)} \right]^{\frac{1}{2}}. \tag{4.6}$$

At a fixed value of ω the field is maximum when $\omega L = Z_0$, the characteristic impedance of the coax. This analysis neglects the loss in the coil, which if it were comparable to the coax loss would shift the optimum inductance higher.

This coil inductance is the same one that maximizes the received signal to noise. This can be understood by examining the equivalent circuit as "seen" by the coil looking into the coax. The coil will see a capacitance C, in parallel with a resistance R (again, if we impedance match at the top), forming an equivalent tank circuit. The fact that the coax loss is much larger than the coil loss means that this resistor is much larger than the resistor which represents the coil loss itself. For a fixed sample and coil volume, and fixed frequency, the power delivered by the spins will be proportional to $\omega L/R$ where R is the equivalent resistance. Since this is equal to $1/\omega CR$, the power delivered by the spins is determined by the component Q of the equivalent impedance presented by the coax. We can use the coax impedance equations to determine $1/\omega CR$ as a function of $1/\omega C$ and find that there is a maximum at $1/\omega C = Z_0$. For a coax whose loss dominates the loss of the coil, the maximum signal power is generated by the spins when $\omega L = Z_0$. Since we've already impedance matched at the top of the cryostat this gives the best S/N.

4.4.3   Odds and Ends

Developing the impedance matching network requires that one be able to measure the impedance that the coax and coil present at the top of the cryostat. Bridges for just this purpose exist, however they are reasonably expensive. One can assemble an inexpensive bridge to measure the phase and amplitude of the reflection coefficient. This is just the *cw* spectrometer as described in the electronics section. It is calibrated by substituting a short circuit, an open circuit, and a 50 Ω termination (for leakage subtraction) on the unknown or tank circuit side of the "magic tee." This calibration tells you the phase and amplitude sensitivity of your measurement bridge, with which one can determine the impedance of any element.

The coax transformed impedance of the coil at the top of the cryostat may be inductive or capacitive, so that the tuning elements necessary to impedance match may include inductors as well as capacitors. One needs to use very high Q elements. Air core capacitors are easy to obtain, and fixed value air core inductors can be purchased inexpensively (about $2) whose Q's are very high ($\sim$ 100). These inductors can be tuned somewhat by the addition of a small length of coax. Variable air core inductors are also available, although they are more expensive and somewhat bulky.

Presently our favorite coax installation consists of a good commercial hard-line (such as UT-85-SS or UT-20-SS from Uniform Tubes) from room temperature down to the one degree pot, heat sinking at 4 K and if possible between 4 K and room temperature. (A few 3 foot UT-85-SS coaxes in a vacuum tube can boil off a liter a day of bath.) At the pot, the inner can be heat sunk with a short run of "stripline" made of double sided copper laminate on thin plastic

(.005" or so) where the ground plane is thermally grounded. A short (0.5"), thin (.020"-.030") strip can be easily made using commercial masking pens and etching solution, providing effective heat sinking for the inner; and is short enough to avoid significant reflections for frequencies below a few hundred MHz. From there down we use a superconducting coax made by fitting a monofilamentary superconducting wire inside a thin teflon tube and pulling the whole works through a niobium capillary. Stripping the CuNi cladding off of all but the ends of the inner improves the rf properties of the coax and helps reduce the problem of heat leaks down the inner. This coax can be made to be 50 Ω by careful choice of wire size or by uniformly flattening the entire coax to a thickness determined by trial and error. The Cooner coax mentioned in the recipes section has terrible rf properties: the stainless inner and outer version has 15 dB of loss in a ten foot length at 10 MHz.

Isolation of the preamp from the transmitter pulse is another problem in doing pulsed NMR. The easiest solution for a single coil spectrometer is to separate the receiver from the tank by a quarter wavelength cable with crossed diodes to ground at the input of the receiver. The diodes limit the voltage across the receiver during a pulse, and the cable transforms the diode low "on resistance" to a high resistance at the tank. Another solution which can have additional benefits in a low temperature experiment is to use a crossed coil geometry. One can generally obtain 40 dB or better isolation by using orthogonal transmitter and receiver coils. Then you have two separate tank circuits with the receiver being nearly decoupled from the transmitter pulse. It is possible to significantly reduce the amount of sample heating from the joule heat of the transmitter coil, provided one can heat sink the transmitter coil to somewhere other than at the sample. One could heat sink the transmitter coil to the mixing chamber, for example, while the sample coil is on the demagnetization stage. If reducing the amount of heating during the transmitter pulse is crucial, the choice of sample holder material becomes important. Using separate coils one can lower the sample heating by orders of magnitude; then the dielectric loss of some cell materials (such as Stycast 1266) can become the dominant heat source. In such cases, the use of even lower loss materials such as quartz or Stycast Series 35 (which can be bought as bar stock) should be considered.

4.4.4   Electronics

The rest of this chapter will describe two NMR receivers in use in our laboratory as well as some of the electronic design criteria. A valuable source of information about rf equipment are manufacturer catalogues. Besides containing the specifications for their equipment, these sometimes contain explanations of rf jargon and even more complicated design considerations.

Figure 4.33 depicts a simple cw spectrometer designed for field sweep measurements. The source must have both frequency and amplitude stability as the spectrometer is essentially a bridge configuration sensitive to small changes. The "Magic T" takes a signal from the source and splits it into two arms, one of

Source

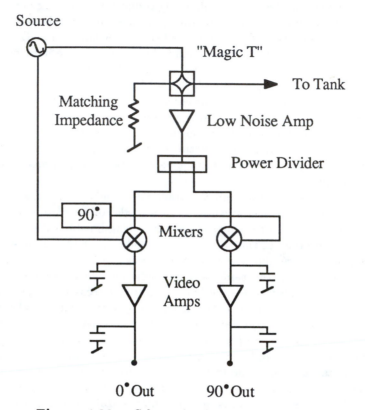

Figure 4.33. *Schematic of a cw spectrometer.*

which has a 180° phase shift. One of these arms is connected to the tank circuit (which has been tuned to 50 Ω); the other to an rf termination. Half of the power reflected from the arms is summed and appears at the output. Because of the 180° phase shift, the amount of drive signal that gets to the output is due to the *difference* in the power reflected from the tank and from the termination side.

The signal then is due to the changes in the reflected power caused by variations in the tank impedance as the field is swept through resonance. Since the rf levels are very small, the signal is amplified before being mixed down. Half of the noise power from the room temperature termination reaches the amplifier, so a front end with a noise figure much lower than 1.8 dB will result in a minimal decrease in the noise. There should be enough gain at this stage so noise from subsequent sections is negligible. The amplified signal is then split into two equal parts and then mixed down to "dc." The local oscillator drives for the mixers should differ by 90° (often just a length of coax suffices) and should have a variable phase shifter in the line before being split so that the two outputs correspond to the absorption and dispersion of the spin resonance. The filters attenuate the rf leakage in the mixers; and the low frequency amps are used because of the low maximum linear output level of mixers (typically −10 dBm, about 71 mV rms across 50 Ω).

Perhaps the largest problem in doing cw NMR in cryostats are the effects of thermal drifts on the tank tuning. If faced with a particularly unstable baseline one might consider using a derivative spectrometer where an audio frequency component is added to the ramped field. Lock-in detection of the spectrometer output at this frequency then yields a derivative spectrum, substantially reducing drift effect.

The electronics used in our present pulsed spectrometer are shown in Figure 4.34. Although this setup is typically used between 5 and 10 MHz, it can be used over a much broader frequency range. The transmitter section is basically a reasonably powered source capable of turning on and off quickly with a high on-off ratio. The long spin relaxation times encountered at low temperatures can require that this on-off isolation approaches values as high as 180 dB. The signal from a source with good frequency stability is split; one arm supplies a cw signal to the rf gates. Gates are typically double balanced mixers that have been optimized for isolation. The gates in our setup are driven by a standard pulse generator and have an on-off ratio of about 80 dB. This ratio can be improved by the insertion of a pair of series crossed diodes. The diodes offer a high resistance for small signals, but require that the gates can pass a large enough voltage to turn them "on". The diodes' degree of isolation is typically limited by their junction capacitance. The diodes for our transmitter were chosen for their low capacitance and low turn on voltage. The variable attenuator is for convenience in changing the output of the power amplifier. Further isolation is gained by a set of series crossed diodes, each pair of which is shunted by a 1 KΩ resistor to ground. 1 KΩ is a compromise between the isolation determined by the voltage division between the junction capacitance and the resistor, and the effects of the resistor when the pulse is on. The diodes we use (although reasonably expensive as diodes go), have very low capacitance combined with exceptional ability to carry large currents (they have 6 pfs./pair and drop only about 1 volt for a current of 1 amp). One needs to be careful here because the wrong diodes can use up substantial amounts of transmitter power. For powers around a few watts the 2835s might be more suitable. Measured isolation of these diodes (tabulated below) is very close to what one would calculate from their capacitances.

The receiver circuit in Figure 4.34 is a good example of what can easily be assembled from commercially available electronics. This receiver has been designed for very fast recovery times from feedthrough of the transmitter pulse (∼ 1μsec.), and for a reasonable dynamic range (> 60 dB for a 1 MHz bandwidth). A key element in the short recovery time is use of the Avantek UTL 1002 signal limiters which keep the rf gain stages from saturating. Additionally, the Merrimac Gate is used to prevent the amplifier on our scope from being blasted, as well as to prevent the filter capacitors from being charged to a large voltage. Substitution of these circuit elements with simple crossed diodes to ground would result in a

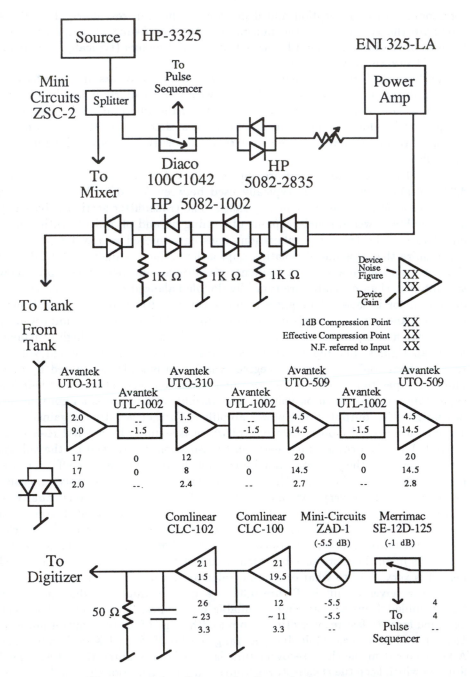

Figure 4.34. *Schematic of a pulsed spectrometer.*

Table 4.2 *Measured diode isolation.*

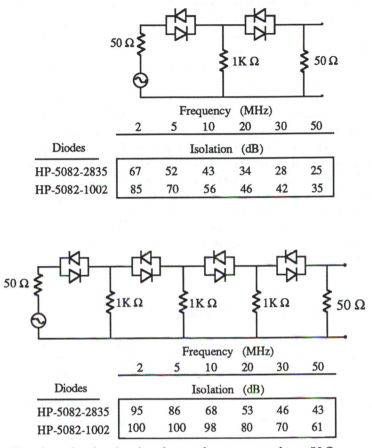

Diodes	Frequency (MHz)					
	2	5	10	20	30	50
	Isolation (dB)					
HP-5082-2835	67	52	43	34	28	25
HP-5082-1002	85	70	56	46	42	35

Diodes	Frequency (MHz)					
	2	5	10	20	30	50
	Isolation (dB)					
HP-5082-2835	95	86	68	53	46	43
HP-5082-1002	100	100	98	80	70	61

Isolations listed are for the circuits above when compared to a 50 Ω source driving a 50 Ω load.

reasonable savings in cost ($\sim$ 40%) with a modest reduction in recovery times.

Choice of the amplifiers to be cascaded involves consideration of gain, noise figure, and the 1 dB compression point of each stage, as well as that for the cascaded set before any stage. There should be enough gain in each stage so that the noise contributed by subsequent stages is negligible. Amplifier manufacturers specify an amplifier's noise figure, which in a 50 Ω system is defined as:

$$\text{Noise Figure} = 10 \cdot log_{10}\left\{\frac{(V_{\text{DEV}})^2 + (V_{50})^2}{(V_{50})^2}\right\} \tag{4.7}$$

where V_{DEV} is the device's equivalent input noise voltage and V_{50} is the noise voltage of a 50 Ω resistor ($\sim .9nV/\sqrt{Hz}$). The effect of a device with loss rather than gain can be included by adding its loss in dB to the noise figure of the next stage. Figure 4.34 shows both the noise figure of each stage as well as the noise figure referred to the input for all stages up to and including a given stage.

A benchmark for the range of linear operation is the "output 1 dB compression point" which is the output level in dBm at which the amplifiers gain is reduced by 1 db. ($\sim$ 10% voltage reduction). One can compare each device's own 1 dB compression point with the effective compression point of the previous stage increased by the gain of the device. For unequal values, the lower of these will then be the effective 1 dB compression point. As is clear from the figure the earliest gain stages have higher compression points than necessary in order to enhance their fast recovery from the tipping pulse feedthrough. The later stages are reasonably well matched for gain compression. Filtering is a very simple 3 pole R-C filter with each pole separated by an active device. A more complicated filter such as a 3 or 4 pole Bessel is often useful. A terrific reference for filter design is Zverev's *Handbook of Filter Synthesis.* Our final effective noise figure combined with our total gain results in a $2.9 \mu V / \sqrt{Hz}$ noise at the output. Since our final output compression point is 23 dBm this gives us a 61 dB dynamic range for a 1 MHz bandwidth.

4.4.5   Sources of Useful NMR Hardware

This list includes most of our favorite rf manufacturers. There are generally many manufacturers for each component you'll want to buy; these are the companies from whom we have purchased equipment that we have found to be reliable.

- Anzac, 80 Cambridge St., Burlington, MA 01803 (617) 273-3333. A good source for amplifiers which can be used at low frequencies (below 5 MHz) as well as "Magic T's" and VSWR bridges.
- Avantek, 3175 Bowers Ave., Santa Clara, CA 95051 (408) 496-6710. Good selection of connecterized amplifiers for > 5 MHz work. Their high output level amps have very good pulse recovery.
- Barker & Williamson, Bristol, PA High Q air core inductors.
- Comlinear Corporation, 4800 Wheator Drive, Ft. Collins, CO 80525 (303) 226-0500. Terrific video amps. Frequency responses from D.C. up to 100 or 200 MHz, and their amps have exceptional pulse response and dynamic range.
- Daico Industries Inc., 2351 East Del Amo Blvd., Compton, CA 90220 (213) 631-1143. R.F. switches (gates) with very high isolation and good switching speed. Prices are high for single pieces but become competitive in quantities.
- E.N.I., 100 Highpower Road, Rochester, NY 14623 (716) 427-8300. Our preferred source for reliable broad band power amplifiers. We've had bad luck with other manufacturers' amps oscillating under unusual load impedances.
- Hewlett-Packard, Microwave Semiconductor Division. Their diodes are great for isolation purposes; they have low capacitances, low turn on voltages, and

good current carrying capability. H.P. makes lots of general rf and microwave equipment of very good quality, but they are not inexpensive.

- J. F. W. Industries Inc., P. O. Box 336, Beech Grove, IN 46107 (317) 887-1340. High quality, reasonably priced feedthrough attenuators and terminators.
- Merrimac, P. O. Box 986, 41 Farfield Place, West Caldwell, NJ 07007 (201) 575-1300. Good rf switches (gates) and broadband 90° power splitters.
- Mini-Circuits, P. O. Box 166, Brooklyn, NY 11235 (718) 934-4500. An inexpensive supplier of quality mixers, splitters, and rf transformers.
- Pasternak Enterprises, P. O. Box 16759, Irvine, CA 92713 (714) 261-1920. An inexpensive supplier of all sorts of rf connectors and adaptors.
- Trontech Inc., 63 Shark River Road, Neptune, NJ 07753 (201) 922-8585. Their amplifiers make great cw front ends.
- Weinschel Engineering, One Weinschel Lane, Gaithersberg, MD 20877 (301) 948-3434. High quality variable and switched attenuators, also high power attenuators.
- Hermetric feedthroughs can be purchased from Pomona and Omni-Spectra.

4.5   Ultrasound

by David Thompson

Ultrasonics are sound waves whose frequencies lie above the range of human hearing, that is above about 20 kHz. Their use in physics can be divided into two parts: low amplitude waves that do not change the medium in which they propagate; and high energy waves in which the sound induces changes in the medium. In the first category, which is by far the more common one, you can include measurements of sound velocity and attenuation coefficients to determine elastic moduli and relaxation times, while the second might include measuring saturation energies of transitions.

In low temperature work ultrasound includes zeroth and first sound which are both density waves.

4.5.1   Why Use Ultrasound?

For a given characteristic frequency you have the choice of two wave probes: sound and light. Electromagnetic radiation requires an electric or magnetic multipole moment in order to couple to a system, while ultrasound will couple to mass distributions. For example, you can, with some difficulty, study the Cooper pairs in a superconductor either with microwaves or ultrasound, while the Cooper pairs in ^{3}He have been most usefully studied by ultrasound (often zeroth sound). Sound and light also have different momenta associated with the same energy phonon or photon which can make ultrasound preferable to light in some cases.

4.5.2   Production of Ultrasound

Ultrasound is usually produced by converting the oscillations of an electromagnetic field into oscillations of a solid. Most commonly piezoelectric crystals, such as quartz, are used, although other methods such as electrostriction, magnetostriction, and mechanical generation are possible (Blitz, 1967).

The fundamental frequency of a crystal is inversely proportional to its length and in the pursuit of higher frequencies you may require crystals whose size is such as to make them too fragile. However you can drive the crystal at higher harmonics thus obtaining higher frequencies with large crystals. Unfortunately, the efficiency of exciting the nth harmonic is inversely proportional to n^2. The piezoelectric effect occurs only when opposite charges appear on the faces of the crystal and so only the odd harmonics of the crystal can be excited. (See Figure 4.35.)

For work with electrically conducting samples you can use electromagnetic, or Lorentz force, transduction. Here a radio frequency coil is placed close to the plane face of the sample (Dobbs, 1973; Frost, 1979):

This induces electronic eddy currents and when a static, solenoidal magnetic field of a few kilogauss is applied to the sample from an external magnet, a

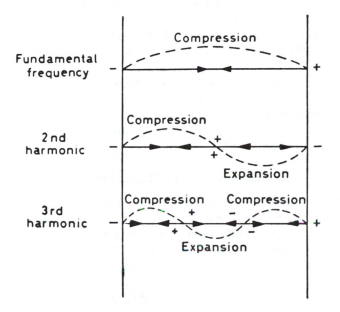

Figure 4.35.   *Distribution of pressure and charge for a quartz crystal oscillating at the fundamental frequency and at the second and third harmonics. [Bergman, L., Ultrasonics, (Bell, London), 1938.]*

shear wave is generated from the metal surface within the electromagnetic skin depth of the eddy currents.

This scheme has the great advantage of needing no mechanical bonds between the transducer and the sample.

4.5.3   Types of Experiments

For a solid sample, a single transducer is mounted on one face and is used as the generator and detector of the ultrasound. An ultrasonic pulse is sent into the sample. It is then detected each time that it is reflected back from the opposite face of the sample. This requires a large acoustic mismatch between the crystal and the sample so that a number of echoes are observed. This may, however, result in distortion of the echoes (Kittinger and Rehwald, 1977). From the transit time you determine the sound velocity and from the amplitude of the signal you determine the attenuation. In liquid ^{3}He experiments, separate transducers are used to generate and detect the ultrasound and they serve as the faces of the sample.

For a comprehensive discussion of ultrasonic velocity and attenuation techniques see, for example, Papadakis (1976).

You can also do acoustic nuclear magnetic resonance, frequently abbreviated as NAR (Bolef, 1966; Sundfors *et al.*, 1984) and acoustic paramagnetic resonance which are analogues of NMR and EPR. The difference is that the r.f. field is replaced by an ultrasound field. Spin flips occur when the ultrasound frequency matches the resonance frequency.

One place in which NAR beats NMR hands down is in the study of single metallic crystals. The skin depth kills NMR but the ultrasonic field used in NAR has no problems penetrating the entire crystal. A recent development in this field is the use of a SQUID to detect the induced changes in magnetization. (Pickens, 1984; Pickens *et al.*, 1984 a,b).

4.5.4   Choice and Preparation of a Transducer

The most common choice of transducer is a quartz crystal. By cutting the crystal at different angles to its axes, referred to as X-cuts, Y-cuts, AC-cuts, etc., you, or rather the manufacturer, can make generators of either longitudinal or transverse waves with different coefficients of thermal expansion. The temperature dependence of quartz's piezoelectric coefficients is small. One drawback with quartz is that to obtain a fundamental frequency of 20 MHz you need a crystal about 0.15 mm thick. Such thin crystals require careful handling both in general and specifically in cooling down.

Quartz has a very large mechanical Q factor, of order 1000 for coupling to air, which gives very high signal to noise ratios. However it does mean that a great deal of care must be taken to ensure that the frequency of the driving pulse exactly matches the mechanical resonance frequency of the transducer.

For higher frequency work with solids, thin films of cadmium sulphide (de Klerk and Kelly, 1965) and zinc oxide (Foster, 1984), which are piezoelectric, have been vapor deposited directly onto the sample and used as transducers.

If transducers of unusual shape are required, such as to aid in focusing ultrasound, then ceramic materials such as barium titanate can be used and the transducers built up out of a mosaic (Anana'eva, 1959).

For work below 1 K lithium niobate transducers have been used (Beamish *et al.*, 1983). They are available commercially and have electro-mechanical coupling factors about five times as large as quartz. This gives you larger ultrasound amplitudes for a given input voltage. It also results in a broader resonance curve and so the tuning of the drive circuit can be less accurate. The price you pay for this is a lower signal to noise ratio than you would obtain with quartz transducers (Elbaum, 1985). At Cornell our usual supplier of transducers is Valpey Fisher.

4.5.5   Transducer-Sample Bonding

In bonding quartz transducers to solids, you want a low mechanical loss bonding material that can withstand the strain produced in cooling due to the different thermal contraction coefficients of the transducer and the sample. The use of a low freezing point bond allows you to avoid much of the thermal contraction from room temperature. Byer (1967) claims that 1-pentene, which freezes at 138 K, is preferable to propane and dry natural gas. Yap (1973) describes an elaborate routine for bonding alkali halides to quartz with 1-pentene. In essence, this scheme involves making the bond at 160 K in a nitrogen dewar and then slowly cooling first to 77 K and then to 4 K. Thin bonds were found to have less mechanical loss and to have a better chance of surviving the cool down. This technique, owing to its complexity and requirements of specially designed equipment, is only used as a last resort when nothing else will hold your transducer on.

In bonding lithium niobate transducers to vycor glass Beamish *et al.* (1983) used Dow Corning 1000 poise silicone oil. The silicone is diluted with trichloroethylene so that when it is put on the surface it spreads out to cover it. The bond is assembled and quenched in liquid nitrogen. It is then rapidly transferred to the cryostat and cooled from there.

Another common bonding agent is nonaqueous stop-cock grease. If you are having trouble with your transducers popping off you should try smaller diameter ones.

4.5.6 Sample Preparation

The sample needs to have flat parallel faces to avoid interference effects across the face of the transducer. For a liquid sample such as ^{3}He, the sample cell must have flat parallel faces. The criterion of flatness is that the surface roughness be much less than the wavelength of the ultrasound. This is usually achieved by polishing the surfaces to optical flatness, a more stringent condition. Viertl (1973) describes fully the polishing of crystals to produce flat parallel faces. (See Section 4.5.9.) The requirement of the faces being parallel is that the separation of opposite faces changes by less than a wavelength across the faces. Again this can be done using optical methods.

Figure 4.36 shows the sample cell used by Giannetta and co-workers (Giannetta, 1980; Giannetta *et al.*, 1980; Polturak *et al.*, 1981) for zero sound studies in ^{3}He. The version currently in use at Cornell is in magnetic fields up to 9.4 T and consequently the brass spacer has been replaced by a quartz one. The distance between the two transducers is 0.769 cm and 0.650 cm for two different cells. This leads to times of flight for the pulses of about 20 to 30 microseconds. Both the brass spacer and the epoxy cell have holes drilled in them to create contact with the ^{3}He bath. The quartz crystals are gold plated and electrical contact is made through the brass spacer via copper cups. Leads are soldered to the copper cups and twisted into pairs with the ground leads before being taken out of the top of the cell through epoxy vacuum feedthroughs. The transmitter and receiver lines are brought to the top of the cryostat through small coaxial cables, using superconducting copper clad inner conductors.

For the high field cell (de Vegvar, 1985) the connections are shown in Figure 4.37. The electrode on the bottom face of the transducer is wrapped around the sides and the connections both soldered on to the same face. To reduce spurious transmission of sound, the inner walls of the quartz spacer have staggered

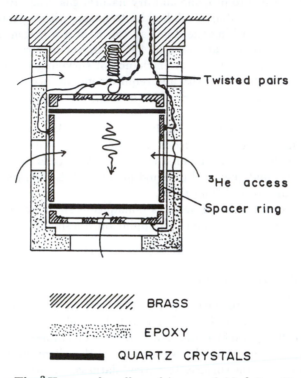

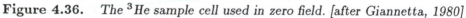

Figure 4.36.   *The ^{3}He sample cell used in zero field. [after Giannetta, 1980]*

grooves cut perpendicular to the axis of the cell. The quartz spacer was bought from Valpey along with the transducers.

4.5.7   Circuit Design

We will give here an example of a circuit design for ultrasound experiments based on a design by Giannetta (1980) and modified by de Vegvar (1985) using separate transmitting and receiving transducers to study the collective mode oscillations of ^{3}He. The transducers are quartz crystals with a 10 MHz fundamental.

The similarity between ultrasonic experiments and NMR should be noted here. In many set-ups the only significant difference is that at the bottom of the cryostat the ultrasound experiment has transducers and the NMR experiment has coils.

Figure 4.38 shows the equivalent electrical circuit of the quartz transducers and Figure 4.39 the block diagram of the electrical circuit. The master oscillator, which provides both the driving signal for the pulse amplifiers and a continuous reference signal, is stabilized to $\Delta f/f = 10^{-7}$ by a phase-locked loop synchronizer. The saturation amplifier is used if you are worried about instabilities in

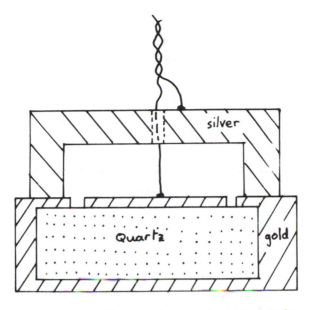

Figure 4.37.   *The electrical connections to the high field ³He transducers.*

the amplitude of the pulses. The pulses are amplified to saturation and then attenuated to the desired size. Since driving the amplifier to saturation is likely to make it very unhappy, you should check the stability of the pulses before you install it. In the current set-up at Cornell the output of the pulse generator is sufficiently stable that we do not use a saturation amplifier. The crossed diodes are a block to DC and low amplitude leakage from the transmitter to the cryostat. The variable capacitive-inductive tank circuit is tuned so that the impedance of the source matches that of the transmitter transducer. (You must include the impedance of the connections between the top of the cryostat and the transducer in these calculations.) Of course you do not have to calculate the impedance of the equivalent circuit a magic tee will do the job for you.

The whole of the receiver arm of the circuit has 50 Ω resistance throughout. This is to minimize signal losses due to reflections at impedance mismatches. The output tank circuit is important in matching the transducer to 50 Ω to maximize the coupling between the transducer and the receiver. The signal is split into two equal parts in order to use vectorial phase detection. Each signal is fed into the R port of a double balance mixer. The reference signals are obtained by splitting the original reference signal in two and phase shifting one part by 90° with a trombone phase shifter. These signals are fed to the L ports of the mixers. After mixing, the signals are put through low pass filters to remove any unwanted higher frequencies (such as harmonics and sum frequencies) generated by the electronics before being fed into the recorder. In order for the mixers to

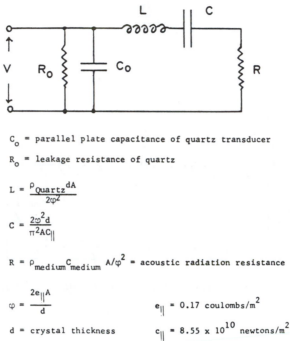

C_o = parallel plate capacitance of quartz transducer

R_o = leakage resistance of quartz

$$L = \frac{\rho_{Quartz} dA}{2\varphi^2}$$

$$C = \frac{2\omega^2 d}{\pi^2 A C_{||}}$$

$R = \rho_{medium} C_{medium}\ A/\varphi^2$ = acoustic radiation resistance

$$\varphi = \frac{2e_{||}A}{d}$$ $e_{||}$ = 0.17 coulombs/m^2

d = crystal thickness $c_{||}$ = 8.55 x 10^{10} newtons/m^2

A = crystal area ρ_Q = 2650 kg/m^3

Figure 4.38.   *Equivalent electrical circuit for a quartz transducer.*
[Kinsler, L. E. and A. R. Frey, Fundamentals of Acoustics, (John Wiley & Sons, NY), 1962.]

be run by the reference signals, they should be at least 10 dB greater than the r.f. signal.

The oscilloscope connected to the recorder is used in tuning the circuit to resonance. This has to be done at the operating conditions since the loading on the transducer, and hence its resonant frequency, varies with pressure and temperature. The tuning technique is to feed the two signals to the x and y inputs of the scope and to zero beat them by adjusting the driving frequency.

The coherent interference due to the receiving transducer picking up the initial transmitter r.f. pulse feedthrough was determined at 6 mK where the attenuation due to the ^{3}He was strong enough to block any received sound pulse. It was assumed that this interference was temperature independent and stable on a time scale of the order of 24 hours and it was then subtracted from the signals. The important point here is that the interference signal is completely coherent with the sound signal.

The incoming signals are integrated, with respect to time, and then a baseline

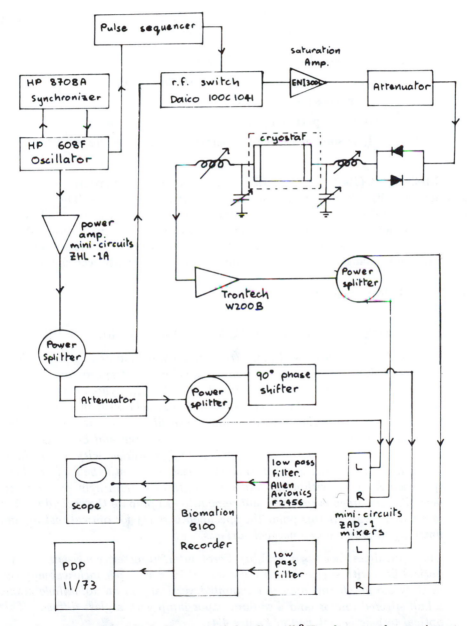

Figure 4.39. *Block diagram of the Cornell ^{3}He ultrasound experiment.*

and the coherent interference subtracted off. You get the attenuation coefficient and the velocity from the amplitude and phase of the signal:

$$\alpha(T) - \alpha(T_0) = -\frac{1}{L} \ln \frac{A(T)}{A(T_0)} \qquad (4.8)$$

$$\frac{c(T) - c(T_0)}{c(T_0)} = -\frac{\phi(T) - \phi(T_0)}{2\pi} \frac{\lambda}{L} \qquad (4.9)$$

where : α is the absorption coefficient;

A is the integrated amplitude of the signal;

c is the sound velocity;

L is the path length;

T_0 is some reference temperature, T_c for the ^{3}He work.

Gibson *et al.* (1981) have described a heterodyne pulsed spectrometer suitable for use with ultrasonics in the range 200 kHz to 157 MHz. The group at Northwestern (Mast, 1982; Mast *et al.*, 1981) have also done cw ultrasound experiments in which the electrical impedance of the transducer is monitored. When the attenuation in the sample changes, the mismatch between the sample and the transducer changes and hence the impedance of the transducer changes.

4.5.8   Polishing of Crystals

The following is taken from J. R. M. Viertl's thesis (1973):

A cleaved or a string-sawed piece (of crystal) was mounted on a polishing jig (Figure 4.40) consisting of a stainless steel cylinder approximately 4 in. (10 cm) long and 4 in. outer diameter. The wall of the cylinder was approximately 1 in. (2.5 cm) thick. The specimen was held with mounting pitch against a movable spring-loaded piston which fit snugly into the cylinder. The mounting pitch consists of resin, hard black pitch and beeswax in approximately equal proportions. The polishing jig is worked with its specimen in 3 μm garnet powder and kerosine on a glass plate until the surface of the specimen is flat to within one-half wavelength of sodium light over its surface. The specimen is removed and remounted to prepare the second surface in a similar way. At this point the specimen is ready for optical testing and finishing to produce two parallel surfaces.

The optical test set-up is a Fabry-Perot interferometer, consisting of two optical flats with surfaces with a flatness of 10^{-6} inch per inch and approximately 2.5 in. (6 cm) diameter mounted vertically in an adjustable stage, a half-silvered mirror, and a sodium vapor lamp with a glass diffuser. This optical test jig is sketched in Figure 4.41.

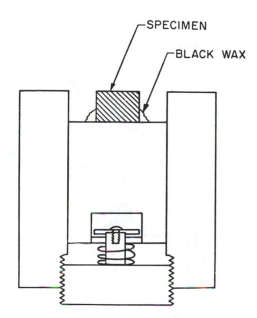

Figure 4.40. *Crystal polishing jig used to produce flat parallel faces on crystals. [After Viertl, 1973].*

The specimen is cleaned briefly in kerosine and dried with tissue paper. The specimen is placed on the lower optical flat of the Fabry-Perot interferometer and pressed lightly with a cotton tip until Newton interference fringes are seen between the lower optical flat and the specimen, showing that the specimen is in good contact with the optical flat. The specimen is moved adjacent to the upper optical flat. For parallel optical flats the Fabry-Perot fringe system consists of concentric circles with the same order visible and stationary as the observer moves across the viewing field. The Newton fringe system produced between the top of the specimen and the upper optical flat form a contour map of the upper surface with an interval of one-half wavelength of sodium light. This allows the determination of the degree of parallelism and flatness. Once the locations of the high spots have been found, they are polished away by hand with 3 μm powder in kerosine on a cotton tip. The testing procedure is continued until the surface contours show that the specimen is flat and parallel over more than 50% of its central area and to within one-tenth the acoustic wavelength to be propagated. The acoustic wave is propagated within this central area.

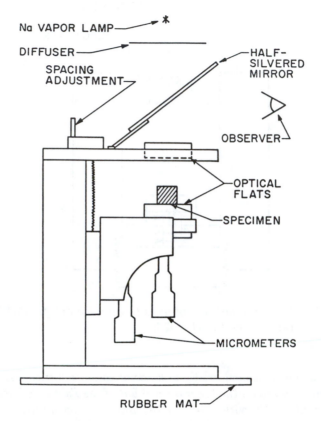

Figure 4.41. *Optical test jig used for finishing crystals with flat parallel faces. [after Viertl, 1973]*

4.5.9   Manufacturers

- Allen Avionics, Inc., 224 East Second Street, Mineola, NY 11501.
- Biomation, Gould Inc., Instrument Division, Biomation Operation, 4600 Old Ironsides Drive, Santa Clara, CA 95050.
- Daico Industries, 2351 East Del Amo Blvd., Compton, CA 90220.
- Dow Corning Corporation, Midland, MI 48640.
- ENI, 3000 Winton Road South, Rochester, NY 14623.
- Hewlett-Packard, 1820 Embarcadero Road, Palo Alto, CA 94303.
- Mini-circuits, P.O.Box 166, Brooklyn, NY 11235.
- Trontech, Inc., 63 Shark River Rd., Neptune, NJ 07753.
- Valpey Fisher Corp., 75 South St., Hopkinton, MA 01748.

4.6   High Frequency Methods

by Bryan W. Statt

Experiments requiring the use of high frequency signals at low temperatures are in need of techniques not found in run of the mill cryostats. I will describe some useful low temperature techniques for high frequency work. For convenience the relevant part of the electromagnetic spectrum will be broken up into three bands:

- Radio Frequency—1 MHz to 1 GHz
- Microwave—1 GHz to 100 GHz
- Optical—100 GHz to . . .

This division is completely arbitrary and should not be taken literally. For example, RF circuit techniques using discrete components are used in 40 GHz amplifiers. Waveguides, typically thought of as being used only for microwaves, can be used in any of the above three bands. Optical components such as Fresnel zone plates and parabolic reflectors are used as lenses in satellite TV receivers at 4 GHz. Use your imagination in choosing the technique best suited for your application.

A low temperature cryostat is naturally suited as an environment for low noise devices. Liquid helium cooled amplifiers and detectors can provide state of the art noise figures. Take advantage of this fact when designing your detection schemes. Another possibility is mixing signals in the cryostat. Converting a high frequency signal to a low IF signal provides the advantage of piping a lower frequency signal out of the cryostat.

Although I have cut off the spectrum at the optical frequencies, many experiments use higher parts of the spectrum. X-ray diffraction can be done at low temperatures; for example, Heald and Simmons (1977) have used X-ray diffraction to study solid ^{3}He down to 60 mK. Beryllium and aluminized mylar windows were used to get the X-rays in and out of the cryostat. Nuclear orientation studies involve even higher frequencies (gamma rays). In this case no particular effort goes into the "window" design as the gamma rays pass through most cryogenic construction materials with the greatest of ease.

Techniques associated with each band will be discussed first from the point of view of getting a signal into and out of the cryostat. I will then point out some miscellaneous techniques useful at low temperatures. I will assume you have a working knowledge of electromagnetism.

4.6.1   RF Techniques

Coaxial cables are by far the most popular transmission lines used to transport RF signals into and out of the cryostat. In addition to the usual considerations in choosing a particular cable, such as characteristic impedance and attenuation, other properties are now of interest. Thermal conductivity, thermal contraction and vacuum integrity are important factors. Maintaining a low VSWR (voltage standing wave ratio) requires care to be taken in the use of connectors and vacuum seals. This is especially important at high frequencies where the 1–2 meters of cable in the cryostat are many wavelengths long. I will now discuss each of the above concerns in greater detail.

Attenuation is usually caused by resistive losses in the inner and outer conductors. At microwave frequencies dielectric losses also become important. You might expect that the attenuation can be reduced by lowering the temperature. RF currents flow within a few skin depths of the surface of the conductors. The skin depth, δ, is proportional to $1/\omega\sigma$. As the temperature is lowered the conductivity increases thereby reducing the attenuation. This continues until the electron mean free path approaches the skin depth. At this point the current flows in the anomalous skin depth which is now temperature independent. Further cooling does not serve to reduce the attenuation. As an example, consider a Copper coax with a Residual Resistance Ratio of 100 between 300 K and 4 K. For all frequencies greater than 3 MHz the anomalous skin depth region is reached before $T = 4$ K. See *Principles of the Theory of Solids* by Ziman for a detailed discussion of the anomalous skin depth.

Superconductors can also be used to reduce resistive losses if the corresponding photon energy is less than the gap energy. As shown in Figure 4.42 the surface resistance ratio for superconducting to normal metal can be significantly reduced. Another advantage to using superconductors is their very low thermal conductivity compared to normal metals. This makes their use at liquid helium temperatures attractive. Two distinct disadvantages of superconductors are the presence of large magnetic fields and the difficulty of making good electrical contact (e.g., cannot solder to NbTi).

Thermal conduction between the top of the cryostat at 300 K and the innards at liquid helium temperatures through the coax must be minimized. Fortunately, the high frequency currents flow only within a few skin depths of the surface. This allows you to gold or silver plate the inner conductor which can now be made of a relatively poor thermal conductor. As the current density is much higher on the inner rather than outer conductor, it may only be necessary to plate the inner conductor. For example, you can use a stainless steel outer conductor with a silver plated Be-Cu inner conductor. The most commonly used inner conductor is copperweld—iron with copper cladding. If magnetic materials pose a problem, copperweld should be avoided. Stainless steel (300 series) and Cu-Ni are somewhat less magnetic but may also cause problems. The physical size of the coax is a compromise between minimizing thermal conduction with a small cross sectional area and minimizing the attenuation with a large inner conductor diameter.

The choice of dielectric for the coax is largely determined by its loss tangent. Teflon is the most common dielectric used in commercially available coax. Thermal contraction tends to pull the teflon away from the end of the coax causing an impedance mismatch. This problem can be remedied by slightly crimping the

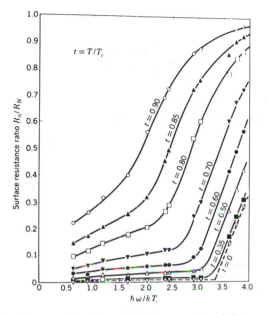

Figure 4.42. *Isotherms of surface resistance ratio of aluminum versus photon frequency. [Biondi, M. A. and M. P. Garfunkel, Phys. Rev. Lett. 2, 143 (1959).]*

coax at several places along its length. A convenient tool for doing this is easily constructed from a tube cutter. Machine a rounded Al wheel to replace the sharp cutting wheel. Use the "tube crimper" as you would the original tool to slightly crimp the coax. The slight impedance mismatch caused by these crimps is probably less than that of a section of coax with the teflon pulled completely away.

Heat sinking the coax at various stages in the cryostat is a must. The outer conductor presents no problem but not so with the inner conductor. A thick piece of teflon between the inner and outer conductors provides an excellent thermal insulator. Two possible solutions are the following: If a $\lambda/4$ line is not too large you can attach one to the coax, in parallel, shorted at the end to be heat sunk. This provides good thermal conductivity but leaves an open circuit at the top of the coax (see Figure 4.43a). A $\lambda/4$ line is, however, somewhat narrow band. Another method is to remove the teflon over a short distance and replace it with a better thermal conductor such as Stycast 2850FT as illustrated in Figure 4.43b. Minimizing the VSWR requires a step change in the outer conductor radius to match the change in dielectric constant.

This method also serves as a vacuum seal which is necessary if the outer conductor is to be used as a vacuum line. If the coax is to be placed inside a

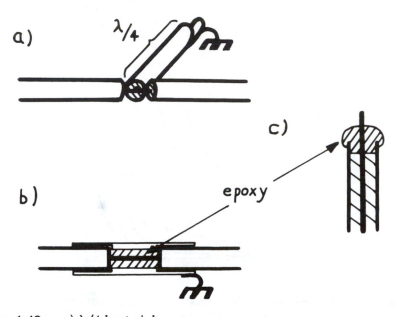

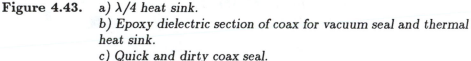

Figure 4.43.   *a) λ/4 heat sink.*
b) Epoxy dielectric section of coax for vacuum seal and thermal heat sink.
c) Quick and dirty coax seal.

vacuum line then a hermetically sealed bulkhead feedthrough can be used. Some of these commercial feedthroughs can also be used at low temperatures with the rubber O-ring replaced by an In O-ring. If impedance mismatch is of no great concern Figure 4.43c illustrates a quick and dirty way to make the end of a coax vacuum tight for either room or low temperature use.

4.6.2   Typical Types of Coaxial Cable

Stainless Steel Multi Wire Inner and Braided Outer Conductor Good for low frequencies, also quite lossy. Small, easily bent and heat sunk. Available from Cooner Sales Co.

Semi-rigid and Rigid Coax Several sizes and char. impedance available. Many metals (SS, Cu, Be-Cu, etc.), including superconductors, for inner and outer conductors as well as many choices of dielectric including spline teflon. Available from Uniform Tubes, Inc.

Homemade Coax (L. Friedman) Outer conductor 30 mil Cu-Ni tubing. Inner conductor NbTi or wire of your choice. Dielectric is teflon tubing. To assemble: Stretch teflon tubing until diameter is small enough to fit into Cu-Ni tubing. Insert wire inner conductor into teflon, greased so that wire sticks to teflon, and epoxy a blob at each end to hold the teflon in place. Pull wire and teflon into outer conductor. Allow teflon to relax after insertion to a tight fit.

Connectors:

- SMA and others for semi-rigid and rigid coax are available from Omni Spectra.
- Micro dot connectors for small semi-rigid and flexible coax are available from Malco.

Installation of coaxial cable is in general much easier during the initial construction of the cryostat rather than at a later date. If a commercial unit (i.e., dilution refrigerator) is being purchased it is a wise move to have the manufacturer install a lifetime supply of coax, wires, etc. Oxford Instruments will install a vacuum dielectric coax which can be disassembled, allowing you to replace the inner conductor if desired.

4.6.3   RF Odds and Ends

A useful point to consider is the possibility of thermally separating the RF part of the apparatus from the sample. As an example consider an NMR experiment with a gaseous sample. Figure 4.44 illustrates how the gas, contained in a pyrex tube heat sunk to the mixing chamber of a dilution fridge, is thermally separated from the NMR coil (loop gap resonator) which is heat sunk to the still. Note that the capacitive coupling can also serve to provide thermal isolation. Not only does this technique provide for thermal isolation between stages but RF power dissipated in the resonator is not dumped into the sample. In this example the tuning and coupling is adjusted via a clever arrangement of strings, pulleys and springs. Moving the coupling coax up and down is also a possibility. In this case the coax can be vacuum sealed at the top of the cryostat with bellows or perhaps a double set of O-rings.

Certain components can be cooled to low temperatures without self destructing. For example some stripline directional couplers can operate at low temperatures. Experimentation may be necessary to find suitable components. Cooling of mixers and detectors should be done with caution. Construction techniques adequate at room temperature may not survive cool down such as epoxies used to hold semiconductor elements in place.

Modifying a small (6 inches or so) length of commercial semi-rigid coax may be desirable, for example, to change the inner conductor. You can do this by immersing the coax in liquid nitrogen and removing the inner conductor and dielectric after thermal contraction takes place. When warmed back up, the inner conductor can be removed and replaced. Reverse this procedure to reassemble.

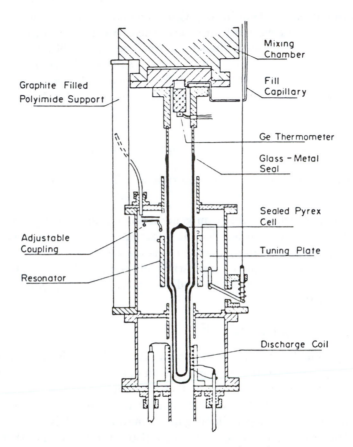

Figure 4.44. *Apparatus used by Hardy et al. (1984) to study hyperfine resonance of H on liquid helium surfaces. H_2 and He are sealed
in a glass cell; H is formed by an rf discharge. [Hardy, W. N.,
M. Morrow, R. Jochemsen and A. J. Berlinsky, Physica **109** +
110B, 1964 (1982).]*

4.6.4   Microwave Techniques

Waveguide and coaxial cable make good transmission lines for microwave
signals. The previous discussion of coaxial cables applies equally well to microwaves. Metallic waveguide of rectangular or circular cross section is most
common. Commercially available waveguide typically covers the bands from
8.2 GHz to 325 GHz and is made of copper, brass, coin silver (higher bands) and
stainless steel. Two companies that sell waveguide and flanges are Alpha Ind.
and Systron-Donner. The thermal isolation provided by the stainless steel waveguide sometimes produces unacceptable attenuation. Plating the inside of the
waveguide is a solution, but keep in mind that a micron of copper is roughly

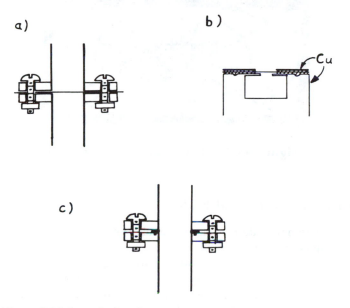

Figure 4.45. *a) Mylar window between two waveguide flanges.*
b) Low temperature epoxied mylar window covering resonator coupling hole.
c) In or Pb 0-ring seal between two waveguide flanges. O-ring groove optional.

equivalent to 10 mil of stainless steel below 1 K.

Heat sinking contacts are easily provided for metallic waveguides. Styrofoam blocks can be used near the top of the waveguide to absorb room temperature radiation. A sharp bend or two, heat sunk at 4 K, will also do. Vacuum integrity can be maintained largely by treating the waveguide as you would a pumping line, with the necessary respect paid to the inner surfaces. Vacuum seals are most easily made at room temperature with a piece of mylar (about 1 mil thick), lightly greased with vacuum grease, sandwiched between two waveguide flanges (Figure 4.45a). If nylon bolts are used this can also provide a DC block. Low temperature seals can also be made by epoxying mylar to metal. The surface of the mylar to be epoxied must be roughened and pressed in place on both sides during curing so that the stresses from thermal contraction will not break the seal (Figure 4.45b).

Dielectric waveguide can also be used. This has the advantage of low thermal conductivity but is harder to make good thermal contact to for heatsinking while maintaining satisfactory microwave characteristics.

Figure 4.46.   *Cryogenic waveguide spider.*

4.6.5   Microwave Odds and Ends

Waveguide components can also be cooled to low temperatures such as directional couplers and low pass filters. If a vacuum tight connection is desired, an In or Pb O-ring can be used between two flanges as shown in Figure 4.45c. Beware: the squashed In or Pb must come out to the edge of the waveguide else a TM_{010} cavity will be created causing a notch filter at its resonant frequency. This procedure is recommended over the soldering of components together since you can then reuse them at a later date and there is no risk of damaging the component during the soldering operation.

You can also fabricate your own components. For example, you may need an adapter with specific dimensions to fit into your cryostat. Electroforming Cu around an Al or stainless steel mandrel allows you to make quite a variety of components. If the mandrel is tapered, stainless steel can be used. When electroforming is complete, you can pull the mandrel out. Al is recommended when the mandrel is not tapered as it can be etched away with NaOH without damaging the Cu substantially.

High frequency microwave plumbing, especially sub-millimeter work, is usually plagued with VSWR problems. This is because of the stringent tolerance requirements when using waveguide of small dimensions (e.g., 75–110 GHz band waveguide is 50×100 mil). As most microwave spectroscopists know, "waveguide spiders" like to set up shop at misaligned flanges and other imperfections (Figure 4.46) which cause large reflections. Unfortunately these beasts are not afraid of the cold! A signal from room temperature to the cryostat can travel many wavelengths before reaching its destination. Reflections at the top and bottom can cause severe standing wave patterns. If sufficient power is available this can be remedied by using a lossy waveguide, with say 20 dB of attenuation along its length to damp out reflections. This can usually be done in the transmitter line but is not advisable in the receiver arm as it decreases signal to noise. This is another reason to do your detection in the cryostat.

Tuning of resonant cavities, detectors and the like is most conveniently done at room temperature. Most of the relevant properties however change with tem-

perature which then changes the tuning. If this imprecise tuning is acceptable it
will save you a lot of work involved in the installation of strings, pulleys, moving
rods and the like. Methods of tuning cavity frequencies, couplings, etc. are the
same as those used at room temperature with the exception that the mechanical
apparatus used must not conduct or generate too much heat in the cryostat. One
possible compromise, such as in the tuning of a narrow band detector, is to set up
a rig in a test cryostat in which the detector is tuned at the operating temperature. The tuning parts can now be fixed in place, the tuning mechanism removed
and the now permanently tuned detector mounted in the final apparatus. This
method is particularly useful in a crowded dilution fridge.

State of the art noise figures for GaAs microwave amplifiers and semiconductor mixers and detectors are obtained with cryogenically cooled devices. These
types of amplifiers are discussed in Section 4.3. As I mentioned earlier, it is
natural to incorporate these devices into your cryostat rather than to transport
signals from the cryostat to room temperature devices. Cooled devices are also
used by physicists not of the low temperature persuasion. Radio astronomers often cool their detectors and amplifiers with closed cycle refrigerators at 20 K. IR
spectroscopists sometimes cool their detectors in small 1–2 liter dewars, suitable
for a day's run.

Superconducting cavities provide very high Q's and frequency stability. For
example, Mann and Blair (1983) have achieved Q's of about 10^8 at 9 GHz.
In their case the cavity is part of an ultra-low phase noise oscillator used in a
gravitational wave detector.

4.6.6   Optical Techniques

Optical signals are often incoherent as opposed to coherent RF and microwave signals. This allows for more flexibility in the methods used to pipe
signals to and from the cryostat. An extension of the waveguide is the brass
light pipe for transporting incoherent FIR (Ohlmann et al. 1958). This is in fact
a highly over-moded waveguide. Optical fibers can also be used to transport
both coherent and incoherent visible light (low or over-moded dielectric waveguide). Solid optical fiber pipe has been used to view images at low temperatures
(Williams and Packard 1980). Inexpensive flexible plastic light fibers are suitable for illumination, etc. The expensive flexible coherent (image) optical fiber
bundles are held together at the ends with a metallic ferrule, which when cooled
would probably sever the glass fibers. Finally, it can "all be done with mirrors."
Light beams deflected with the usual optical components have the advantage of
no material connection from one stage of the cryostat to another. Note that any
light absorbed is, however, dumped into the cryostat.

The above methods are largely useful for bringing the light in from the
top of the cryostat. Another option is to shine it in through windows in the
side (or bottom) of the dewar. Optical dewars of this type can be purchased
commercially (see list on next page). This saves the trouble of making vacuum
tight windows for use at low temperatures as well as aligning them for low

temperature operation. Mylar windows can be made in the manner discussed earlier (Silvera 1970) or quartz windows can be epoxied to flanges or sealed with an In O-ring. Precaution must be taken in order not to over stress the window upon cool down. For example, using an invar (low thermal expansion coefficient) clamp to hold the window against an In O-ring will reduce the stress. This can be important if the light polarization is to be preserved as the stress induced birefringence can rotate the polarization vector many cycles.

4.6.7   Commercial Suppliers

Useful Components

- Cooner Sales Co. Inc., 9186 Independence Ave., Chatsworth, CA 91311, 213-882-8311
- Uniform Tubes, Inc., Collegeville, PA 19426, 215-539-0700
- Omni Spectra Connector Division, 140 Fourth Ave., Waltham, MA 02154, 617-890-4750
- Malco, Microdot Connector Grp., 12G Progress Dr., Montgomeryville, PA 18936, 215-628-9800
- Alpha Industries, Inc., 20 Sylvan Rd., Woburn, MA 01801, 617-935-5150
- Systron-Donner Corp., Microwave Components Sales Dept., 14844 Oxnard Str., Van Nuys, CA 91409, 213-786-1760

Optical dewars:

- Janis Research Co. Inc., 20 Spencer St., Stoneham, MA 02180, 617-438-3221
- Oxford Instruments North America Inc., 3A Alfred Circle, Bedford, MA 01730, 617-275-4350
- Kontes Martin, 1916 Greenleaf St., Evanston, IL 60202, 312-475-0707

4.7   Pressure Measurements

by Jeffrey S. Souris and Timo T. Tommila

Undoubtedly the parameter most often used to characterize a volume of gas, regardless of its density, is pressure. Even in describing very high vacuum, where more appropriate measures might be molecular mean free paths or monolayer formation times, its usage persists. The amount of literature devoted to the measurement of pressure is staggering: hundreds of methods have been devised, tested, and refined. Yet relatively few of these techniques find practical application and widespread use.

In this paper we discuss those methods most commonly used to measure total pressure in low temperature studies. For a majority of these measurements, the pressure gauge is operated at either room temperature or very low temperature. Measurements made at room temperature are relatively straightforward; they rely on well established methods and commercially available gauges. At low temperatures, however, pressure measurements are not nearly so commonplace and frequently require greater expertise and more esoteric devices, often built by the investigator.

We therefore begin our discussion with two separate treatments of pressure measurement, based on the environment in which they are to be carried out. A description of various calibration methods then follows and we conclude with a few miscellaneous notes. Our presentation is not meant to be exhaustive, but to familiarize you with the fundamentals of current practice. For greater detail regarding specific devices and techniques, the lists of references and suppliers appended should prove helpful.

4.7.1   Room Temperature Pressure Measurement

The room temperature portion of a cryostat usually consists of a number of gas handling systems. Some are associated directly with the maintenance of low temperatures while others, such as that of a ^{3}He melting-curve pressure gauge's, with the experimental area itself. For a majority of the pressure measurements done in these places, great precision is not required. Very often you are only interested in monitoring the operation of a gas handling system, or in following pressure changes as occur when the system is being cleaned. In these instances, Bourdon, thermoconduction, and semiconductor strain gauges are usually sufficient. If still greater sensitivity is desired, capacitance, ionization, and piezoelectric gauges may be used though at considerably greater cost.

A great deal of literature concerning each of these gauges in their many varied forms has been published over the last two decades; and we shall not attempt the herculean task of restating it here. Instead our intent is to provide a brief and general summary as to how these commonly used gauges work, and when they might best be employed. More complete presentations may be found

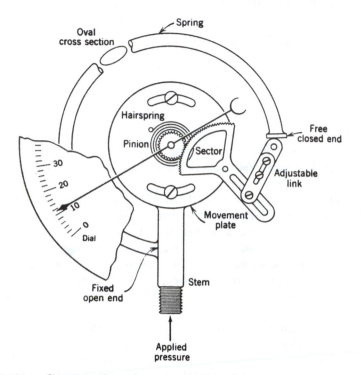

Figure 4.47. *Construction of a common Bourdon tube pressure gauge. [Benedict, R. P., Fundamentals of Temperature, Pressure and Flow Measurement, (John Wiley & Sons, NY) 1977.]*

in the discussions of Beavis (1969), Benedict (1977), Berman (1985), Denison (1979), and Leck (1964).

Bourdon Gauges The simplest Bourdon gauge design is that built from a tube of oval cross-section, bent in a circular arc. As shown in Figure 4.47, one end of the tube is fixed and open so as to receive the applied pressure. The other end is sealed and connected via a mechanical linkage to the pointer of the gauge. When pressure is applied, the oval cross section of the tube becomes more circular, thereby changing the tube's overall length and so moving the pointer.

Gauges incorporating metallic tubes or corrugated bellows and capsules, often refered to as diaphragm gauges, show dynamic ranges of 10^3 and can measure pressures down to .5 Pa, with uncertainties of about .5% of the full scale reading. More elaborate gauges, with uncertainties of .01%, employ back-to-back Bourdon tubes as well as tubes of helical and twisted geometries that "unwind" under applied pressure. Substituting more elastic materials in tube construction, such as quartz, offers greatly improved performance. These springy devices exhibit dynamic ranges of 10^5 and have smallest measurable pressures near 10^{-5} Pa. An example of such a gauge, marketed by Mensor Inc., uses a mirror, mounted along a helical Bourdon tube of quartz, to optically measure the angular rotation of the tube during pressure changes.

The disadvantages in using Bourdon gauges are their slow response and somewhat high sensitivity to shock and vibration. In addition, if helium is a major constituent of the gas and you are using a glass tube gauge, helium diffusion through the tube's walls will result in erroneously low pressure readings. For this application a metallic diaphragm gauge should be used.

Capacitance Gauges Capacitance pressure gauges often employ a metallic diaphragm as one plate of a capacitor as shown in Figure 4.48. Under the application of pressure the diaphragm moves with respect to the other plate, changing their relative separation. The diaphragm's deflection is detected by the unbalancing of an a.c. capacitance bridge, the voltage required to rebalance the bridge being directly proportional to pressure. Capacitance gauges have excellent frequency response, good linearity, and a large signal-to-noise ratio permitting the high amplification of response signal needed to detect small electrode deflections. Many commercially available gauges are designed for near room temperature operation and can be used for pressure measurements from 10^{-3} Pa to 10 MPa with accuracies of .02%. They come in a wide variety of configurations that include both differential and absolute pressure sensing devices. MKS Instruments offers a gauge, the MKS Baratron, which comes with a temperature controlled, differential pressure head. By using various pressure heads you can select the range of pressure measurement with optimal sensitivity.

Semiconductor Strain Gauges A few years ago, semiconductor pressure gauges were little more than a novelty; poor in performance compared to the available alternatives. More recently though, a number of devices such as those of Sensym's LX1800 series have found their way into the lab, replacing the ordinary Bourdon gauges used in gas handling systems of 50 Pa to 3 MPa. Pressure measurements made with these gauges commonly have accuracies of 1% of the full scale reading.

At the heart of the device is a diffused silicon, piezoresistive strain gauge. Using standard IC fabrication techniques, a silicon diaphragm is formed onto which a resistor network is then diffused. When pressure is applied the diaphragm deflects, causing its integral resistors to stretch or compress. The resulting change in resistance is measured by means of a Wheatstone bridge circuit located adjacent to the diaphragm on the chip. Most transducers also include a preamplifier, allowing you to use an ordinary digital voltmeter for pressure readings. They are relatively immune to vibration, show little hysteresis, and have excellent long term stability. Their principal drawback is that changes in ambient temperature can adversely affect the device's null, span, and sensitivity. Most manufacturers include temperature compensation circuitry on the chip to greatly reduce these effects.

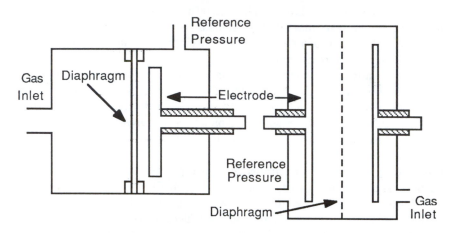

Figure 4.48. *Capacitance gauges. The design on the left is 'one-sided' and is suitable for protection of electrodes from damaging gases. The design shown on the right is a more sensitive twin electrode design.*

Piezoelectric Pressure Transducers Piezoelectric devices offer a simple and reliable way to make secondary pressure standards and to monitor pressures at room temperature in the range of 1 kPa to 100 MPa, generally with uncertainties of .5%. They are largely insensitive to vibration and electromagnetic interference, and have excellent reproducibility.

Although these transducers can be found in a variety of forms and crystal compositions, the more sensitive gauges often incorporate a crystalline quartz, oscillating beam whose resonant frequency varies with applied loads. For the device shown in Figure 4.49, four metallic electrodes have been vacuum deposited onto the beam such that diagonally opposed electrodes are electrically connected. The beam is forced into flexural vibration by an oscillator circuit which tunes itself to the beam's resonant frequency. Under compression the response frequency decreases, while with tension it increases. To maintain the characteristically high Q (of order 10^4) of the transducer, silicon rubber isolators are used as mechanical, low-pass filters in both the sensing and mounting arms.

Thermoconduction Gauges Thermoconduction gauges operate at pressures where energy transport, between sensing elements at different temperatures, occurs by gaseous conduction. At lower pressures, radiation and thermal conduction along surfaces dominate energy transfer. At higher pressures, molecular interaction within the gas leads to convection.

For each species of gas present, its molecular weight and accommodation coefficient principally determine how effectively thermal conduction occurs. Generally, more complex molecules accommodate more efficiently than simpler mol-

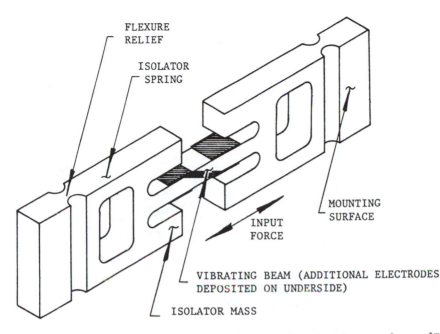

FLEXURE
RELIEF

ISOLATOR
SPRING

MOUNTING
SURFACE

INPUT
FORCE

VIBRATING BEAM (ADDITIONAL ELECTRODES
DEPOSITED ON UNDERSIDE)

ISOLATOR MASS

Figure 4.49.   *A Paroscientific quartz beam, piezoelectric transducer. [Paro-scientific Inc., catalog, Paroscientific Inc., 4500-148th Ave. NE, Redmond, WA 98052.]*

ecules, while the converse holds in regard to energy transport. The composition of the gauge sampled gas must therefore be known beforehand, especially if hydrogen or helium is present, and the gauge calibrated accordingly. Most manufacturers calibrate their thermoconduction gauges for air.

One of the simplest thermoconduction gauges is the Pirani gauge. A transition metal wire serves as the sensing element. The wire is both heated and its resistance measured by means of a Wheatstone bridge circuit. Although variations of this gauge have been used to measure pressures down to 10^{-6} Pa, most commercially available Pirani gauges operate between 1 Pa and 10^{-4} Pa. As with all thermoconduction gauges, to obtain maximum sensitivity the gauge should be operated in a cool, static environment.

An alternative design and perhaps the most ubiquitous of thermoconduction gauges is the thermocouple gauge. These devices employ a thermocouple to measure the temperature of a heated wire. The gauge may be operated in either constant current or constant voltage modes, though the former is somewhat more common. Most thermocouple gauges have dynamic ranges of about 10^3, and they can be used to measure pressures as low as 10^{-2} Pa to within a few percent. With the careful maintenance of temperature, pressures down to 10^{-4} Pa may be measured, although then aging effects of the thermocouple junction may not

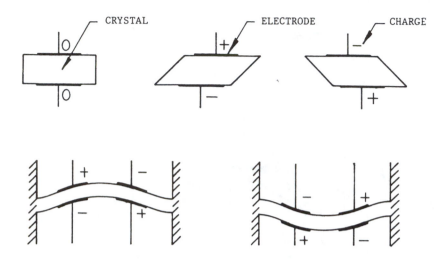

Figure 4.50.   *Illustration of the change in electrode charge with quartz beam deflection. [Paroscientific Inc., catalog, Paroscientific Inc., 4500-148th Ave. NE, Redmond, WA 98052.]*

be negligible. A greater output and somewhat higher sensitivity may be gained by connecting several thermocouples in series (a thermopile gauge), but does not provide any appreciably larger range of pressure measurement.r

Ionization Gauges At present the most commonly used ionization gauges are variations of the Bayard-Alpert gauge. Two basic configurations are commercially available: The "chaste tube" (10^{-1} to 10^{-6} Pa), in which the gauge elements are enclosed in a glass tube, and the "nude tube" (10^{-1} to 10^{-9} Pa), which uses no enveloping tube at all. The latter is built on a ceramic base and may be heated to temperatures in excess of 450° C; a consideration in systems that require heat-cleaning.

The operation and internal structure of both nude and chaste tube varieties are primarily the same. For the nude tube gauge of Figure 4.51, electrons are emitted from a heated filament and accelerated by a positively charged grid, consequently ionizing surrounding residual gas molecules. The ions are then drawn to a negatively charged collector where they may be counted in the form of a current. The nude tube gauge's ability to measure lower pressures reflects its more efficient, closed grid structure.

Since the ionization gauge measures, effectively, the density of the residual gas, concurrent temperature measurements are necessary to accurately determine the pressure. More often than not, though, ionization gauges are used as relative pressure gauges and at constant temperatures, so temperature/density effects are of little problem in that sense. To ensure accurate readings at low pres-

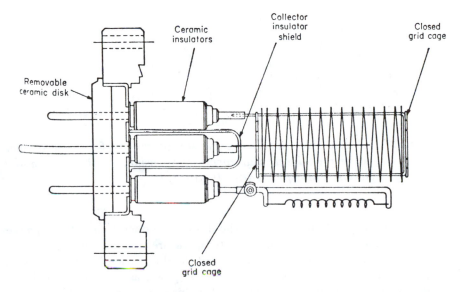

Figure 4.51. *A bakeable nude-tube version of the Bayard-Alpert ionization gauge. [Redhead, P. A., J. P. Hobson, and E. V. Kornelsen, Physical Basis of Ultra High Vacuum, Chapman & Hall, Ltd., London) 1968.]*

sures, you must "outgas" the grid by heating it electrically a few minutes before measurements are to be taken and should use a LN_2 trap near the gauge's entrance. Most ionization gauges should not be operated at pressures greater than .1 Pa, although chaste tube designs have reportedly survived accidental exposure to atmosphere. A very complete and authoritative survey of more sensitive variations of the Bayard-Alpert gauge is given in Berman (1985).

4.7.2 Low Temperature Pressure Measurement

For pressure measurements at temperatures around that of liquid helium, you will often be concerned with one of two pressure regimes: high pressures of up to 30 MPa, as in performing temperature calibration via the ^{3}He-melting curve, and low pressures of less than 100 Pa, as in measuring the pressure of spin polarized, atomic hydrogen. In addition, some experiments may require the measurement of ultra high pressures of up to a hundred gigapascals at low temperatures.

Nearly all commercially made gauges are designed for room temperature use; they cannot be adapted to function reliably, with the power efficiency and size restrictions required, in the low temperature setting. To make these pressure measurements you will most likely need to construct and calibrate your own gauge.

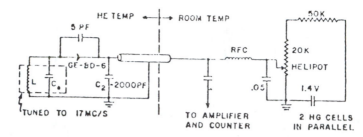

Figure 4.52.   *A sensitive capacitance measuring method for a low temperature pressure measurements. [Boghosian, C., H. Meyer and J. Reves, Phys. Rev. **146**, 110 (1966).]*

In the following section we discuss the design and operation of those gauges most often used at these pressures, in low temperature environments. Since most of these devices are electrically operated, you should provide good thermal anchoring of their leads and use as little excitation power as possible, to minimize joule heating and the formation of temperature/pressure gradients. Capacitance type gauges have seen extensive use in the lower two ranges while at ultra high pressures more exotic means are necessary.

Capacitance Gauges The measurement of gauge capacitance is usually made by means of a capacitance bridge as described in Section 4.1. Another suitable method is the use of tunnel diode or FET oscillator circuits, as first reported by Boghosian *et. al* (1966) and Kelm (1968). The tunnel diode circuit is shown in Figure 4.52.

These techniques require the gauge to be the frequency determining element in an oscillator circuit and use a frequency counter to observe the electrode's frequency of oscillation. For many low temperature applications it is necessary to place the entire tunnel diode/FET oscillator circuit into a liquid helium bath to obtain the good stability, reduced thermal noise, and improved frequency response sought. Occasionally it is convenient to keep the circuit inside the cryostat's vacuum can, but this necessitates a good thermal link to either the liquid helium bath or to the refrigerator as these devices dissipate moderately large amounts of power.

At higher pressures, up to 30 MPa, the first highly sensitive capacitance pressure gauges for low temperature use were constructed by Straty and Adams (1969), primarily for studying liquid and solid helium. Figure 4.53 shows a cross-sectional view of a similar gauge used by Greywall and Busch (1982) to measure pressure along the ^{3}He-melting curve. These gauges are constructed mainly of Be-Cu which has been heat treated for three hours at 320° C in vacuum. Gauges with nylon chambers, while not as sensitive or stable, are suitable for NMR or magnetic thermometry using a paramagnetic salt in the sample chamber. The relative sensitivity, dp/p, can be as fine as 10^{-8} for the Be-Cu cells. Straty and

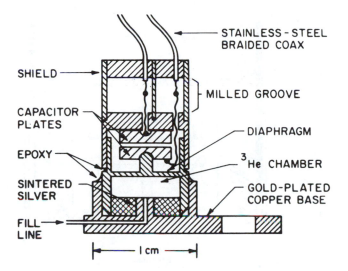

Figure 4.53. *Cross-section of a higher pressure capacitance gauge used in* ^{3}*He melting curve thermometry. [Greywall, D. S. and P. A. Busch, J. Low Temp. Phys.* **46**, *451 (1982).]*

Adams (1969) give a detailed description of construction and design considerations.

Still greater resolution may be obtained by using twin diaphragms, each of which is attached to an electrode. Greenberg *et al.* (1982) report, for the device shown in Figure 4.54, a factor of two improvement in sensitivity over that of single diaphragm, gauge designs. The gauge is built in two symmetrical halves, and subsequently epoxied together at room pressure. Because of the entrapped air, however, the device should not be used above liquid nitrogen temperatures. For higher temperature applications, evacuation of the cell's interior and use of more rigid, less sensitive diaphragms are required.

In the low pressure region a number of sensitive capacitance pressure gauges have been built using very thin foils as one plate of the capacitor. Matthey *et al.* (1981) used Mylar and Kapton foils with good success to measure the pressure of spin polarized, atomic hydrogen between 10^{-3} and 100 Pa, at temperatures as low as 100 mK. A similar Mylar foil gauge, built by Yurke (1983) and capable of resolving 8×10^{-5} Pa/$\sqrt{\text{Hz}}$, is shown in Figure 4.55.

To facilitate repair of this gauge's diaphragm should it rupture, the back electrode and diaphragm are each mounted on separately removable, annular flanges. Construction of the back electrode begins with a brass disk being epoxied onto one of these flanges with Stycast 1266. Once the epoxy has hardened, the disk is lathed flat and sanded with successively finer abrasives. Its surface is then polished with a diamond paste until optically flat, as checked with a simple

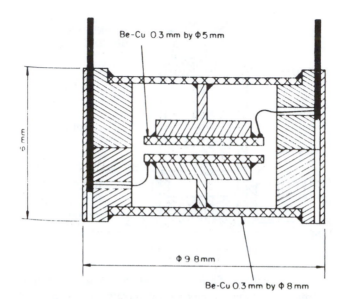

Figure 4.54.   *A twin diaphragm Be-Cu strain gauge for low temperature ^{3}He melting pressure measurements. [Greenberg, A. S., G. Gvernler, M. Bernier and G. Frossati, Cryogenics **22**, 144 (1982).]*

interferrometer. Surface smoothness is essential since to achieve high sensitivity and linearity, the spacing between the back plate and the foil should be as small and as uniform as possible. The gap between the stationary electrode and the diaphragm is maintained by an annular piece of Mylar, whose thickness is the same as that of the diaphragm.

The diaphragm is made of a 6 μm thick Mylar sheet, aluminized on the surface facing the back electrode. In preparation for mounting, the Mylar is first stretched over and cemented to a temporary support ring so as to yield a uniform tension throughout the foil. An extremely thin layer of pumped on Stycast 1266 is then spread across the outer area of the diaphragm flange's top surface; a groove in the flange, dividing its surface into two concentric rings, serves as a sink for the epoxy flowing inward during mounting. The temporary ring with its Mylar covering is then placed atop the diaphragm flange, allowing its weight to maintain the foil's tautness as the epoxy hardens. The excess Mylar is subsequently trimmed away, with a small tab left for electrical connection.

The choice of proper membrane thickness is not too critical so long as the diaphragm's diameter is much larger than its thickness. To achieve the same range of capacitance/pressure measurement, thin foil structures generally require larger electrode separations than devices built with thicker foils. While thin foils may allow for quicker response and higher resolution, moderate voltage swings or

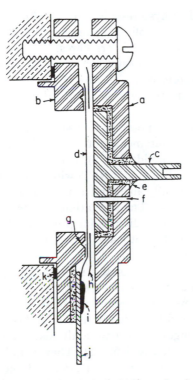

Figure 4.55.   *Cross-section of Yurke's (1983) mylar foil capacitance gauge for*
measuring low pressures within the cryostat.
(a) back electrode flange, (b) diaphragm flange, (c) back electrode, (d) mylar diaphragm, (e) epoxy, (f) vent hole, (g) glue
catcher, (h) mylar spacer, (i) silver paint, (j) binding post making electrical contact with aluminized coating of mylar, (k) indium O-ring

sudden changes in ambient pressure can very easily and permanently deform the
diaphragm. Too thick a foil, however, may not produce a suitable deflection for
commonly encountered pressures. Through trial and error, foils with thicknesses
of 5 to 40 μm have shown the best overall performance.

It is best to perform the epoxying in a clean-room, as then the entrapment
of airborne particulates within the epoxy can be minimized. Thinner foil diaphragms require extra care in their construction as they are very susceptible to
the nonuniform stresses often introduced during mounting. To guarantee reliable
operation once the pressure gauge is assembled and installed, care must also be
taken so that pressure difference across the foil is never greater than a couple
hundred pascal.

As with most capacitance pressure gauges, these gauges show hysteresis after the initial cool down. The hysteresis is a result of unevenly distributed stresses left in the membrane during cooling. To remedy this problem the gauge must be repeatedly driven with high and low pressures. You should follow the decrease in hysteresis during these cycles to establish when the gauge is ready for use. The degree to which the hysteresis can ultimately be eliminated depends upon the metal/membrane used and the process of manufacture.

Other Methods It is also possible to use a superconducting filament as a pressure transducer for measurements in the range 5×10^{-3} to 5 Pa. Such a device, as built by Cesnak and Schmidt (1983), can be constructed by removing the copper cladding from a multifilament NbTi wire and cutting all the filaments but one. The wire is then thermally anchored at both ends to 4.2 K and driven locally normal by small current pulse. The relaxation time to reach the superconducting state depends upon the surrounding vapor pressure. Usually the time difference between current pulses must be at least few seconds, making continuous pressure measurement impossible.

At higher pressures, conventional strain gauges have also been examined for their use in cryogenic applications. Cerutti *et al.* (1983) tested commercially available strain gauges over the pressure range of 0.5-3.5 MPa, making measurements at temperatures of 5 to 27 K, 77 K, and 295 K, and in magnetic fields of 0 to 6 T. The strain gauges studied were all absolute bridge-connected devices with resistances between 400 and 600 Ω. They found under these conditions, discrepancies of up to 19% between their standard and the manufacturer's calibration. Upon recalibration, however, uncertainties were kept to below 1% of the device's full scale reading.

Ultra High Pressure Gauges In the ultra high pressure regime, the measurement of pressure is frequently made either by using strain gauge techniques or by tracking known phase transitions of various materials as discussed in Bradley (1969). The superconducting transition in metals, as characterized by variations in the a.c. suceptibility of the metal, can also be used to measure high pressures. Berton *et al.* (1979) measured pressures of 1 to 2 GPa by observing the superconducting transition of tin between temperatures of 1 and 20 K; the precision of their pressure measurements being limited only by the definition of dT_c/dP (2%).

A common technique for measuring still higher pressures, in the range of 10 to 50 GPa, is to examine changes in the X-ray diffraction pattern of a crystalline powder which has been mixed with the sample. To ensure uniform pressure throughout the sample and powder, it is necessary that the powder have a small yielding strength with respect to shearing strength so that the powder can flow plastically between the sample and its retaining structure. Typically these powders consist of NaCl or CsI as they are relatively inert, have diffraction patterns that contain only a few lines thus minimizing interference effects, and are transparent at optical wavelengths. Other optical high pressure methods, such as observing the shift in Cr-ruby luminescence spectra with changes in pressure, may be found in Jarayaman (1983).

4.7.3   Room Temperature Calibration Techniques

A number of the devices described so far do not actually sense force directly; instead they reflect gas density in terms of thermal conductivity and ionization potential. And in gauges where the transducer does experience an applied force and is displaced accordingly, pressure is not usually reproduced mechanically, but deduced electronically from changes in current, voltage, capacitance, or resistance. The use of these indirect-reading devices as pressure gauges depends upon their calibration against pressure standards and the environments in which they are employed.

In general, the room temperature techniques used for establishing "known" pressures fall into two broad catagories: static expansion and dynamic flow. The static expansion method, most commonly used because of its relative ease of implementation, is best suited for the calibration of devices which operate between 10^5 Pa and 10^{-3} Pa. The dynamic flow technique covers a wider range, with lowest practical calibrations around 10^{-8} Pa, but its test chamber is somewhat more difficult to construct and can result in very large errors if the chamber's geometry is not thoroughly considered.

To calibrate gauges for high pressure work, the most widely accepted methods employ a version of deadweight, free piston gauge. It is a self-contained system, requiring no other gauges to serve as pressure standards for comparison, and may be used to calibrate devices from 70 Pa up to 3 GPa with great precision.

Static Expansion   The static method relies upon the isothermal expansion of a small, relatively high pressure, volume of gas (P_0, V_0) into a larger evacuated volume (P_1, V_1), as shown in Figure 4.56. The initial measurment of P_0 is made with a reference gauge. Once the two volumes have been opened to one another and the system has reached equilibrium, the expansion volume's calculated pressure

$$P_1 = \frac{P_0 V_0}{V_0 + V_1} \tag{4.8}$$

is compared to the test gauge reading. V_1, typically 10^3 times greater than V_0, is then evacuated and the expansion/evacuation process repeated several times yielding a set of calibration points. After the n*th* repetition, the expansion volume's pressure should be

$$P_n = \frac{P_0 V_0^n}{(V_0 + V_1)^n} \tag{4.9}$$

provided that after each cycle V_1 was evacuated to a residual pressure much less than P_n. Gas adsorption and desorption at the walls, the cumulative effect of inaccurately determined volumes, and gauge pumping restrict the utility of this method to pressure above 10^{-3} Pa.

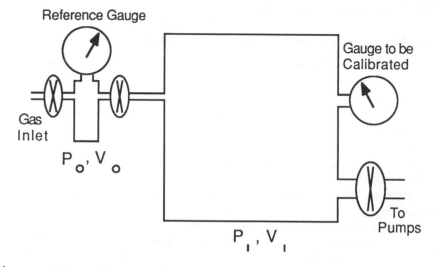

Figure 4.56. *Schematic of apparatus used in static expansion calibrations.*

The test chamber is constructed of stainless steel, to permit high temperature cleaning, and ideally has electropolished inner surfaces. Even with these precautions, gases such as H_2, CO, CO_2, and O_2 cannot be used for lower-end pressure calibrations because of their high adsorption characteristics. The chamber is evacuated between cycles by a well baffled diffusion pump, generally through a LN_2 trap. Likewise the reference standard and gauges to be calibrated employ LN_2 traps so as to prevent their contamination with mercury or hydrocarbon vapors that may be present. For greater detail concerning the choice of materials and test chamber geometry, see Berman (1985), Denison (1979), and Messer (1977).

Dynamic Flow The most widely accepted methods of calibrating devices below 10^{-3} Pa rely upon deducing the pressure reductions that result from the impeded flow of a gas through a series of calculable conductances. In its simplest form, the apparatus consists of a reference chamber (V_1, P_1) and a calibration chamber (V_2, P_2), connected to one another through an aperture of small known conductance C_1 as shown in Figure 4.57. The calibration chamber in turn is connected, via a second aperture of conductance C_2, to a high speed, trapped diffusion pump. Under steady-state conditions, the rate of gas flow through V_2 is

$$Q = C_1(P_1 - P_2) = C_2(P_2 - P_e) \qquad (4.10)$$

where P_e is the equilibrium pressure at the diffusion pump's entrance. The pressure in the calibration chamber then is

$$P_2 = \frac{C_1 P_1 + C_2 P_e}{C_1 + C_2}. \qquad (4.11)$$

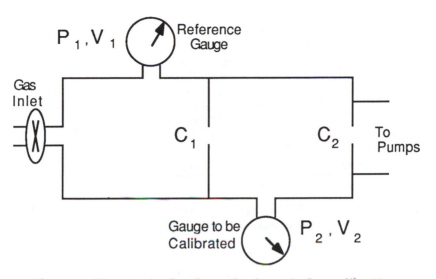

Figure 4.57. *A simple scheme for dynamic flow calibrations.*

The measurement of P_e can be circumvented to some degree provided that the pumping speed is large compared to the conductance C_2: as a rule of thumb, $P_2 \geq 100 P_e$. Equation 4.11 then simplifies to

$$P_2 = \frac{C_1 P_1}{C_1 + C_2}. \tag{4.12}$$

Often, however, P_e's contribution is not negligible and a gauge similar to the one being calibrated is installed in the pumping line. The ratio of their indicated pressures may then be used in estimating P_e. To permit very low pressure calibrations the ratio C_2/C_1 must be very large, since P_1 must remain within the reference gauge's range of accepted response.

Although calibrations down to 10^{-10} Pa can be achieved, difficulties in evacuation, calculation of conductances, gauge pumping and, to a lesser degree, adsorption/desorption at the walls, limit the application of the dynamic flow technique to pressures above 10^{-8} Pa. For the construction of this apparatus and more sophisticated designs, which include greater numbers of apertures and differential pumping of stages, see the discussions in Beavis (1969), Berman (1985), Denison (1979), and Redhead (1968).

Deadweight Tester The calibration of high pressure gauges, in the pressure range of 70 Pa to 3 GPa, is most often done by using a deadweight tester similar to the one shown in Figure 4.58. Various weights of known mass are loaded onto one end of a precision machined piston. Pressure is then applied to the other end of the piston, enabling it to float freely within the cylinder. To prevent

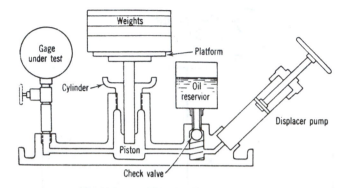

Figure 4.58.   *Design of deadweight, free-piston tester, commonly used for high pressure calibrations. [Benedict, 1977 (same as Fig. 4.47).]*

widening of the gap between the cylinder and piston with increasing pressure, some deadweight testers employ a secondary pressure, external to the cylinder.

Friction is essentially eliminated by allowing the outward flow of high pressure fluid through the piston/cylinder gap and rotating or oscillating the piston within the cylinder. Because of this leakage, however, the system pressure must be continuously trimmed upward to maintain the piston at a static height. In this state of equilibrium the unknown system pressure, as measured by the gauge undergoing calibration, is then proportional to the force of the piston-weight combination divided by the cross-sectional area of the piston.

Although pressure measurements with accuracies of .01% to .05% of the reading are routinely done, several corrections must first be made. Air buoyancy, local gravity, surface tension of the hydraulic fluid, as well as changes in the piston's cross sectional area with temperature and pressure, if not properly dealt with, can each lead to significant error. Benedict (1977) provides a general summary of these correction procedures.

4.7.4   Calibration Reference Gauges

At present, there are relatively few devices which can be regarded as pressure standards for gauge calibration. The standard must be capable of measuring force per unit area directly. And as the quantity of gas in the gas phase approaches that amount adsorbed onto the test chamber's walls, this task becomes difficult if not meaningless. In this low density regime extension techniques, such as the dynamic flow method just outlined, must be used in conjunction with a pressure standard. The most widely accepted standards in current use are variations of the U-tube manometer and McLeod gauge. A less accepted, though in our opinion practical and less time-consuming alternative for moderate and high pressure calibrations, is the use of high precision secondary standards such as the MKS Baratron at lower pressures, and the Parosci piezoelectric gauge at higher pressures.

U-tube Manometer The simplest calibration device for moderate pressures is a U-tube manometer. One end of the U-tube is exposed to the calibration chamber while the other end is either differentially pumped or sealed at vacuum. Pressure is determined by measuring the difference in heights of a known density liquid in each arm of the tube. In the steady state, the difference between the reference pressure and the unknown pressure is balanced by the weight per unit area of the equivalent displaced manometer liquid: that is

$$\Delta P = \beta \Delta h_e \tag{4.13}$$

where Δh_e is the equivalent manometer fluid height and β the specific weight of the manometer fluid. Corrections such as those for temperature, local gravity, and capillary effects are discussed at length in Benedict (1977).

Commonly used liquids include mercury and various oils of low saturated vapor pressures. With mercury you can calibrate gauges between 10 and 10^5 Pa with relative ease. The major disadvantages with using mercury are that it is poisonous and its relatively high vapor pressure, about .3 Pa at room temperature, requires the use of a liquid nitrogen trap between the vacuum system and the manometer. By using diffusion pump oil, again with a cold trap, you can lower the calibration pressure range roughly by one order of magnitude. Still greater precision can be achieved by measuring the position of liquid levels with a cathetometer.

In addition to standard U-tube manometers, micromanometers may be used as low pressure reference standards; extending the capabilities of calibration another three orders of magnitude lower than that of the U-tube. An authoritative survey of these devices is provided in Brombacher (1970).

McLeod Gauge The McLeod gauge is a refined verson of the standard U-tube manometer in which the gas, whose pressure is sought, is compressed from a larger volume into a small capillary tube as shown in Figure 4.59. The mercury level is raised, by applying pressure with external pumps, so as to isolate a portion of the gas in A and B. Once the mercury level in E reaches the same height as the sealed end of B, h is measured and related to calibration by:

$$PV = \alpha h(h + P) \tag{4.14}$$

where P is the pressure of the gas isolated in A and B, V is the combined volume of A and B, and α is the cross-sectional area of the capillary B. If $P \ll h$, as is usually the case, then $P = \alpha h^2 / V$. Therefore, the larger the ratio of the initially trapped gas's volume to its compressed volume, the greater will be the magnification of the pressure P and the manometer reading h.

Typically, McLeod gauges are used for calibration of devices that measure pressures of 10^{-3} to 10 Pa. Calibration at very low pressures by this method

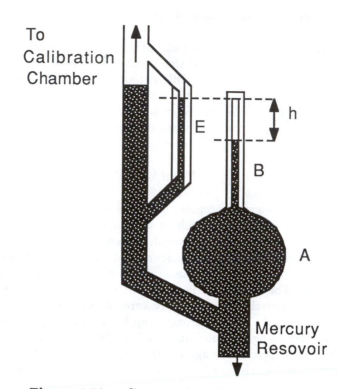

To
Calibration
Chamber

E

B

A

h

Mercury
Resovoir

Figure 4.59.   *Cross-section of a McLeod gauge.*

is limited because of complications arising from temperature gradients in the oil, gas adsorption, inaccuracies of the volume measurement, and vibrational disturbances of the liquid surfaces.

4.7.5   Low Temperature Calibration Techniques

Most low temperature pressure gauges require *in situ* calibration because of changes in geometry that occur as the gauge is cooled or heated. A few of the more practical approaches routinely taken in these calibrations are presented below. Often it is worthwhile to employ at least two of these techniques, as the precision with which any one may be done is somewhat variable.

Vapor Pressure Calibration To calibrate a low pressure gauge within an operating cryostat, you should not use a room temperature calibration device because of the thermomolecular effect associated with unknown temperature gradients and complicated tube structures. However, the temperature dependence of the saturated vapor pressures of ^{4}He and ^{3}He is well known and may be used in conjunction with precise temperature measurement to determine pressures accurately inside the cryostat.

Since the vapor pressure curve is an extremely sensitive function of temperature, the first task is to make a careful calibration of temperature as discussed in Chapter 5. If, for example, the temperature calibration included a small temperature offset, a plot of pressure versus the reciprocal of capacitance would show the correct linear behavior but with an incorrect slope that would most likely go unnoticed.

Next the gauge sampling volume is filled with either ^{3}He or ^{4}He. It is important that there always be sufficient liquid to produce a saturated vapor and that you are measuring the lowest temperature in the volume, since that is the temperature which determines the ambient's vapor pressure.

Finally the temperature is regulated, by precisely throttled pumping and/or carefully controlled refrigerator heating or cooling, and pressure measurements taken with the suspected gauge. You should monitor both temperature and pressure changes during each successive step in the temperature sweep to establish when the system has reached equilibrium and a calibration point may be taken. By using both ^{4}He and ^{3}He you can cover different pressures within the same temperature range. This gives a method of determining errors in pressure calibration although even then some offset in the temperature calibration may be hidden, yielding imprecise pressures.

Electrostatic Calibration Another practical technique, especially for calibrating a low pressure capacitance gauge already mounted in the cryostat, is the electrostatic method. If a d.c. voltage (V) is applied to the electrodes of the gauge, the resulting electrostatic force simulates a pressure of

$$P_s = \frac{1}{2}\frac{C^2V^2}{A^2\epsilon} \tag{4.15}$$

where C is the gauge's capacitance, A the area of one electrode plate, and ϵ the dielectric constant of the residual gas. The values of C, V, and A can be measured to a high degree of precision. For many parallel plate gauges, though, the electrode's separation is rarely uniform. As a result, unevenly distributed forces are exerted on the diaphragm and the capacitance measured is different from that found in Equation 4.15. For small diaphragm deflections, however, the gauge's capacitance may be rewritten as

$$C = C_o + \left[\frac{\partial C}{\partial P}\right]_{\substack{P=0 \\ V=0}}(P + fP_e) \tag{4.16}$$

in which f is a geometric factor relating gas pressure response to voltage response via Equation 4.15, and P is the gas's pressure. Yurke (1983) found f=2.06, for the foil gauge described in Section 4.7.3, by comparing electrostatic and vapor pressure calibrations to one another. For the ideal parallel plate gauge, f should be unity.

To calibrate a gauge, its capacitance is measured as a function of applied voltage and fit to a second order polynomial. Higher order fits should also be

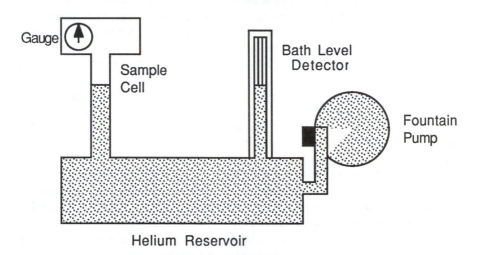

Figure 4.60.   *A liquid ^{4}He manometer for calibrating gauges within the cryo-stat.*

tried, to appraise the correctness of using of Equations 4.15 and 4.16. If the electrodes have a permanent voltage difference in normal operation, the applied voltage is either added to or subtracted from it, depending upon the polarities involved. The plot of simulated pressure versus capacitance may then be obtained from Equations 4.15 and 4.16. In general, and especially at higher pressures where electrode deflections are no longer small, you should compare your electrostatic results with those obtained by vapor pressure calibration, as an estimate of errors. This is also the easiest way to determine the $\partial C/\partial P$ term in Equation 4.16.

L^{4}He Manometer Within the cryostat, a column of liquid helium can function as a pressure head to calibrate gauges inside the sample cell. A liquid ^{4}He filled reservoir is used to connect the sample cell to a bath level detector, forming the U-tube shown in Figure 4.60. The height of the liquid is determined by measuring the capacitance across two coaxially mounted, copper tubes. Superfluid film flow, as regulated by a resistor heated fountain pump attached to the reservoir, controls fluid height within the apparatus. Since the dielectric constant of liquid ^{4}He is 1.05, practical capacitances allow pressures from 10^{-1} to 10^2 Pa to be easily measured. During gauge calibration, however, you must take care to maintain constant temperature throughout the manometer, so as to prevent the formation of pressure gradients which may result from thermomechanical fountain effects.

4.7.6   Additional Notes

In measuring low pressures inside an operating cryostat with a room temperature pressure gauge, the themomolecular effect must be taken into consideration. If the pressure is so low that gas flow occurs in the molecular flow region, the equilibrium state pressure will vary with temperature since the steady state condition requires the number of particles flowing in both directions be equal and the flow rate be temperature dependent. Usually this pressure difference is nearly impossible to calculate because of the complicated tube geometries and complex temperature distributions often present. However, if the pressure is higher so that the flow is characterized as viscous, then the system will have constant pressure everywhere. The transition between molecular and viscous flow occurs when the gas's mean free path λ is such that $d/3 < \lambda < d/100$, for a system's characteristic dimension d. To a rough approximation then, the pressure at which both molecular and viscous flow occur simultaneously is $P_{trans}(Pa) \sim .67/d(cm)$.

If you are trying to measure the vapor pressure of liquid ^{4}He below the lambda-point with a room temperature gauge, the situation is a bit more complicated. The superfluid ^{4}He film flows toward higher temperatures and evaporates at the point where the film flow is not substantial enough to maintain a saturated vapor pressure over itself. This means that the pressure measured at room temperature corresponds to the vapor pressure of the superfluid ^{4}He at the temperature and location at which the film evaporates. Generally this pressure is different from that sought within the colder experimental area. Internal structures, such as baffled tubing, can be used to lessen these film flows considerably, but you should still be mindful of where evaporation is taking place when interpreting pressure readings.

Finally, the importance of the gauge's location cannot be overemphasized. Even with the most precisely calibrated gauge, improper placement can often result in highly misleading readings. Usually the vacuum system is not in static equilibrium but instead undergoing dynamic gas flow. Sources of concern include: gas jetting into the gauge, pressure gradients near pumping ports, backstreaming of oil and hydrocarbon vapors, and pumping effects of the gauge and its mounting. With the intelligent location and adequate baffling of gauges, and the strategic use of LN_2 traps, you can minimize these hazards and your time spent later correcting for them.

4.7.7   Manufacturers

Bourdon Gauges
- Mensor, Inc., P.O. Box 55642, Houston, TX 77055.

Capacitance Gauges
- Datametrics Inc., 340 Fordham Rd., Wilmington, MA 01887.
- MKS Instruments, Inc., 22 Third Ave., Burlington, MA 01803.
- Setra Systems, One Strathmore Rd., Natick, MA 01760.

Semiconductor Strain Gauges
- Sensym Inc., 1255 Reamwood Ave., Sunnyvale, CA 94089.

Piezoelectric Transducers

- Paroscientific, Inc., 4500 - 148th Ave. NE, Redmond, WA 98052.

Thermoconduction and Ionization Gauges

- Varian, 6489 Ridings Rd., Syracuse, NY 13206.

Conventional Strain Gauges

- BLH Electronics, 42 Fourth Ave., Waltham, MA 02254.

Deadweight Testers

- Harwood Engineering Co., South Street, Walpole, MA 02081.

Other Related Sources of Note

- Alcatel Vacuum Products, 60 Sharp Street, Hingham, MA 02343.
- Balzers, 8 Sagamore Park Road, Hudson, NH 03051.
- Edwards High Vacuum Inc., 3279 Grand Island Boulevard, Grand Island, NY 14072.
- Granville-Phillips Company, 5675 East Arapahoe Avenue, Boulder, CO 80303.
- Huntington Mechanical Laboratories Inc., 1040 L'Avenida, Mountain View, CA 94043.
- Key High Vacuum Products Inc., Flowerfield Industrial Park, St. James, NY 11780.
- Leybold-Heraeus Vacuum Systems Inc., P.O. Box 483, 120 Post Road, Enfield, CT 06082.

4.8   Electromagnetic Compatibility

by John S. Denker

4.8.1   Introduction, or Zen and the Art of Grounding and Shielding

These notes are meant to give you a running start into the EMC business: to indicate useful ways of looking at things, to warn you of pitfalls, and to cultivate your intuition about orders-of-magnitude. I will treat a couple of elementary examples, but fortunately the techniques discussed should suffice for nearly all the measurements we make. There are additional sneaky tricks, useful in special situations, which you may find in the references. I have made an effort to include at every turn typical values for the voltages, currents, and impedances involved. These are of course highly variable; don't get excited if your situation differs by a couple orders of magnitude.

The reason that grounding and shielding is often compared to necromancy is that interference processes result from nonidealities in the circuit elements, from the breakdown of cherished assumptions about how things work. It ain't the things you are puzzled about that cause problems—it is the things you are confident of, wrongly. Therefore, you must acquire a feeling for which approximations can still be trusted, and which cannot.

Let me illustrate this with a koan (a Zen riddle to stimulate meditation):

What is the stray-field potential difference from one side of the lab to the other, typically?

Some people guess a thousand volts, or a couple of volts . . . A few millivolts is closer to the truth, but the point of the koan is that the voltage in question is NOT a potential! A potential is a function of position, independent of the path taken to get there. In the presence of a changing magnetic flux the voltage depends not only on position, but also on the shape of the circuit (Figure 4.61).

This causes no end of trouble.

Let me conclude this introduction by dispelling a few more common misconceptions.

60 HERTZ $\neq$ INEVITABLE

- As Morrison points out, it is possible to shield things so well that you can attach an oscilloscope to a man's heart, where microvolts of interference would be intolerable, and microamps of pickup would be lethal. The problems we deal with in H-corridor are pretty simple by comparison.

EMC $\neq$ BLACK MAGIC

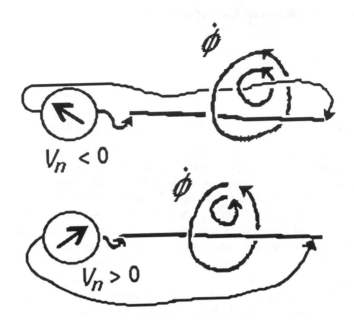

Figure 4.61. *Voltage is not the same as Potential.*

- The principal sources of interference are quite well known. They can be measured, and they can be systematically designed out of the apparatus.

WIRES $\neq$ EQUIPOTENTIAL

- Ordinary wires have a non-negligible resistance, and an even larger inductive impedance. Coax cables are extremely resistive.

TOPOLOGICAL CIRCUIT DIAGRAM $\neq$ CIRCUIT

- It is easy to wire up two circuits which are topologically equivalent, one of which works, one of which doesn't.

GROUND $\neq$ GROUND

- Good ole terra firma is not an equipotential. If you see a circuit "grounded" at more than one point, you should be very suspicious of any voltages measured with respect to "ground". Which ground?

4.8.2   Resistive Couplings

Suppose you were working on a movie called "Jaws" and you needed a robot shark. You would hook up a computer, a long cable, some power amplifiers, motors, etc. (Figure 4.62).

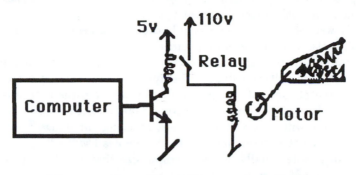

Figure 4.62.   *Circuit Diagram for a Shark*

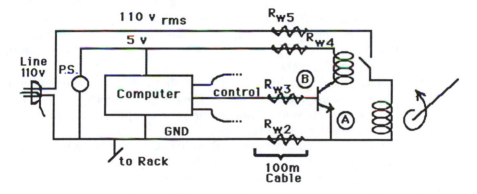

Figure 4.63.   *Actual Layout.*

TTL signals come out of the computer, turning on the transistor, which closes the relay, which turns on the motor. This circuit is obviously reasonable and correct as far as it goes. Now, having been warned that topological circuit diagrams can be misleading, I have drawn a much more detailed semi-scale layout diagram (Figure 4.63).

For each wire in the cable, I have shown a squiggle to represent the inevitable parasitic impedance of the wire. Query: what happens when you turn on this circuit? You may guess that for some reason it doesn't work properly, but that is the least of it! The most important thing to notice is that one side of the power line is attached to the rack. The cold side of the line is typically only a volt or so from ground, but if you turn the plug around, you get a hot rack! You can imagine the scene: salt water, innocent bystanders, 110 volts . . .

Nobody was killed, but they certainly could have been. If you don't remember anything else from these notes, remember this: 110 volts kills. No matter how you ground things, make sure that even if there is an insulation breakdown or other fault, high voltage will never be applied to instrument cases, knobs, or racks. Note: the purpose of the third prong on power cords is to provide a low impedance path to ground, so that if some fault tries to apply voltage to the chassis; A) there will be a voltage divider, reducing the danger; and B) it will draw enough current to blow the fuse. In these notes I will advocate using fewer grounds than people usually use, so I emphasize that you have to think twice about the safety before snipping the ground wire.

On a couple of occasions my technicians have tried to bump me off by the hot-instrument method . . . . There is nothing like a big blue SSPAAKKK!! to make you grateful for that third prong.

The other interesting thing about the circuit of Figure 4.63 is this: even if you plug it in the "right" way around, when you turn it on, you blow out the computer! Let's analyse why. Assume the cable consists of 100 meters of #10 copper wire. According to the almanac, that has a resistance of 3 mΩs per meter, or 0.3 Ωs total. The motors draw 20 amps rms when they start. Ohm's Law gives a drop of 6 volts rms across R_{w2}. At the negative peak, the voltage at node A will be -8.5 volts, and the voltage at node B will be one diode drop above that, i.e., -7.8 volts. Now if you apply negative 7 or 8 volts to the output of a computer, you will blow every chip in sight.

At this point, you decide to redesign your shark. It's a multi-megabuck production; you can afford a new computer. The design in Figure 4.64 is a vast improvement. It won't zap your technician or your computer. The relay now isolates the 110 v power loop from the logic.

Unfortunately, this circuit doesn't work either. Can you see why? At this point, the filming is a year behind schedule, largely because of problems with the shark, and the producer REALLY wants it fixed.

The problem is that as more and more channels are turned on, the current in R_{w3} increases, and since the left side of R_{w3} is grounded, the voltage at node E increases until the transistor cuts off. This is a truly obnoxious bug, since it depends on many things, and you can't find anything wrong with any particular channel.

At this point, since the beast almost works, you might be tempted to improve things by using bigger wires, smaller relays, more tolerant circuitry, etc.; but I would hate to have to guarantee the infallibility of the resulting device. I prefer the solution shown in Figure 4.65, which is: A) elegant; and B) manifestly correct.

We can identify three separate current loops in the cable: the logic loop (wires 3 and 4), the relay loop (wires 2 and 5), and the 110 v power loop (wires 1 and 6). The current in R_{w3} is small, so Ohm's Law says the drop in wire 3 is small, and node E is essentially zero volts, relative to the ground at node G. We expect some drop in R_{w2}, so the voltage at node G' will be a volt or so

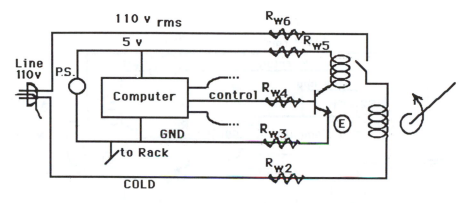

Figure 4.64.   *Immensely Better Design.*

NEGATIVE with respect to ground. Strange but true. The trick is clear now. The logic loop doesn't mind if it develops a couple dozen millivolts of drop. The relay loop doesn't mind if it develops a volt or so of drop. The 110 v power loop doesn't mind if it develops many volts of drop. On the other hand if one loop interacts with another one lower in the heirarchy, you have a mess.

Another koan:

What is the best way to ground node G′ in Figure 4.65

DON'T. The drop in R_{w2} is inevitable. Yield to it. If you raise node G′ to ground, you will raise node E a volt or so. Relay-loop currents can now flow through R_{w3} and complete their circuit via the ground strap from G to G′. The circuit then has all the problems of its predecessor, Figure 4.64. We will leave the shark now, but let me point out that this tale is not all wet. In our lab we have a case where a COMPUTER controls the RELAYS on a ratio transformer, connected to a SQUID. You may surmise that the design bears more than a passing resemblance to Figure 4.65.

SUMMARY OF SECTION 4.8.2

- 110 v kills.
- Draw to scale (no topological idealizations).
- Show power, ground, parasitic resistance, everything.
- $V = IR$ (e.g., wire = equipotential iff no current flows).

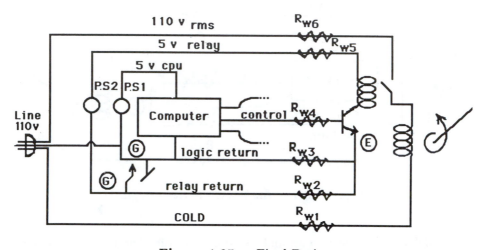

Figure 4.65. *Final Design.*

- There is a RESISTIVE COUPLING if two currents flow in the same R.
- A floating supply is a neat way of separating current loops.

4.8.3   Capacitive Noise Injectors

Parasitic capacitance can couple your circuit to all sorts of nasty noise sources. For example, fluorescent lights and extension cords have in them voltages of hundreds of volts, setting up fields throughout the room. If you randomly string a cable around the lab, you will get an open circuit voltage of 20 to 100 volts peak-to-peak, roughly independent of the length of cable. This couples to the cable through a series capacitance of roughly 75 picofarads per meter, which is an impedance of 35 M Ω-meters at 60 Hertz. The short-circuit is a few microamps per meter.

Since 35 M Ωs is a pretty big resistance, we can model this noise injector as a constant current source. Conclusion: there is an INEVITABLE noise of a few microamps per meter injected into every cable in the lab. What effect does this have? Consider Figure 4.66.

The distributed parasitic impedance of the ohmmeter lead is R_{w2}, at best 10 m Ωs per meter. The lower noise injector forces current out of the ground and into the cable. The current flows through R_{w2} and into the ground, completing its circuit. The resulting noise voltage at node B is a few dozen nanovolts. Big

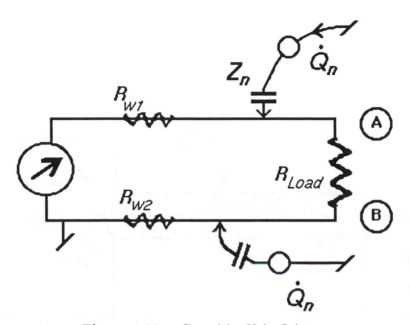

Figure 4.66.   *Capacitive Noise Injector.*

deal! Another noise current is injected into the upper ohmmeter lead. It cannot flow through the high impedance of the meter, so it returns to ground via R_{load} and R_{w2}. This produces a differential-mode noise voltage (from A to B) of $V_n = R_{load} \times \dot{Q}_n$. If R_{load} is kΩs or MΩs, V_n will be millivolts or even volts. This is clearly intolerable.

Since the noise current is more or less inevitable, the strategy is to let it flow, somewhere, just not through delicate high-impedance elements such as R_{load}. Coax cable is universally used for this type of measurement, as shown in Figure 4.67.

All the noise current flows in R_{w2}, producing a negligible noise voltage at node B. Note that the cable capacitance C_{12} (inner to outer) is not small: 100 pf per meter, which is comparable to Zn, the impedance of the noise injector. So why doesn't the noise voltage on the outer leap across C_{12} and then (horrors!) flow through R_{load}? Well, it does, but the voltage on the outer is really tiny. It is reduced from the open-circuit stray-field noise voltage by a divider ratio Z_n/R_{w2}, so we are about a billion times better off in Figure 4.67 compared to Figure 4.66.

That is the EE's way of looking at it. The physicist's way of looking at it is Figure 4.68.

Capacitance is defined such that $Q = CV$. Object 1 has a self capacitance: $Q_1 = C_{11}V_1$. There is a mutual capacitance from 1 to 2: $Q_2 = C_{12}V_1$. You can visualize mutual capacitance as electric field lines starting on object 1 and

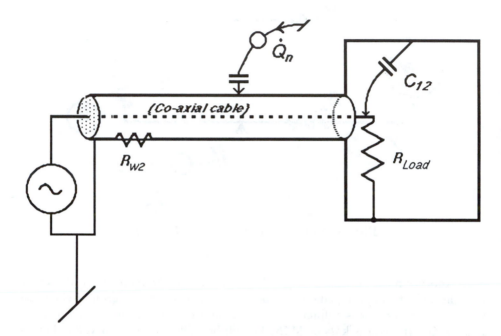

Figure 4.67.   *Coaxial $\dot{Q}$ Shielding.*

landing on object 2. Interposing a shield at constant voltage V_3 greatly reduces the value of C_{21}. This means that object 1 (the noise source) will have much less effect on object 2 (the vulnerable node of your circuit). This tells us how to make a good shield: 1) Make sure no lines of force can get past the shield and terminate on the vulnerable node; and 2) hold the shield at a constant voltage.

Sometimes for one reason or another, the thing we are measuring, R_{load}, has one side grounded. As will be discussed at length in the next section, we don't want two grounds, so we must float everything else. By the way, every BNC connector exposes the outer conductor, so you must always prevent these from touching ground and each other, which is a pain. This in itself is a good reason for using twinax rather than coax.

Figure 4.69 shows a generic experiment, consisting of a load (R_{load}), a signal source (upper instrument), and a scope (lower instrument). In the upper instrument, the circuitry is enclosed in a separate $\dot{Q}$ shielding box, within the chassis.

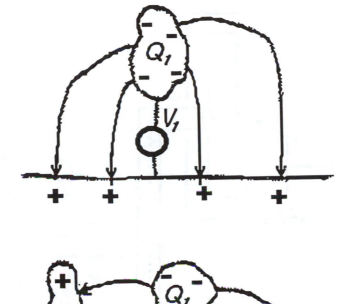

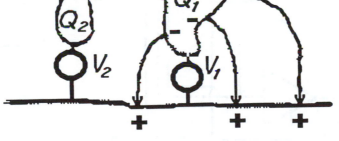

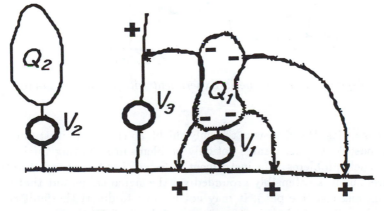

Figure 4.68.   *Mutual Capacitance.*

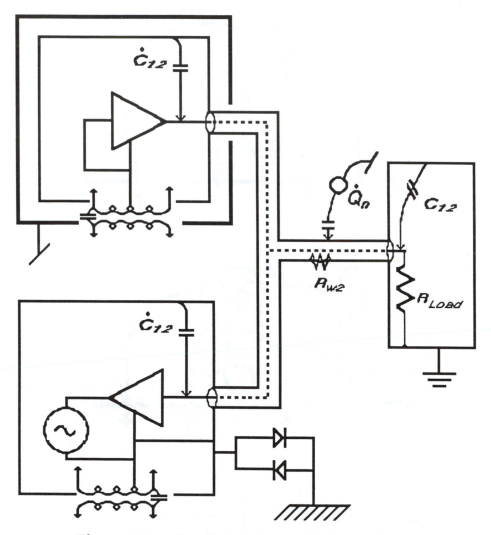

Figure 4.69.   *Generic Experiment with Grounded Load.*

In the scope, alas, the chassis is the only shield, so the whole instrument, knobs
and all, must float. This creates not only problems in mounting the instrument,
but also safety problems. The instrument is designed to be grounded to the rack.
In Figure 4.69 it is ostensibly grounded via the signal cable, but that is a long,
uncertain, and resistive path. It may not be able to divert the fault current or
blow the fuse. It might be wise to put crossed diodes from the chassis to ground,
big ones, 30 amps at least.

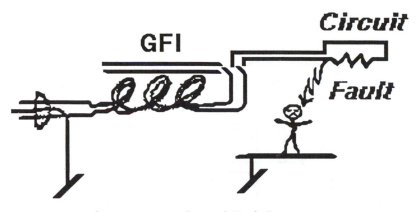

Figure 4.70. *Ground Fault Interrupter.*

There exists such a thing as a ground fault interrupter (GFI), shown schematically in Figure 4.70. Normally, power current flows out the hot side of the line and is returned via the cold side of the line. You should never come in contact with either, so to make you part of that loop would require two faults. Far more likely is that current would flow out the hot side of the line, through you , and be returned via terra firma. The GFI measures the differential current between the hot and cold lines, and if there is an imbalance it trips the relay, cutting all power.

If you ever need to run a two-wire extension cord through a swamp, I recommend using a GFI. On the other hand—

- GFI protection does not extend across transformers.
- GFI will prevent fires and electrical burns, but it is not clear that it acts fast enough to prevent electrocution in all cases.
- The threshold on most GFIs is set shockingly high (to keep Tek scopes from tripping them).

If all instruments were battery operated, or otherwise truly floating with respect to rack ground, Figure 4.69 would work just as nicely as Figure 4.67. Unfortunately, there is another class of capacitive noise injector—power supply transformers, as indicated in the figure.

Some transformers just consist of gobs of wire laid around each other, so that in addition to the desired transformer action (mutual inductance), there can be a huge parasitic interwinding capacitance. Most instruments, and even the open-frame power supplies in the Cornell LASSP stockroom, aren't too bad, but Tektronix scopes are notorious for this. An interwinding capacitance of 15000 pF, with an effective open-circuit noise injector voltage of 30 volts pp, yields a 60 Hz noise current of roughly 175 microamps pp. This is about 100 times worse

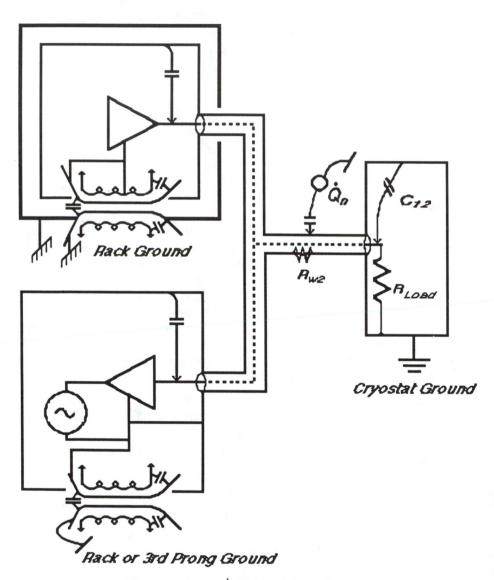

Figure 4.71. $\dot{Q}$ Shielded Transformers.

than the stray-field effects discussed in the previous section, so even the IR drops in R_{w2} can become significant, a microvolt or so.

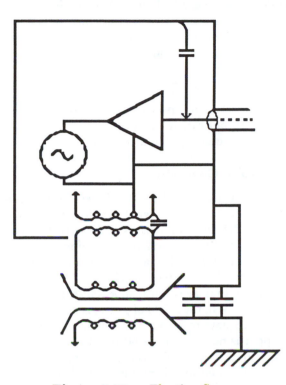

Figure 4.72.   *Floating Scope.*

It is possible to reduce the interwinding coupling by at least 5 orders of magnitude by controlling the geometry of the windings and by putting a shield around each coil (see Figure 4.71). The shield on the primary is returned to chassis ground (third prong) via a low-impedance path. The shield on the secondary is returned to circuit common via a low impedance path. There is no interaction between these (complete!) loops and the rest of the circuitry. Look in Morrison [see reference at end of chapter] for details. Larry Friedman once trenchantly summarized Morrison by saying, "Design everything with triax and ultrashielded supplies, and you'll be OK." Well, what if you're stuck with a cryostat that's not wired quite right, and a Tek scope? How do you make a scope float?

A large glass of mouthwash and two scoops of ice cream.

Don't—float everything else instead.

Don't—get a signal transformer or a guarded preamplifier and hook the scope to the out put of that—see next section.

Use an external shielded power transformer, Figure 4.72.

Suitable transformers are available from Xentec, Signal, TRW-UTC, Topaz, and probably N other vendors. Effective interwinding capacitance is 10^{-1} to 10^{-4} pF; cost is on the order of 100 bucks. In general you need one for each unshielded instrument hooked to a delicate signal.

SUMMARY OF SECTION 4.8.3

- Capacitive noise injectors behave as small constant current sources.
- They can be shorted out rather easily.
- Power transformers require special attention.

4.8.4   Inductive Noise Generators

We turn now to parasitic mutual INDUCTANCES, which are another route for noise entering your circuit. This is what people often refer to as the dreaded ground-loop. In fact, ANY loop will act as a one-turn transformer or antenna, and any changing flux, $\dot{\phi}$, through the loop will induce a spurious non-potential voltage around it. Typical 60 Hz fields in the H corridor are $3mV/m^2$. Near some transformers the level can be 10 times this size.

You may be thinking that we can just short out this $\dot{\phi}$, just as we shorted out the $\dot{Q}$ in the last section. Well, the shield of a coax is great for carrying a screening charge, but it is not adequate to set up a screening flux. For one thing, the skin depth in purest copper is nearly 1 cm at 60 Hz (1 mm for iron). Magnetic fields freely penetrate anything thinner. To keep flux from reaching your circuit, you would need to put it in a solid metal can made of 1/2 inch thick iron, or several inches of copper thickness. The correct model for this form of noise injector is a good constant voltage source (low Z). The effective impedance is immeasurably small, $\ll \mu\Omega$. Suppose we put a loop of wire on the floor: 3 millivolts across $30m\Omega s = 0.1$ amps. That surely seems ridiculous, but that's what you get.

When faced with 5 mV of 60 Hz noise in the circuit of Figure 4.73, most people instinctively say, "Oh, it's not grounded properly," and make the ground strap bigger. In fact, since ϕ is fixed, doing that only increases the current in the loop, increasing the drop in R_{w2}. The correct strategy is just the opposite. Cut away the ground strap entirely! Then both ends of R_{load} will be hopping up and down 5 mV with respect to local cryostat ground, but that does not produce any observable current or voltage at the meter. Any net flux that gets between the leads of the voltmeter will produce noise, but the effective sideways area between conductors in a coax or twisted pair is nil, so by cutting open the obvious ground loops we eliminate the noise. What can you do if you are stuck with more than one ground? The best and easiest thing to do is use an isolation transformer in the signal coax, as suggested in Figure 4.74.

The ground-loop currents cannot flow across it, so there can be no IR drops in your conductors. The DIFFERENTIAL voltage on one coil is transformed across to the other. The $\dot{\phi}$ contributes a common mode voltage, so it subtracts out. The limitation to this technique is that the core of the transformer will saturate at low frequencies and/or high power levels, although if you have an audio signal

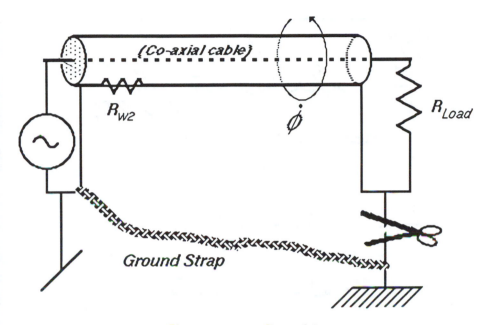

Figure 4.73.   *Ground Loop.*

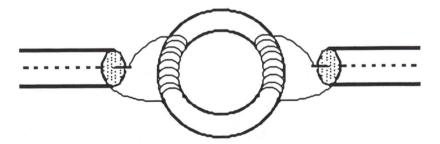

Figure 4.74.   *Isolation Transformer.*

large enough to saturate a Geoformer, you probably aren't too worried about 5 mV of pickup. (Geoformers are available from Triad; at IF and RF you can wind your own or check Mini Circuits Lab.)

If your signal is at a very low frequency, just replace the transformer in Figure 4.74 with a guarded amplifier. That is basically a precision subtractor

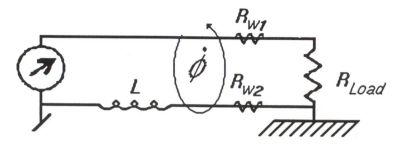

Figure 4.75.   *Stopping Ground Current.*

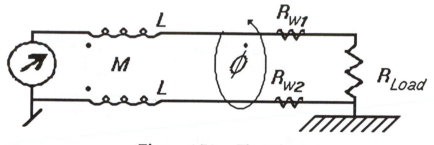

Figure 4.76.   *The Balun*

circuit with all the shielding done right. For 10 bucks you can get a chip with accuracy of 1 part in a couple million (125 dB CMRR at DC, 75 dB CMRR at 10 KHz).

If you need real bidirectional DC continuity, there is one more trick. The idea is indicated in Figure 4.75. Since we know the voltage drop, f-dot, is inevitable let's just collect it all in one place, using an inductor. To be effective at that, the impedance of the self-inductance must be large compared to R_{w2}, i.e.,

$$2\pi 60\ Hz >>> \omega_{Balun} = R_{w2}/L. \qquad (4.17)$$

But hang on a minute. This defeats the whole purpose. The noise voltage appears across L, but that is in series with the measuring circuit! The trick is this: since all the voltage drop is concentrated across L, we can subtract it out using a transformer, as shown in Figure 4.76.

This looks similar to Figure 4.74, but really it isn't. It's called a balun, and is simpler to make than it looks. You just loop a coax bodily through a ferrite bagel N times. This clearly produces a 1:1 turns ratio. Also note that differential currents (flowing down the inner and returning via the outer) induce

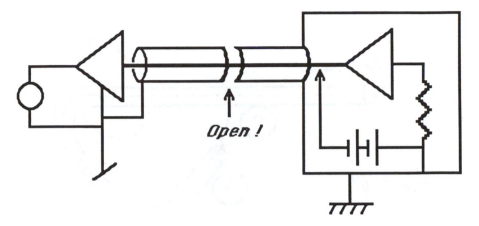

Figure 4.77. *Floating Preamplifier.*

no flux in the ferrite, so there is NO NET INDUCTANCE seen by your signal. EE's would say that the mutual inductance cancels the two self inductances. It is a wonderful thing, but ordinary coax is so resistive that you would need an enormous inductance to achieve useful noise reduction at 60 Hz. Note that at high enough frequencies, the self-inductance of the coax itself acts like a one-turn balun, even without the core.

One of my favorite tricks for breaking ground-loops is to use a battery-powered floating preamplifier. The ground-difference voltage still appears across your cable, but if the gain of the preamp is big enough, you don't care.

A final trick used for isolating circuits is the employment of OPTO-ISOLA-TORS. They provide excellent isolation but are terribly nonlinear. These devices are great for digital circuitry, if you ever need to isolate your logic lines. There exist ways to linearize optos using feedback and matched pairs, but transformers are usually more suitable.

4.8.5   Measuring Techniques

It is not entirely trivial to observe and quantify ground-loops and other noise injectors. You can prevent them by design, usually, but if you inherit an apparatus, it can be hard to diagnose its ills. Remember that a scope all by itself is an instant ground-loop looking for a place to happen, so using a scope to find ground loops can be rather comical. Here are the methods I used to measure the 60 Hz fields in my lab.

To observe the capacitive noise injector, I draped a 4 meter cable around the room. It is hard to measure its open-circuit voltage since it has an impedance of 35 M Ω-meters and the scope has an input impedance of only 1 M Ω. What you

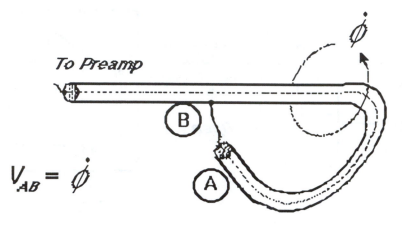

Figure 4.78.   *Split-Loop Antenna*

need is a 100X scope probe. The low temperature group owns one, but I couldn't find it, so I made one by sticking 90 M Ωs in series with a standard 10X probe.

To estimate the effective impedance of the injector, I just put various little capacitors across the probe until I observed a factor of two in attentuation. I measured the short-circuit current directly with a current probe. I used one that works on the wrap-around ammeter principle. It requires a bit of thought to use it, because 60 Hz is below its nominal passband and its sensitivity is pretty poor even in-band. There is a nice one based on the Hall effect which is good down to DC, but I did not have one.

It was gratifying that the separate measurements of V, I, and Z were consistent.

To measure the inductive effects, I used what is called a split-shield antenna. See Figure 4.78.

This device is generally useful. It separates a magnetic field from an electric field. The end of the inner is connected to the outer at point B. $V_{ab} = \dot{\phi}$. The outer screens out the electric field, especially if you put a little box around the joint, grounded to ONE of the outers.

I made no attempt to measure an impedance for $\dot{\phi}$, because even in the absence of skin effects, you can show that the short-circuit current you observe is limited by the self-inductance of the observer, not by anything pertaining to the source.

In class I showed various thingamajigs I made to facilitate inserting the current probe into the circuit. A cute trick for measuring a differential voltage, when you don't have a guarded amplifier you trust, is to put a resistor from A to B and measure the resulting current with the current probe.

4.8.6   Some Useful References

There are two books which have rather interesting discussions about grounding problems.

- Ralph Morrison, *Grounding and Shielding Techniques in Instrumentation, 2nd Edition*, John Wiley & Sons, New York, 1977. This is a clear and sensible analysis of the problem but the book is somewhat narrowly focused.
- Henry W. Ott, *Noise Reduction Techniques in Electronic Systems*, John Wiley & Sons, New York, 1976. This contains a lot of good ideas, but also a lot of confusing excess discussion.

Chapter 5

Thermometry

Eric L. Ziercher, Kenneth I. Blum,
and Yue Hu

5.1 Primary Thermometers

by Eric L. Ziercher

Thermometry is a central concern of low temperature physics. Physical quantities are often studied versus temperature, and a common place to question experimental results is in the thermometry. I will focus on the user's viewpoint. A major region of the discussion will be the superfluid ^{3}He regime (i.e., below 3 mK).

A reasonable question at this point is: "What is a primary thermometer?" The answer: it is a thermometer that is based on a well understood physical principle and is reliable. In other words, it gives us the absolute temperature and gives repeatable results.

Several temperature scales for the superfluid ^{3}He regime have been proposed; four of major interest are:

- The Helsinki scale: based on nuclear orientation thermometry.
- The La Jolla scale: based on noise thermometry.
- The Cornell scale: based on ^{3}He melting curve thermometry.
- The Greywall scale: based on the properties of liquid ^{3}He.

All these scales give values for $T_c(P)$ for superfluid ^{3}He, thus intercomparison is easily done. The thermometers used in the first two scales were unable to function in the superfluid ^{3}He regime, so the primary thermometer was used to calibrate a secondary thermometer which was used at lower temperatures. These scales should be viewed as an evolving set of knowledge. Additional work has been done on them since their original publication and the values for $T_c(P)$ have changed.

Thermometry for these scales is discussed below. I will focus, primarily, on melting curve thermometry. In the higher temperature regions (i.e., above .2 K),

^{3}He and ^{4}He vapor pressure thermometry is used as described in Betts (1976). Additional details of the first three temperature scales from a user's viewpoint are given by Richardson (1977). Greywall's (1986) analysis is the most recent and is probably the closest to the 'truth'. It is the one we employ in our work at Cornell in 1987. An overview of low temperature thermometry is given by Hudson *et al.* (1975) and Soulen (1982).

5.1.1   Nuclear Orientation Thermometry

The most commonly used type of nuclear orientation thermometry is based on the anisotropic γ-ray decay from radioactive nuclei. Nuclear energy levels are split by a magnetic field into $(2I + 1)$ hyperfine levels, where I is the spin quantum number of the nucleus. Each sublevel will have its own radiation pattern with independent variable θ (the angle from the magnetic field direction). The radiation pattern from all the levels is described by a function $W(\theta, T)$ where T is the absolute temperature and enters due to the differing occupations of the sublevels. At the temperature extremes $W(\theta, T)$ is independent of T: at high temperatures all levels are equally populated and at low temperatures only the lowest level is occupied. Thus, the choice of a suitable nucleus is based on the temperature region of interest and a good knowledge of its decay (which allows calculation of the $W(\theta, T)$ function). Two commonly used nuclei are ^{54}Mn and ^{60}Co.

There are several major factors that govern the experimental application of this technique. A metallic host is used to provide conduction electrons which facilitate the spin-lattice relaxation of the nuclei. The host should also be ferromagnetic so that a uniform local field for the nuclei is achieved. A .1 T to 1 T magnetic field is typically used to saturate the host, but a properly shaped ^{59}Co host provides an orientation axis that does not require an external field. Another concern of sample selection is heating effects. A trade off exists for short sampling times for a "hot" source versus a low heat leak for a "cool" source. Heat leaks of 10-1000 pW are typical with counting times on the order of minutes (see Berglund *et al.* 1972). Finally, γ-detectors and counters are required, thus a lab that had this equipment readily available might easily use this thermometry scheme.

More details of this method may be found Berglund *et al.* (1972), Weyhmann (1974), and Marshak (1983); overviews are given in Betts (1976) and Lounasmaa (1974). The National Bureau of Standards has established a temperature scale based on a comparison of nuclear orientation thermometry and noise thermometry (Soulen and Marshak, 1980). These two thermometers were found to agree to within .5% from 10 to 50 mK. They were used to calibrate a CMN thermometer and germanium resistance thermometer, which then represent their NBS-CTS-1 temperature scale which spans a 10 to 500 mK range. This scale is made portable through their SRM768 device (Soulen 1982) which can be bought from NBS (NBS Spec Pub 260-62 (1969), device stock number 002-002-02047-8). This device consists of five superconducting fixed point references that span a

15 to 208 mK range. Users should note that this device must be used in fields of less than 1 μT. Each device is individually calibrated and NBS publishes new calibration temperatures for the device as the NBS temperature scale evolves.

5.1.2   Noise Thermometry

Background and Theory Nyquist's Theorem states that due to the Brownian motion of conduction electrons, one will detect a randomly fluctuating voltage (Johnson noise) across a resistor. The average of the voltage squared is given by $\langle V^2 \rangle = 4k_B T R \Delta f$ where k_B is Boltzmann's constant, T is the absolute temperature, R is the resistance, and Δf is the bandwidth of the measurement electronics. Thus, measurements of the fluctuations yield the absolute temperature of the resistor. In practice this method has several difficulties:

- As noted in the equation above, one is dealing with an averaged quantity and often long averages are needed to get good temperature resolution.
- The system transfer function must be well characterized. In particular, the bandwidth and gain must be accurately known.
- One must eliminate non-thermal noise (such as $1/f$ noise or external interference) from the final results.
- Uncertainties in the system sometime require a calibration against a fixed point or against a 'high temperature thermometer' (such as ^{3}He vapor pressure).

Current Based Devices (Webb, Giffard, and Wheatley 1973) The system arrangement is shown in Figure 5.1. The current noise of the resistor creates a fluctuating magnetic flux in the SQUID magnetometer which is operated in flux locked loop mode. A noise current is measured:

$$\langle I^2(f) \rangle = \frac{4k_B T}{R} \frac{\Delta f}{1 + (2\pi f \tau)^2} \tag{5.1}$$

where f is the frequency at which the measurements are done and $\tau = L/R$. Integration over all frequencies gives $\langle I^2 \rangle = kT/R\tau$ which can be compared to experiment. Webb et $al.$ (1973) found that the noise due to other sources had a different frequency spectrum than the Johnson noise. The extraneous noise was isolated and was found to correspond to a temperature of 50 μK. Statistical calculations gave a 2500 sec sampling time for a 1% temperature precision. Calibration of this thermometer against a CMN thermometer showed the expected linear dependence until the temperature dropped below 15 mK. Below 15 mK the noise thermometer showed a higher temperature than the CMN which was consistent with a 4×10^{-14} watt heat leak into the resistive sensor.

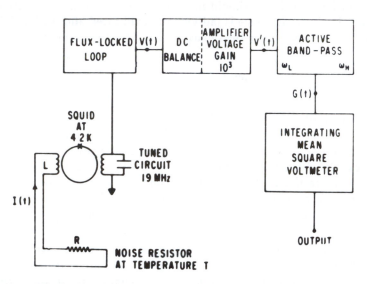

Figure 5.1.   *Block diagram of current based scheme of noise thermometry. [Webb, R. A., R. P. Giffard and J. C. Wheatley, J. Low Temp. Phys. 13, 383 (1973).]*

Voltage Based Device (Soulen 1978, Soulen and Marshak 1980) In this method the Josephson junction is resistively shunted and the voltage fluctuations are read via a frequency method (see Figure 5.2). The junction is given a small dc bias I_0 and a total voltage of $I_0 R + V_N(t)$ is seen; V_N is the noise voltage. The junction oscillates at a frequency $f = (I_0 R + V_N)/\phi_0$; $\phi_0 = h/2e$. The frequency is repeatedly measured and the mean square deviation of the measurements is given by $\sigma^2 = (2k_B T R / \tau \phi_0^2)$; τ is the gate time. From this equation one obtains the absolute temperature T. Extraneous noise occurs since the frequency counter can only count an integral number of cycles. This gives a lower temperature bound of $\Delta T_{min} = \phi_0^2/24k_B R\tau$. Using fast period counters, ΔT_{min} can be made quite small and a measurement can be done in a few minutes. The rms scatter of values of T is given by $\Delta T/T = (2/N)^{1/2}$ where N is the number of measurements of frequency done to find σ^2. For $\Delta T/T = 1\%$, (e.g., a temperature resolution of 20 μK at $T = 2$ mK) and $\tau = .01$ sec one finds $N = 20000$ and a measurement time of $N\tau = 200$ seconds (Soulen 1986).

DC SQUID Noise Thermometry (Roukes *et al.*1984) A design for a highly sensitive dc SQUID noise thermometer is being developed. A figure of merit for SQUID noise thermometry is: $T_N t_{meas} \propto \varepsilon_c/\sigma^2$ where T_N is the noise temperature and t_{meas} is the measurement time. Using $\varepsilon_c \sim 10h$ for state of the art coupled energy sensitivity and $\sigma = .01$ for a 1% temperature precision one finds

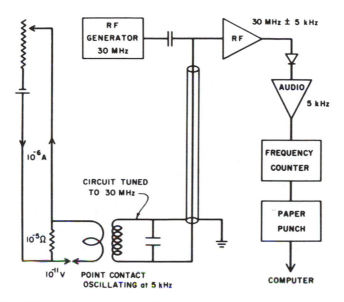

Figure 5.2.   *Block diagram for voltage based noise thermometry. [Kamper, R. A. and J. E. Zimmerman, J. Appl. Phys. 42, 132 (1971).]*

$T_N \, t_{meas} = 5 \, \mu K$ sec. Thus it should be possible to equal the noise temperature of the system described by Webb *et al.* (1973) using a much shorter measurement time.

5.1.3   Melting Curve Thermometry

Background A melting curve thermometer (MCT) is a thermometer based on the properties of the ^{3}He melting curve (see Figure 5.3). Note that in the two temperature regions 0 to T_{min} and T_{min} to large T there is a unique temperature associated with each pressure. Thus, the temperature is determined by measuring the pressure of a sample of ^{3}He that is on the liquid-solid coexistence curve. Since very sensitive pressure transducers can be made, it is reasonable to use this phenomenon as the basis for a thermometer. The melting curve is a basic property of ^{3}He and is only slightly affected by the presence of ^{4}He impurities (the pressure at T_{min} will be decreased slightly but T_A will not be affected, see Richardson 1977). Thus, an advantage to this method is that the results are largely insensitive to sample preparation. Disadvantages are sensitivity to magnetic field (operation in up to 5 kg fields is reasonable) and moderately long time constants (on the order of minutes). The pressure vs. temperature characteristic of the melting curve can be determined from thermodynamics and basic measurements.

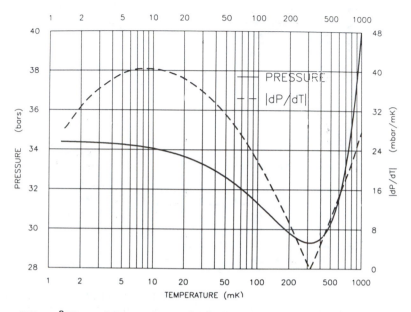

Figure 5.3. ^{3}*He melting curve and its derivative. Generated from the spline fit in Table 5.1 and the data of Grilly (1971).*

Construction One of the standard designs for the MCT sensor is found in Greywall and Busch 1982, as illustrated in Figure 4.53. Another variation, used at Cornell, is the one designed by Peter Gammel shown in Figure 5.4. The pressure is sensed by a capacitive strain gauge of the Straty-Adams design (1969). A reference capacitor can be heat sunk to the refrigerator for stability. A typical gas handling system is shown in Figure 5.5 and its usage is described below.

Calibration and Usage During the cool down the MCT can be filled and used as an ideal gas thermometer to show how cooling is progressing. At 1 K the MCT diaphragm should be exercised to reduce hysteresis. This consists of cycling the MCT pressure between 29 and 35 bar about ten times at about a bar/minute rate. Calibration of the pressure vs. capacitance characteristic of the thermometer is done next. A dead weight tester is a common high precision pressure standard to use for this. Another standard which is slightly less precise but much quicker to use is a Paroscientific quartz pressure transducer (Paroscientific Inc., 4500 148th Ave. N.E., Redmond, WA 98052).

The final step is to use T vs. P melting curve data so that one finally obtains a temperature vs. capacitance calibration. Since $T(P)$ data are evolving, it is wise to keep data in terms of pressure so that it can be reanalyzed at a later date with new $T(P)$ data. Smoothed fits to the region $T < T_{min}$ are given in Tables 5.1, 5.2 and 5.3. Careful construction and calibration can yield resolutions of .1 μK

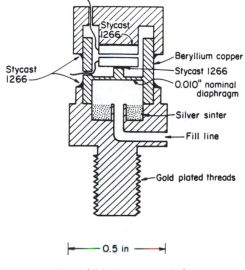

Figure 5.4.   *Melting Curve Thermometer transducer of Peter Gammel.*

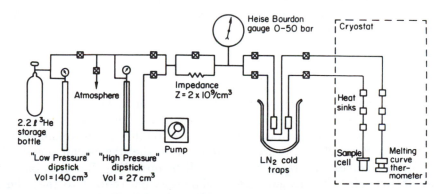

Figure 5.5.   *Gas handling system for sample cell and melting curve thermometer. [after Sagan, 1984]*

and accuracies, ignoring errors in the fixed point (see the discussion below), of a few μK (Greywall 1985, Sagan 1984).

Table 5.1.   *Spline Fit for T(P) Melting Curve Data [after Wildes, 1985].*

P_0(Bar)	A	B	C	D
29.45000	254.8733	-223.6473	368.3388	-528.6911
29.60000	227.8293	-148.8319	130.4253	-115.7982
29.80000	202.3534	-110.5575	60.9461	-35.1547
30.10000	173.7222	-83.4816	29.3070	-10.0690
30.60000	138.0495	-61.7263	4.2035	-3.0948
31.30000	100.7393	-46.3908	7.7044	-1.3063
32.00000	71.5928	-37.5248	4.9612	-0.6685
33.00000	38.3608	-29.6077	2.9558	-0.2243
33.50000	24.2678	-26.8201	2.6194	-0.9323
33.90000	13.8996	-25.1721	1.5008	-1.8569
34.15000	7.6713	-24.7699	0.1081	-19.4619
34.25000	5.1756	-25.3322	-5.7313	-56.4450
34.31500	3.4896	-26.7923	-16.7359	-163.0204
34.35000	2.5241	-28.5637	-33.8585	-1038.3926
34.39400				

Note 1:   See Note 1 under Table 5.2.

Note 2:   Evaluation of spline fit. Read down the P_0 column until you reach a value greater than the pressure you wish to convert, then move back one row.

$\Delta P \equiv P - P_0$, $T(P)=A+\Delta P \times (B+\Delta P \times (C+\Delta P \times D))$. $T(P)$ is given in mK.

Note 3:   The range of validity is from 1.2 to 230 mK. Errors are within ± 10 μK for $T < 230$mK and within ± 1.5 μK for $T < 14$ mK.

Once the calibration is complete, the MCT is pressurized to some value based on your desired operating range. Referring to Figure 5.6, the horizontal lines refer to pressures in bars, and the ratio a/b gives the fraction of the volume that is solid. One must select a pressure that remains in the liquid-solid coexistence region in the temperature range that you will be operating. However, one also wishes to minimize the amount of solid since thermal contact to the solid is poor. If too low a pressure is used the MCT will "fall off the melting curve." For example, if 30 bar was used and one cooled below .15 K, all the solid in the cell would melt and the cooling trace on the bridge output would suddenly flatten. The plugged capillary method is used to seal the MCT at the desired pressure, thus the fill capillary must be heat sunk in several places to ensure solid formation.

Fixed Point Calibration Due to pressure head effects, the pressure vs. ca-

Table 5.2. *Spline Fit for P(T) Melting Curve Data from Wildes 1985*

T_0(mK)	A	B	C	D
1.0000	34.397390	-2.92616×10^{-2}	4.98414×10^{-3}	-6.26180×10^{-3}
1.5000	34.383222	-2.89738×10^{-2}	-4.40857×10^{-3}	1.00537×10^{-3}
2.2786	34.358467	-3.40102×10^{-2}	-2.06033×10^{-3}	4.40024×10^{-4}
3.0000	34.333024	-3.62960×10^{-2}	-1.10798×10^{-3}	1.13416×10^{-4}
5.3830	34.241771	-3.96445×10^{-2}	-2.97159×10^{-4}	4.33486×10^{-5}
7.5000	34.156925	-4.03198×10^{-2}	-2.18574×10^{-5}	7.45226×10^{-6}
12.5000	33.955711	-3.99795×10^{-2}	8.99266×10^{-5}	1.45851×10^{-6}
22.7857	33.555596	-3.76667×10^{-2}	1.34932×10^{-4}	-5.42809×10^{-7}
29.1200	33.322279	-3.60226×10^{-2}	1.24617×10^{-4}	-2.24052×10^{-7}
74.3582	31.926967	-2.61233×10^{-2}	9.42097×10^{-5}	-1.86438×10^{-7}
99.4888	31.327011	-2.17414×10^{-2}	8.01538×10^{-5}	-1.39847×10^{-7}
136.6138	30.623179	-1.63682×10^{-2}	6.45784×10^{-5}	-9.44628×10^{-8}
177.2302	30.058565	-1.15899×10^{-2}	5.30682×10^{-5}	-6.19892×10^{-8}
231.6965	29.574723	-6.36068×10^{-3}	4.29392×10^{-5}	-3.95604×10^{-8}
256.2500	29.443847	-4.32362×10^{-3}	4.00252×10^{-5}	-5.51849×10^{-8}
330.0000				

Note 1: Meaning of errors in Tables 5.1 and 5.2. These splines are fitted to the data of Halperin (1978) and Greywall and Bush (1982). The quoted errors are due to and comparable to the scatter in the original data. Using the spline fit is superior to interpolating the data due to the smoothing nature of the spline.

Note 2: Evaluation of the spline fit. Read down the T_0 column until you reach a value greater than the temperature you wish to convert, then move back one row.

$\Delta T \equiv T - T_0$ $P(T) = A + \Delta T \times (B + \Delta T \times (C + \Delta T \times D))$. P is given in bars. The fit and its first derivative are smooth; its second derivative is continuous.

Note 3: The range of validity is from 1.2 to 330 mK. Errors are within ± 80 μbar for $T < 330$ mK and within ± 60 μbar for $T < 24$ mK.

pacitance calibration will be offset by some small amount. The most common way to handle this is to recalibrate through the use of a fixed point. Four fixed points that can be used are: the A transition, the B transition (warming only), the solid transition, and the melting curve minimum. The signatures of the three transitions are shown in Figure 5.7. Then by using the published fixed point temperature from your favorite reference you can recalibrate and have a functional melting curve thermometer.

T(P) Melting Curve Data The determination of $T(P)$ for the melting curve

Table 5.3.   *Melting Curve Polynomial Fit from Greywall 1986*

$$a_{-3} = -.1965270 \times 10^{-1}$$
$$a_{-1} = -.78803055 \times 10^{-1} \qquad a_{-2} = .61880268 \times 10^{-1}$$
$$a_0 = .13050600$$
$$a_1 = -.43519381 \times 10^{-1}$$
$$a_2 = .13752791 \times 10^{-3}$$
$$a_3 = -.17180436 \times 10^{-6}$$
$$a_4 = -.22093906 \times 10^{-9}$$
$$a_5 = .85450245 \times 10^{-12}$$

$P - P_A = \sum a_i \times T^i$ P is in bars and T in mK. The fit is valid from 1 to 250 mK and the residuals are are less than .25%.

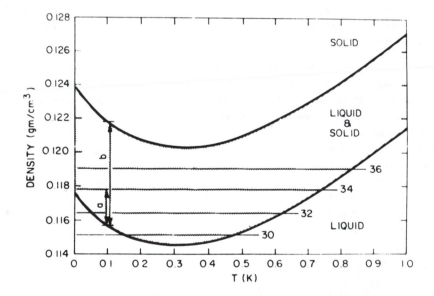

Figure 5.6.   *Density of liquid and solid ^{3}He along the melting curve. [Greywall & Busch, 1982 (same as Fig. 4.53).]*

has been done in several ways (Betts 1976). One method is based on thermodynamics and measurements of latent heat of solidification of ^{3}He (Halperin *et al.* 1978). Using the Clausius-Clapeyron relation the slope of the melting curve is given by $dP/dT = (s_s - s_l)/(v_s - v_l)$ where the differences in molar entropy and

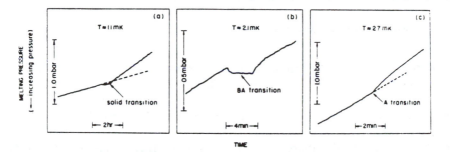

Figure 5.7.   *Signatures of transitions in a melting curve thermometer. Traces show melting pressure for a slow warming trend. [after Greywall, 1986]*

molar volume are between the solid and liquid phases. A servo system is used so that pressure (and thus temperature) in the cell is maintained constant when a heat pulse is injected. During this process Δn moles of liquid are converted to solid and we have $\Delta n = \Delta V/(v_s - v_l)$ where ΔV is the change in sample volume that is measured. The heat input matches the latent heat $\Delta Q = \Delta nT(s_s - s_l)$ and we finally get $T(dP/dT) = \Delta Q/\Delta V$. By integrating the data it is possible to calculate the temperature at any pressure with respect to a reference pressure. One usually uses T_A as a reference and defines $T^* = T/T_A$. T_A is determined with the Clausius-Clapeyron equation and with measurements of $(v_s - v_l)$ and the entropy of liquid ^{3}He, and a knowledge of the behavior of the entropy of solid ^{3}He. Greywall (1985) has renormalized this scale slightly to force it to coincide with the NBS-CTS-1983 scale and to tie in with higher temperature data of Greywall (1982). More recently, Greywall (1986) has published a careful rationalization of a temperature scale based upon his measurements of the melting curve and the thermal properties of liquid and solid ^{3}He. The coefficients of the polynomial of temperature versus melting pressure are listed in Table 5.3. At the present time, this is probably the most reliable temperature scale to use.

Miscellaneous Design and Usage Considerations Epoxy should be kept out of the fringing fields region of the capacitor to eliminate effects due to the temperature dependence of the dielectric constant. Stycast 2850 should not be used at all due to its highly lossy nature. Heat sinks to the fill line are 50 cm lengths of capillaries wrapped and hard soldered to copper cold fingers with 50 to 100 cm lengths separating the heat sinks. Heat leaks can come from the fill line and the signal line. Rough calculations give a 10 pW heat leak for 75 cm of 4 mil I.D. capillary heat sunk to the mixing chamber (Sagan 1984). Another 10 pW can come from the signal line. Local experience shows that a bridge excitation 4 volts RMS at 5 kHZ works fairly well. One might consider the elimination of the reference capacitor by the use of a differential capacitor in the MCT transducer (see Gonano and Adams, 1970, for details on this technique).

5.1.4   Some Useful References

- Betts, D. S., 1976, *Refrigeration and Thermometry below One Kelvin.*
- Greywall, D. S., 1986, *Phys. Rev.* **B33,** 7520.
- Lounasmaa, O. V., 1974, *Experimental Principles and Methods Below 1 mK.*
- Soulen, R. J., "Noise in Physical Systems," 1978, *Proc 5th Intl Conf on Noise,* 249.
- Webb, R. A., R. P. Giffard, and J. C. Wheatley, 1973, *Journ Low Temp Phys* **13,** 383.

5.2   Magnetic Thermometry

by Kenneth I. Blum

What, art thou mad? art thou mad? Is not the truth the truth?

—Falstaff

5.2.1   CMN Thermometry

There was a time in the not so distant past when it was believed, or perhaps hoped, that the electronic paramagnetic susceptibility of certain salts would provide a nearly ideal thermometer at very low temperatures—a thermometer which above all could be understood from theoretical first principles. Although this has proved not to be the case, the cerous magnesium nitrate (CMN) and lanthanum diluted CMN (LCMN) thermometers remain very suitable devices for many purposes down to the low millikelvin regime.

The main advantages of this type of thermometer are its sensitivity and its accuracy for work on liquid ^{3}He. Using bridges, tenths of a microkelvin at millikelvin temperatures can be resolved. These salts enjoy an anomalously low Kapitza boundary resistance with liquid ^{3}He, ensuring good thermal contact. To exploit fully the sensitivity of this thermometer, it must be carefully calibrated throughout the entire region of interest. The physical source of the high sensitivity at low temperatures is the Curie Law $1/T$ dependence of the susceptibility of an ideal paramagnet. Unfortunately, CMN and LCMN salt crystals, powdered for high surface area and good thermal contact, are hardly ideal single crystal noninteracting paramagnets. CMN thermometers often exhibit Curie-Weiss-like behavior from several degrees Kelvin down to tens of millikelvin, and similarly for LCMN down to perhaps 1/2 mK. Below these temperatures the seams begin to come apart; as the ordering temperature is approached it is advisable to calibrate against the NMR thermometers discussed below or the melting curve thermometer described in Section 5.1. (See Table 5.3 for Greywall's (1986) polynomial fit of pressure versus temperature.) Still more upsetting, the LCMN thermometers sometimes have sample and run dependent low temperature behavior. On the other hand, some groups have found that the superfluid ^{3}He transition temperature, measured with LCMN at various pressures, reproduces to a few μK over long periods of time (Parpia, 1987, private communication). Do not trust extensions of high temperature calibrations into the microkelvin region. Only trouble will result.

The formula for Cerous Magnesium Nitrate is $(Mg(H_2O)_6)_3 \, (Ce(NO_3)_6)_2$: $6H_2O$. The Ce^{+++} ions are paramagnetic with a ground state characterized by S=1/2, L=3, J_{tot}=5/2. Lanthanum has a 1s ground state and is thus not paramagnetic, though it may be weakly diamagnetic. If ninety to ninety-five percent of the cerium ions are replaced by lanthanum, a weaker electronic spin-

spin interaction and depressed ordering temperature result.

To use these thermometers for extremely low temperature applications the problem of fitting calibration curves arises. Higher order corrections to the Curie-Weiss expression should be tried first. This may be a good approach to take, but things will not always be this straightforward. A convenient way to examine data for Curie-Weiss-like behavior or deviations from it is to plot $1/\chi T$ versus $1/T$, emphasizing the low temperature end:

$$\frac{1}{(\chi - \chi_\infty)T} = \frac{1}{C} - \frac{\Delta}{C} \cdot \frac{1}{T}. \tag{5.2.}$$

Usually Δ is slightly negative; $\Delta \simeq -0.12 \; mK$ is a common value. The background susceptibility, χ_∞, is generally left as an additional fitting parameter. You will find that the value obtained in the fit differs from the measured susceptibility at 1 K, for instance. The material in the environment around the salt thermometer contributes to the background susceptibility and does not necessarily obey a Curie Law.

If you wish to make a thermometer from scratch (check with Biomagnetic Technologies Inc. and Oxford Instruments for commercially available systems) here are some guidelines. You can obtain $Mg(NO_3)_2$, $Ce(NO_3)_3$, and $La(NO_3)_3$ from any convenient chemical supply house. For CMN mix three molecules of $Mg(NO_3)_2$ with two molecules of $Ce(NO_3)_3$. Add enough water to form a saturated solution. Let the mixture stand at room temperature until crystals form. A common practice to improve the purity of the crystals is to discard the last of the liquid, redissolve the crystals, let them recrystallize, perhaps repeating the process many times. If you want a single crystal, the solution must be held at nearly constant temperature during the evaporation. For a powder, evaporate at room temperature, grind with a mortar and pestle, and separate with a sieve. Powder smaller than $45\mu m$ is generally used. This powder, which coheres at about a fifty percent packing fraction, can be packed into a right circular cylindrical pill. It might be best to pack the pill in the thermometer cell itself.

A cautionary word or two about hydration and dehydration: the water vapor pressure of CMN between 25°C and 85°C is well described by Betts (1976, p. 206):

$$lnP = 23.14 - 7046/T \tag{5.3}$$

where T is in Kelvin, and P is in torr. At 25°C this gives a relative humidity of 25%; if the environmental vapor pressure of water drops below this, dehydration will occur. If you pump vigorously on CMN crystals sixteen of the twenty-four waters of hydration pop off, changing the Curie constant, drastically increasing Δ, and possibly ruining the thermometer. One solution to this problem is to put the pill into a small chamber isolated by a large impedance. If the sample chamber is a few cubic centimeters, and is attached to a meter of four or ten mil fill line, dehydration will be retarded. In this situation the sample chamber develops a partial pressure of water vapor consistent with the above equation,

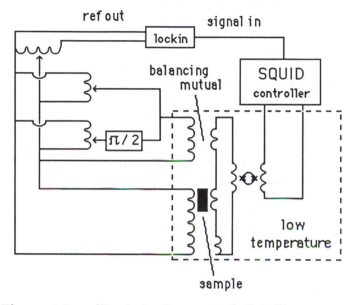

Figure 5.8.   *Circuit for Paramagnetic Salt Thermometer.*

but the number of molecules involved is a small fraction of those retained in the crystals. Diffusion through the impedance, particularly if it is filled with helium gas, is a very slow process, and the salt pill remains stable.

An example of a circuit used to monitor the temperature of an LCMN or CMN thermometer is pictured schematically in Figure 5.8. The sample coil and its counter-wound partner make up an astatic pair. This is the secondary coil which is driven by the primary coil wrapped around it. For an ideal empty astatic pair no emf will be induced by excitations from the primary. Of course, in practice, the cancellation is not exact so there is some small emf produced. This 'off-set' signal will be part of the geometry dependent factors which contribute to the value of χ_∞. Insertion of the sample in one half of the secondary changes the situation. The sinusoidal drive signal induces an ac magnetic field at the sample. The magnetization vector of the sample follows the alternating magnetic field and induces a net signal in the secondary coil. Since the salt sample is paramagnetic, the size of the secondary signal is proportional to χ.

The off-balance signal is due to flux penetrating the SQUID which is detected by the rf probe. You should be able to resolve from 10^{-4} to 10^{-6} of ϕ_0 ($\phi_0 \sim 2 \times 10^7$ $G - cm^2$). Measure the susceptibility by nulling the signal. Change the voltage appearing across the balancing mutual, both in-phase and out-of-phase, by changing each of the transformer ratios. The two phases of the lock-in should be aligned with these two arms of the bridge. The arm in phase with the drive measures the dispersive part of the susceptibility which is large and strongly temperature dependent. The out-of-phase signal should be very small in comparison and roughly constant. When ordering begins to set in at the lowest temperatures the out-of-phase component sometimes reaches as high as one percent of the in-phase component.

An example of a CMN thermometer cell is shown in Figure 5.9. Contact to ^{3}He is provided by an array of holes in the NbTi cap. There is some flux penetration through these holes which can be minimized by making them small compared to their length. To make this cell, cast a bar of Stycast 1269A epoxy from Emerson & Cumming. Machine it down. Drill a hole for the pill and ^{3}He, making sure to indent for the secondary and primary coils. Wind the coils. Fill in the machined indentations with fresh epoxy, and machine the piece into a cylinder to fit inside the superconducting shield. Use tissue paper or Spectramesh (obtained through Fisher Scientific Co. or another standard scientific supply company for about $50/3ft^2) to hold the LCMN pill in place. Spectramesh is a nylon mesh with a range of hole sizes to choose from; $10\mu m$ is good for this application. The mesh can be secured on top of the pill with small drops of Stycast 1266 epoxy. All of the wiring for this thermometer done at temperatures below 4.2 K should consist of NbTi superconducting wires twisted in pairs to reduce flux induced EMFs. Care should be taken to keep all solder joints superconducting as well. The twisted pairs should be slipped inside Nb or NbTi tubes to shield them.

Greywall and Busch (1982) observed some apparent anomalies in the susceptibility of CMN and conjectured that these were attributable to a proximity effect. The copper cladding on superconducting wire can be induced into a superconducting state, leading to a temperature dependent background. They did not have this trouble when copper wire was used for their CMN thermometer, but the excitation had to be dropped to a level where self-heating became acceptably small. Their LCMN thermometer was too insensitive for this remedy and was left unchanged.

A Nb shield would be better than the NbTi one shown in the figure. Type I superconductors are better than Type II superconductors since some flux penetration is possible for the latter between H_{c_1} and H_{c_2}. Also, the NbTi cap, labeled the inner shield, might be nicer as the outer shield. This would present a more uniform cavity, which is important for the nulling measurement since the asymmetric effect may not be independent of the field inside. The amount of epoxy in the secondary should be kept to a minimum because the epoxy is paramagnetic. To maintain the astatic pair cancellation the same amount of epoxy should be contained in each half of the secondary. Because 1266 and 1269 epoxies do not wet Teflon you can wrap the coils around a Teflon bar, pot the wire in epoxy, and later push out the Teflon coil form. In general, you should worry about the paramagnetic properties of materials used in the thermometer. Remember to cool through the superconducting transition of the shield with a mu-metal cylinder around the cryostat in order to reduce the magnetic flux trapped inside the shield.

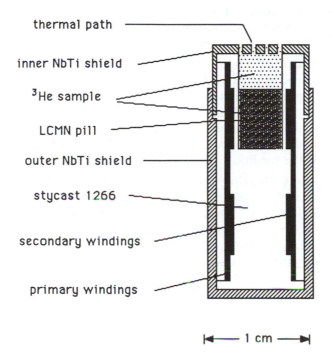

thermal path

inner NbTi shield

^{3}He sample

LCMN pill

outer NbTi shield

stycast 1266

secondary windings

primary windings

|◄——— 1 cm ———►|

Figure 5.9.   *CMN thermometer cell.*

5.2.2   Pt NMR Thermometry

Nuclear paramagnets obey the Curie Law to much lower temperatures than
do electronic paramagnets. Their usefulness extends well into the microkelvin
regime. The magnetic ordering temperature of nuclear dipoles is much lower
than that of electronic dipoles because nuclear dipoles are 1000 times smaller
than electronic dipoles. Unfortunately, the same factor of 1000 gets squared
in the relative signal sensitivity in a susceptibility measurement. The nuclear
signal is a million times smaller. Static measurements of nuclear susceptibility
are consequently quite difficult. There are huge impurity effects which are difficult
to remove. In this section nuclear resonance techniques are discussed, as they are
the most successful solution to this problem. Magnetic resonance is capable of
identifying the signal from a specific nuclear species because every isotope has
a unique resonant frequency. Thus, there are no background signals of the sort
which plague static techniques.

Metal samples are chosen for NMR thermometers because the conduction
electrons keep the spin-lattice relaxation time, T_1, short. The equilibrating mechanism is the hyperfine interaction between the magnetic moments of the nuclei

and conduction electrons. The electronic thermal reservoir is assumed to be maintained at the lattice temperature.

Because the thermometer is a metal several considerations compel the use of finely divided samples—powders or wires:

1. The Skin Effect Oscillatory EM fields decrease in strength inside a conductor with a characteristic length $\delta \sim (resistivity/frequency)^{1/2}$ called the skin depth. If the skin depth is not greater than the conductor size, several difficulties arise. The rf tipping field in the sample is severely attenuated, and so is the NMR signal. In the extreme case, the interior of the metal is not probed. If the resistivity depends on temperature the skin depth will as well. These problems are alleviated if the condition $\delta > sample\ size$ is satisfied.

2. Eddy Current Heating Currents induced in an ordinary conductor by changing magnetic fields lead to Joule heating. Calculation of these currents is extremely complicated in general (see Smythe, 1939). Assuming full field penetration, as the first point demands, the energy dissipated per volume decreases with decreasing sample size. In general, the power dissipated is proportional to the square of the smallest sample dimension transverse to the applied magnetic field.

3. Thermal Contact to ^{3}He Thermal equilibrium times are size dependent. Thermal resistance is inversely proportional to surface area, and heat capacity is proportional to volume. The thermal relaxation time in a lumped elements model is proportional to the product of these two. We can now compare the thermal time constants of many small particles and one large chunk of the same volume. For N small particles the resistance $\sim 1/Na^2$, where a is the characteristic length of the particle. If these particles are now pushed together in three dimensions to form one big chunk, all internal surfaces are lost and the resistance $\sim 1/N^{2/3}a^2$. Clearly, the heat capacities are identical, so the ratio of thermal time constants $\tau_{chunk}/\tau_{powder}$ is just $N^{1/3} = N^{1/3}a/a = A/a$, where A is the characteristic length of the chunk. Thus, for thermal contact as well, small particles are desirable.

There are lower bounds to the size of metal samples to use. Platinum is the favorite choice for an NMR thermometer because of its short T_1, long T_2, and because there is only one Pt isotope with a magnetic moment. The short T_1 permits more rapid sampling rates of the temperature and the long T_2 (or narrower line) permits one to use simpler electronic detection apparatus. But platinum particles exhibit conduction electron density oscillations at their surface when the sample size is less than $\sim 0.5\mu$m (see Yu et al, 1980). This effect broadens the resonance linewidth—i.e., shortens the FID—which leads to difficulties measuring temperatures, as we will see. If wires are used, thermal conductivity along their length becomes a problem if they are too thin.

One type of sample cell for Pt NMR thermometry measures the lattice temperature of a metallic experimental stage. An example is a thermometer developed by the Jülich group (see Buchal et al, 1978). This thermometer is made of

950 25μm Pt wires of 99.999% purity cut to a length of 1.4 cm. In the Jülich model the wires are crimped in a carefully machined platinum holder to ensure good thermal contact. Frequently, silver is used for the wire holder because it can be prepared with excellent thermal conductivity and it has a smaller nuclear heat capacity than platinum. The wires are separated slightly, cleaned, and coated with about 100 Å of SiO_2. This insulation reduces eddy current heating and makes the condition $\delta > sample\ size = 25\mu m$ easy to satisfy. The wire holder is then screwed directly to the stage whose temperature is sought.

Measurement of the temperature of ^{3}He samples is complicated by the large Kapitza resistance it suffers at interfaces. For ^{3}He temperature measurements you could try an arrangement like the one in Figure 5.10. This is a schematic drawing based on a thermometer used by the Ohio State group (Feder, 1979). The ^{3}He is in good thermal contact with the platinum powder. In order to isolate the NMR magnetic field from the rest of the experimental stage the thermometer is placed in an epoxy tower. The rf coil consists of 600 turns of 0.05 mm copper wire with $\sim$ 1 mH inductance. You can dilute the Pt with silica particles until the percolation threshold is reached to minimize eddy current heating. The copper coil form and field confining shield are heat sunk to the mixing chamber and are isolated from the thermometer and stage by graphite spacers. If possible, locate the sample in the region of highest field homogeneity without using a spacer; this eliminates the heat leak to the thermometer from the magnet. There is a significant dielectric loss in Stycast 1266, so it is better to use Emerson & Cumming Stycast 1269A epoxy or some other medium such as 'Q dope' with a low dielectric loss.

To produce a convenient H_0 field for NMR and to confine it to the vicinity of the thermometer you can place a solenoid magnet inside a superconducting shield. The superconducting shield forces most of the lines of induction to complete their loops within its radius. This greatly diminishes the ability of the field to extend to other regions and it increases field homogeneity. Computer calculations to optimize the dimensions of the coil and shield have been published (Smith, 1973); unfortunately, the calculations demonstrate that the goals of field confinement and good field homogeneity conflict. For field confinement the shield should be longer than the solenoid. For good homogeneity the shield should be a little shorter than the coil. The best compromise is usually to have the longer shield in order to confine the magnetic field.

You can also try a different arrangement. Place the superconducting cylinder inside the solenoid, turn the field on at high temperatures, and cool down, allowing the cylinder to trap the field when it goes superconducting. Afterwards, the solenoid field can be turned off. Any other method of producing an H_0 field can be used, and the superconductor will trap the flux and leave it highly homogeneous.

Another possible technique for field homogenization is to wind a coil with different densities of windings over the length in such a way that it produces a more uniform field within. Smith (1973) claims that these various techniques can increase field homogeneity to better than fifty times that of an ordinary solenoid.

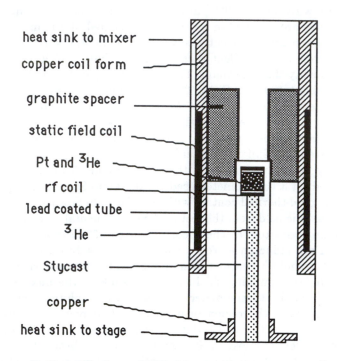

heat sink to mixer

copper coil form

graphite spacer

static field coil

Pt and ^{3}He

rf coil

lead coated tube

^{3}He

Stycast

copper

heat sink to stage

Figure 5.10.   *Platinum NMR Thermometer for Liquid ^{3}He.*

A useful discusssion of the methods for producing homogeneous magnetic fields in coils has been published by Garrett (1951).

Methods of operation are not particularly involved. Tom Gramila has given a discussion for the optimization of coupling to NMR circuits in Section 4.4. The basic technique is quite simple: the net magnetization vector is tipped through a small angle by an rf pulse, and the free induction decay (FID) is observed. χ is then proportional to the initial size of the FID. There is a brief interval following the pulse in which the signal cannot be measured because of the ringing of the tuned tank circuit and the recovery time of the electronics. You should expect to lose the first ten to twenty microseconds of signal. With optimized circuitry the dead time can be cut down to about two microseconds. Since the intrinsic decay time of the platinum signal is half a millisecond, very little signal is lost even with ordinary recovery times. The Q of the circuit must be optimized between the desire for a large signal and the need for a fast recovery of the amplifiers from overload after the transmitter pulse. In practice, a Q of about twenty should not be exceeded. The preferred way to obtain the exponentially decaying envelope is to multiply the FID by the rf drive. What results are the sum and difference bands which in this case are a DC component ($\omega - \omega = 0$) and a high frequency component ($\omega + \omega = 2\omega$). Filter out the high frequency component, leaving the decaying exponential. The initial signal amplitude can readily be obtained from extrapolation of this curve.

In the development of NMR thermometry one of the highlights was the clever use of the Korringa relation for calibration. This law has limited validity, but deserves some mention here. Recall that the conduction electrons mediate the equilibration of the nuclear spins and the lattice. The relaxation time, T_1, should then be proportional to the number of conduction electrons available. But only the electrons near the Fermi surface contribute to the conductivity, and their number is proportional to the electron temperature, $T_{electron}$. Thus,

$$T_1 T_{elec} = K = \text{the Korringa constant for a given metal}$$

is expected to be valid. Knowledge of this constant and a measurement of T_1 suffices to measure the temperature of the sample. The published value of K_{Pt} seems to range from 29.6 to 29.9 msec-K. (See, for example, Aalto, $et\ al$, 1974.) Unfortunately, the Korringa "constant" may not really be constant. Stresses and impurities in the platinum change the value of K by amounts typically less than 10%. In addition, the value of T_1 in any metal will depend upon the strength of magnetic field as the resonant frequency is lowered below a few hundred kHz.

To measure T_1, send in a fairly large tipping pulse which reduces the net magnetization in the z-direction. Follow the recovery along the z-axis by measuring the magnetization in that direction at well defined intervals. Tip the recovering signal through some small angle and observe the FID. This part of the measurement is identical to the susceptibility measurement described above.

Recent work by Greywall (1986) covering the temperature range from 0.6 mK to 5 mK indicates systematic error in earlier thermometry which relied on the Korringa relation. Over the same temperature range, the susceptibility of LCMN is well fit by the Curie-Weiss expression. In Greywall's analysis the temperature scale is set by the plausible requirement of a linear specific heat in normal liquid ^{3}He at these low temperatures. Even for NMR susceptibility measurements, extrapolation to still lower temperatures from the calibration regime is a risky undertaking. Perhaps in a few years we will find the grounds for a deeper trust in magnetic thermometry at extremely low temperatures. In the meantime, as Groucho advises, always examine the dice.

5.3   Other Secondary Thermometers

by Yue Hu

5.3.1   What Can Be Used as Secondary Thermometry?

The answer is simple: anything, as long as it has some physical property varying with temperature. If this statement seems too general, here are some additional features a thermometer ought to have:

- Ease of measurement. You don't want three racks of $10K/piece equipment to monitor your thermometer.
- A short time constant. You don't want to wait forever to find out the temperature.
- Stability and Reproducibility. You don't want to recalibrate it all the time.
- Small size. You don't want it to occupy much of the valuable space in your refrigerator.
- Insensitivity to environmental change such as magnetic fields.
- Good sensitivity in the temperature range of interest.
- Low cost or easy to make.

Your experimental requirements will of course determine how picky you want to be. We will discuss some of the most commonly used secondary thermometers for the helium temperature range in this section.

5.3.2   Carbon Resistors

Carbon resistors are extremely cheap and commonly used in the temperature range from 0.04 K to 300 K. A good review article was written by Anderson (1972). Among all types of carbon resistors, those made by Allen-Bradley Company and Airco Speer Electronics Division of Airco, Inc. have attracted the most attention and are widely reported in the literature. But when you go around shopping for Speer resistors, you will find that they are not manufactured any more, unless some good-hearted friends, who have been around long enough, give you some. At one time Ohmite resistors were manufactured by Allen-Bradley but this is no longer the case. When you see an "Allen-Bradley/Ohmite resistor," you should regard it as a piece of history. So next time you discover some resistors you like, buy as many as you will ever need in your life (they are cheap). You should probably tell the manufacturer that you are going to use them as low temperature thermometers.

Figure 5.11 shows typical R-T dependences of some commercial resistors. As you can see, their increasing resistance at low temperatures limits their use. It is generally true that for a given brand, the smaller the nominal resistance

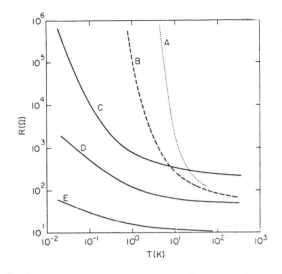

Figure 5.11.   *Resistance versus temperature for several commercial resistors: (A) Thermistor; (B) 68Ω Allen-Bradley; (C) 220Ω Speer; (D) 51Ω Speer; (E) 10Ω Speer. [Van Sciver, S. W. and J. C. Lottin, Rev. Sci. Instr. 54, 762 (1983).]*

at room temperature, the smaller the sensitivity $(dR/R)/(dT/T)$ at any given temperature. Also, the R-T characteristics of a given brand are qualitatively independent of the size of the unit (i.e., 1/8 W, 1/4 W, 1/2 W, etc). There is no guarantee, however, you just have to try them and throw out the bad ones.

Here are some tips on improving the behavior of your resistors before mounting them on your refrigerator: Remove the insulating covering (by using a lathe or sandpaper). This does not improve the thermal contact since most of the thermal impedance is in the carbon core. It does decrease the heat capacity, however, giving you a shorter time constant. If you are still not satisfied, you can cut the resistor into small disks or slabs. For such a disk or slab, the length of the thermal path will be greatly reduced resulting in a better thermal contact. You may use sealing wax to hold the resistor in place while grinding or cutting it; the wax can later be dissolved with alcohol or some other solvent.

Intimate electrical contact between the modified resistor and its leads is necessary to reduce the noise level. Mezhov-Deglin *et al.* (1969) describes a technique for electroplating the small carbon disk with copper and then soldering constantan leads to the copper. You can also use silver epoxy to glue the leads to the electroplated resistor. For slabs, it is a good idea to leave a portion of the original copper leads embedded in the resulting slab, as pointed out by Robchaux *et al.* (1969), and then use In to solder the leads or spot weld them to the copper at each end of the slab. Electrical leads must have a low thermal conductivity to avoid heat leaks. Cu-Ni coated superconducting wire or a resistive alloy such as manganin will usually work satisfactorily.

Now you are ready to mount the resistor onto your refrigerator. The usual procedure is to insert the resistor into a hole in a metal block with a snug fit or bind it with copper foil or fine copper wires. Grease, epoxy or varnish can be used as a thermal agent. The metal is then thermally anchored to the stage of your refrigerator of interest. If the resistor is to be immersed in liquid helium, the thermal contact to copper will probably not improve the thermal response (Van Sciver, et al., 1983).

Thermally grounding the leads can be crucial for obtaining a shorter time constant. This is especially important when the resistor is inserted in a copper hole which has an excess heat capacity. (The time constant can be as long as tens of minutes if the leads do not have a proper heat sink.) The thermal grounding is accomplished by gluing the leads to a metallic heat sink with General Electric 7031 varnish and electrically insulating them from each other with cigarette paper. It can also be done by connecting a capacitor from each lead to ground.

The resistance can be measured by a standard bridge set-up (c.f. Section 4.1) or a potentiometer. The thermal contact between the resistor and its environment becomes worse at low temperatures because of the Kapitza resistance (c.f. Section 3.5). Therefore, the measuring power dissipation should be kept smaller than $10^{-9} T^3$ W (Anderson 1972).

Before you can use your resistor as a thermometer, you must calibrate it to obtain an R-T dependence. An empirical fitting for more than two decades of temperature is difficult because of the lack of a theoretical prediction so any extrapolation is dangerous; i.e., you should calibrate it for the entire temperature range of interest. Here are some expressions you may find useful for R-T fitting:

$$\sqrt{\frac{\log R}{T}} = a + b \log R \tag{5.4}$$

$$\frac{\log R}{T} = a + b \log R + c(\log R)^2 + \frac{d}{T}. \tag{5.5}$$

Or, if you have a good computer, try this:

$$\log T = \sum_i a_i (\log R)^i \tag{5.6}$$

where i runs from 0 to 3 or 6 (Star, Van Dam and Van Baarle 1969). Slight changes in calibration from run to run can be reduced by substituting the reduced resistance $\rho = R/R_F$, where R_F is the resistance at some fixed temperature.

Once calibrated, the thermometer will have a good chance of reproducing itself to within 1% over a year or two if it is not abused. It is well known that heating the resistor with a soldering iron will alter its R-T dependence. Baking a resistor above 420° C will destroy it (Johnson and Anderson, 1971), so handle it with care even though it is cheap. It is always a good idea to thermally cycle the resistor from 300 K to 4.2 K several times before calibration.

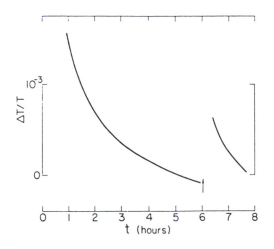

Figure 5.12. *Temporal drift of a 100Ω Ohmite resistor at 1.8 K. At the time indicated by the arrow the resistor was warmed to 4 K and rapidly cooled back to 1.8 K. [Van Sciver, S. W. and J. C. Lottin, Rev. Sci. Instr.* **54**, *762 (1983).]*

Johnson and Anderson (1971) have also found a drifting of $\Delta T/T$ on the order of 10^{-3} over a period of several hours (See Figure 5.12) and more serious drifting at lower temperatures. Their conclusion is that the drifting is intrinsic and not due to the lack of thermal equilibrium. Also, resistors with smaller temperature sensitivities tend to have smaller drift.

It is often desirable to measure temperatures in magnetic fields. Different brands of carbon resistors behave differently in magnetic fields. Figure 5.13 shows $\Delta R/R$ vs. B up to 11 T for an Allen-Bradley resistor where the magnetoresistance is positive.

Figure 5.14 shows some more recent results on a Speer resistor by Naughton *et al.* (1983). They used the expression

$$\frac{\Delta R}{R} = \frac{a(T)\,B^2}{b(T) + B^2} - c(T)\left(1 - e^{-d(T)\,B}\right) \tag{5.7}$$

to fit the experimental results, where $a(T)$, $b(T)$ and $c(T)$ are first order polynomials in temperature T and $d(T)$ is proportional to T^{-2}. The fittings are shown by the dotted lines. The Speer resistors have a smaller magnetoresistance compared to other brands, but are positive at high fields and negative at low fields.

It is clear that the magnetic field dependence is more severe at lower temperatures, requiring the use of some other type of thermometry such as capacitance thermometry which will be discussed later.

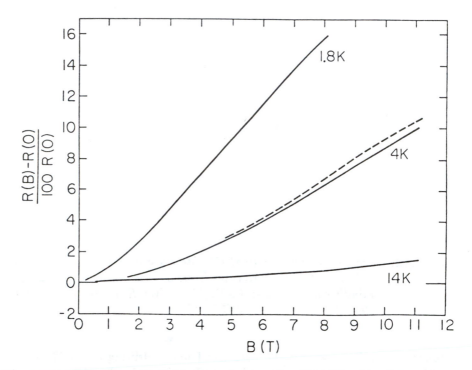

Figure 5.13. *Magnetic field dependence of a 47Ω Allen-Bradley resistance thermometer. Broken line is behavior of 220Ω Allen-Bradley at 4 K. [Neuringer, L. J. and L. G. Rubin, Temperature 4, 1085 (1972).].*

5.3.3   Germanium Resistors

When doped with either gallium (p-type) or arsenic (n-type), the resistance of a single crystal germaniun varies rapidly at temperatures below 10 K. Suitably doped resistors can be used as low temperature thermometers down to 10 mK. Figure 5.15 shows some typical R-T dependences of Ge resistors manufactured by Lake Shore Cryotronics, Inc. Figure 5.16 shows the construction details of a Ge resistor made by the same company. Four gold wires are welded onto the Ge crystal and phosphorus bronze leads are then connected to the gold wires. The crystal is carefully mounted to minimize mechanical stress because it is highly piezoresitive. The unit is encapsulated in a gold-plated copper enclosure. ^{4}He is frequently used as an exchange gas. ^{3}He or N_2 are available as options.

When you order Ge thermometers, you should keep the following in mind: First you must know the temperature range you are interested in. As you can see in Figure 5.15, asking a Ge thermometer to work over a temperature range

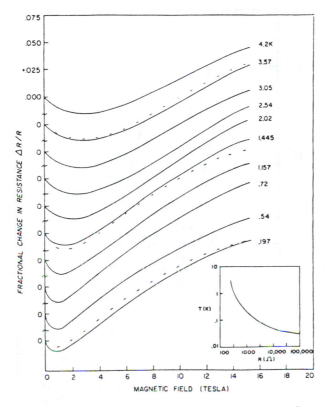

Figure 5.14. *Magnetoresistance of a Speer 220Ω resistor. [Naughton, M. J., S. Dickinson, R. C. Samaratunga, J. S. Brooks, and K. P. Marin, Rev. Sci. Instr. **54**, 1529 (1983).]*

of more than two orders of magnitude is not realistic. Secondly, there is little or no correlation between its resistance at 4 K and its performance at lower temperatures; a modest guarantee can only be made if the resistance ratio at 1.5 K and 4.2 K is known. Only a calibration can ensure the right performance. If you order an uncalibrated thermometer and find out its R-T relation is no good below 1 K, don't be too disappointed. You have to excuse the manufacturer because there is some "black magic" in the crystal-growing—a change of only 2 or 3 parts in 10^{16} of arsenic doping is the difference between a 1000 Ω and a 30 Ω resistor. Lake Shore Cryotronics will sell you uncalibrated Ge thermometors working at temperatures from 0.1 K to 10 K for $125 each. But the one working down to 10 mK is only available calibrated and it costs $1595. (much cheaper than $1600?)

The Ge resistors are very small, as shown in Figure 5.16. The technique for

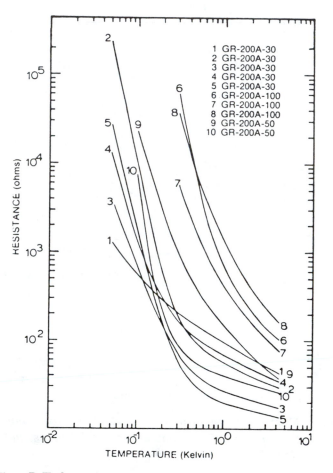

Figure 5.15. *R-T dependences of Ge resistors. [Lake Shore Cryotronics, catalog, Lake Shore Cryotronics, 64 East Walnut St., Westerville, OH 43081.]*

thermally grounding the resistor and the leads is the same for carbon resistors and will not be repeated here. Blakemore (1972) estimated that nearly 70% of the thermal conduction is due to electrical leads so it is more important to thermally anchor the leads than the resistor itself. He also finds that the typical thermal time constant of a Ge thermometer is about 0.05 seconds when it is immersed in liquid helium at 4.2 K, with little dependence on its size.

Fitting $R(T)$ is not trivial (more so than that for carbon resistors). A computer polynomial fitting is often necessary:

$$\log T = \sum_i A_i (\log R)^i \qquad (5.8)$$

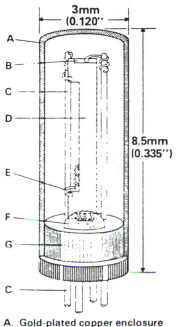

A. Gold-plated copper enclosure
B. Current injection zone
C. Phosphorus-bronze leads 0.20mm (0.008" dia.)
D Sensing element
E. Gold leads 0.05mm (0.002" dia.)
F. Epoxy heat sink
G. Beryllium oxide base

Figure 5.16. *Lead and Thermal Ground Arrangement of a Ge Resistance Thermometer. [Lake Shore Cryotronics, catalog, Lake Shore Cryotronics, 64 East Walnut St., Westerville, OH 43081.]*

where i usually runs from 0 to 8 or 14 for the temperature range 1 to 30 K (Osborne, Flotow and Schreiner 1967).

Besley *et al.* (1978) studied the stabilities of ^{30}Ge thermometers at 20 K by thermally cycling them from 300 K to 20 K more than 90 times. The thermometers showed five types of behavior: stable, drifting, jumping, bimodal and irregular. 90% of the sensors exhibited changes of less than 2 mK, although some bad ones changed more than 20 mK. The instability problem is far less serious at 4.2 K than at 20 K. Lake Shore Cryotronics claims the reproducibility is better than 0.5 mK at helium temperatures.

Despite its excellent reproducibility and high sensitivity, Ge thermometers behave rather poorly in magnetic fields. Because the sensor is made of a single crystal, its magnetoresistance is orientation dependent and it is so large that it

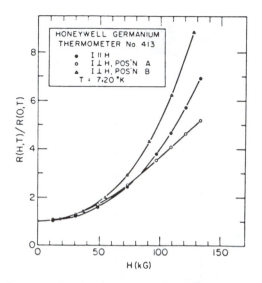

Figure 5.17. *Anisotropy in the magnetoresistance of a single crystal Ge resistor. [H. H. Sample and L. G. Rubin, Cryogenics 17, 597 (1977).]*

is plotted as $R(H,T)/R(0,T)$ instead of $\Delta R/R$. Figure 5.17 shows some results for a Honeywell Ge thermometer at 7.2 K.

The resistance of a Ge thermometer can be measured by a standard bridge set-up (c.f. Section 4.1 on bridges) or a potentiometer. You should be aware that the resistance of a Ge thermometer may be as high as 10^6 Ω and the lead resistance can be 10^5 Ω (because part of the lead is the Ge crystal itself, see Figure 5.16). To obtain an accuracy of 1 part in 10^5, the effective impedance of the potential measuring circuit must be greater than 10^{11} Ω. To minimize the self-heating, White (1979) suggests a measuring current of 100 μA for temperatures above 20 K, 10 μA above 4.2 K, and 1-2 μA for temperatures below 4.2 K.

5.3.4 Capacitance Thermometers

If you are dissatisfied with the performance of resistance thermometers in magnetic fields, try a capacitance thermometer. Lawless (1972) first developed a $SrTiO_3$ glass-ceramic multilayer capacitance thermometer whose C-T characteristics are shown in Figure 5.18. The thermometer is sensitive over three temperature regions: below 0.10 K where $C \propto 1/T$, from 0.10 K to 72 K where C increases smoothly with temperature (including a linear region) and at higher temperatures from 100 K to 300 K. The characteristics at 4.2 K are: $C = 2.5$ to 3.0 nF, $dC/dT \approx 20$ pF/K, $d(\ln C)/dT \approx 0.8\%$/K, and the thermal response time $\tau/T \approx 14$ msec/K. The $SrTiO_3$ thermometer is now available commercially through Lake Shore Cryotronics, Inc.

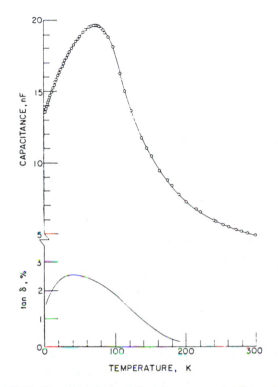

Figure 5.18. *C-T Characteristics of a glass-ceramic Capacitance Thermometer. [Lawless, W. N., Rev. Sci. Instr. **46**, 625 (1975).]*

Lawless did not detect any B-field effect within his experimental accuracy (1 mK) for fields up to 14 T and temperatures down to 1.5 K. Naughton *et al.* (1983) reported a 0.13% change of $\Delta C/C$ (equivalent to 20 mK in temperature) at 19 T and 90 mK. Nevertheless, the capacitance thermometer is still the best sensor to use in powerful magnetic fields. A shortcoming of capacitance thermometers is the time-dependent aging phenomena which is shown in Figure 5.19.

Aging effects can be expected whenever the glass matrix is spatially perturbed, either by thermal expansion (above 77 K), or by electrostrictive coupling to the micro-crystal following a voltage pulse, or by warming or cooling across the 65 K transition temperature. (Lawless, 1975)

The capacitance decays for about 30 minutes before stabilizing so it is highly recommended to wait an hour or so after cooling down to avoid large electrical shocks to the sensor. Swenson (1977) observes an irreproducibility of 0.2 K for different cooling processes from 300 K to 4.2 K and that the thermometer tends

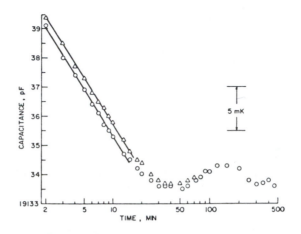

Figure 5.19. *Examples of isothermal time-dependent aging of a capacitance thermometer. Data after rapid cooling to 4.2 K are shown for the same specimen on two different days. [Lawless, W. N., Rev. Sci. Instr. 46, 625 (1975).]*

to become stable and reproducible once it is cold. Even if you do not believe the calibration from a previous cool-down, you can still use it as an excellent null detector to sense any temperature change, especially in a changing magnetic field. Using $\Delta C/\Delta T$ from the old calibration (you can pretty much trust it), ΔT can be calculated.

The capacitance can be measured by a General Radio 1615-A transformer ratio-arm bridge or some homemade bridges (c.f. Section 4.1). Three-lead measurement is highly recommended to reduce the effect of stray capacitance (which may be comparable to the capacitance of the sensor and changes with temperature as well). Here are a few words about the measuring power. The self-heating can be estimated by

$$\dot{Q} = 1.23 \, Cf \tan\delta \, V^2 \text{ (pW)} \tag{5.9}$$

where C is the capacitance in pF, f the frequency in Hz, V the measuring voltage, and $\tan\delta$ the dissipation factor (Lawless 1972). The good news is that the self-heating is in the pW range at helium temperatures and decreases with decreasing temperature which is the opposite for resistance thermometers.

Hartmann and McNelley (1977) fabricated capacitance thermometers from KCl:OH and NaF:OH, which have sensitivities, $d(\ln C)/dT$, one or two orders of magnitude better than those of SrTiO$_3$ thermometers. Their size is a disadvantage, however. ($10 \times 10 \times 3$ mm^3, compared to a 3 mm diameter and 9 mm length of the latter.)

Studies have shown that the dielectric constant of glass, ϵ, has a $-\ln T$ dependence at low temperatures (typically below 0.1 K). This makes another kind of homemade capacitance thermometer possible. Thin glass films are commercially available (e.g., Heraeus-Amersil, Inc. sells silica-fused quartz film of thicknesses down to 0.15 mm). Electrodes can be made by evaporating metal (say Au) films on both sides of the glass. Electrical leads can then be glued on with Ag epoxy. Schickfus and Hunklinger (1976) showed that the change of $\Delta\epsilon/\epsilon$ increases linearly with the dipolar impurity OH^- in the glass. Make sure you buy the glass with a large OH^- content or boil the glass in water vapor at about 200° C for a couple of hours to increase its OH^- concentration.

Mechanical stress caused by thermal contraction will flatten out $\Delta\epsilon/\epsilon$ at low temperatures. When designing the capacitor holder, make sure you leave enough room for the sensor to contract after cool-down. D. D. Osheroff suggests (through a private communication) applying only a small amount of glue (e.g., vacuum grease) to a corner of the capacitor to hold it in place and to leave the other sides free to contract during the cooling process. Frossati et al. (1977) found $\Delta\epsilon/\epsilon$ changes with measuring field strength and frequency, which warns us to calibrate the capacitance thermometer at a fixed frequency and excitation voltage.

5.3.5   Miscellaneous Other Secondary Thermometers

Rhodium-Iron Resistors The resistance of a wire of Rh alloyed with 0.5 atomic percent of Fe decreases monotonically with temperature. Rusley (1975) has investigated the use of such a resistor as an alternative to germanium thermometers in the temperature range from 0.4 K to 20 K. Compared to Ge thermometers, the attractive features of Rh–Fe thermometers are its superior reproducibility, smaller magnetoresistance, and simpler R-T dependence which ensures more accurate interpolations between calibration points. However, thermal contact becomes a problem below 0.4 K for an encapsulated one. Its large size is also a disadvantage.

Carbon Glass Resistors With the hope of improving the stability of carbon resistors to match that of Ge thermometers and retaining the low magnetoresistance, Lawless (1972) developed a process of impregnating pure carbon filaments of controlled sizes into some stable glass matrices. The carbon glass resistor thus obtained has a similar R-T dependence as that of an Allen-Bradley carbon resistor and can be used as a thermometer down to 1 K. Its magnetoresistance has about the same order of magnitude as that of an Allen-Bradley resistor. The resistances of the ith and jth units have a simple relation:

$$R_i = bR_j^m \tag{5.10}$$

where b is a constant independent of temperature and m is very close to unity. If you need many thermometers, carbon glass ones will make the calibration a lot easier. But Besley (1979) found that the stability of carbon glass thermometers is about ten times worse than for Ge ones.

P-N Junction Diodes These are useful for the temperature range from 1–300 K, but need at least 10 μA current to overcome electrical noise which might cause heating problems at helium temperatures. A review article on diode thermometry was written by Swartz and Swartz (1974).

Thermocouples These are not as commonly used in the helium temperature region as the other types of thermometers described above. They are more expensive (mostly due to the voltage measuring apparatus) and very magnetic field dependent. Spurious thermal emf in the leads and low thermal impedance can also cause lots of problems.

Platinum Thermometers A platinum thermometer was used as the International Practical Temperature Scale of 1968 for the temperature range from 13.81 K to 630.74° C, and, therefore, is very widely publicized. People interested in this temperature range are referred to the review article by Tiggelman and Durieux (1972). Platinum thermometers are very stable upon room temperature cycling. A typical one has a sensitivity of 0.008 Ω/K at 15 K, but becomes too insensitive at lower temperatures.

5.3.6   Sources of Calibrated Resistance Thermometers

- Lake Shore Cryotronics, 64 E. Walnut St., Westerville, OH 43081. (614)-891-2243
- Scientific Instruments, Inc., 1101 25th Street, West Palm Beach, FL 33407.

References

Aalto, M. I., H. K. Collan, R. G. Gylling, and K. O. Nores, 1974, *Proceedings of LT-13* **4**, 513.

Abragam, A., 1961, *The Principles of Nuclear Magnetism* (Oxford University Press, Oxford).

Abragam, A., and M. Goldman, 1982, *Nuclear Magnetism: Order and Disorder* (Oxford University Press, Oxford).

Almond, D. P., M. J. Lea, and G. R. Pickett, 1972, *Cryogenics* **12**, 469.

Alvesalo, T. A., T. Haavasoja, M. T. Manninen, and A. T. Soinne, 1980, *Phys. Rev. Lett.* **44**, 1076.

Anana'eva, A. A., 1959, *Sov. Phys. J. Acoustics* **5**, 13.

Anderson, A. C., 1972, *Temperature* **4**, 773.

Anderson, A. C., and W. L. Johnson, 'The Kapitza Resistance Between Copper and ^3He, 1972,' *Journal of Low Temp. Phys.* **7**, 1.

Andres, K., 'Hyperfine Enhanced Nuclear Magnetic Cooling, 1978,' *Cryogenics* **18**, 8.

Andronikashvilli, E., 1946, *J. Phys. (Moscow)* **10**, 201.

Ashcroft, N. W., and N. D. Mermin, 1976, *Solid State Physics* (Holt, Rinehart and Winston, Philadelphia).

Babcock, J., L. Kiely, T. Manley, and W. Wehmann, 1979, *Phys. Rev. Lett.* **43**, 380.

Barone, A., and G. Paterno, 1982, *Physics and Applications of the Josephson Effect* (John Wiley & Sons, New York).

Bassett, W., and T. Takahashi, 'X-ray Diffraction Studies up to 300 kbar,' 1974, in *Advances in High-Pressure Research* **4**, ed. R. H. Wentorf, Jr. (Academic Press, London).

Beamish, J. R., A. Hikata, L. Tell, and C. Elbaum, 1983, *Phys. Rev. Lett.* **50**, 425.

Benedict, R. P., 1977, *Fundamentals of Temperature, Pressure, and Flow Measurements* (John Wiley & Sons, New York).

Bergmann, L., 1938, *Ultrasonics* (trans. H. S. Hatfield) (Bell, London).

Berton, A., J. Chaussy, B. Cornut, J. Odin, J. Paureau, and J. Peyrard, 1979, *Cryogenics* **19**, 543.

322   References

Besley, L. M., 1978, *Rev. Sci. Instr* **49**, 1041.

Besley, L. M., 1979, *Rev. Sci. Instr* **50**, 1626.

Besley, L. M. and H. H. Plumb, 1978, *Rev. Sci. Instr.* **49**, 68.

Betts, D. S., 1976, *Refrigeration and Thermometry Below 1K* (Sussex Univ. Press, Brighton).

Bishop, D. J. and J. D. Reppy, 1980, *Phys. Rev.* **B22**, 5171.

Biondi, M. A., and M. P. Garfunkel, 1959, *Phys. Rev. Lett.* **2**, 143.

Blakemore, J. S., 1972, *Temperature* **4**, 827.

Blitz, J., 1967, *Fundamentals of Ultrasonics 2nd Edition* (Plenum, New York).

Bocko, M. F., 1984, *Rev. Sci. Inst.* **55**, 256.

Boghosian, C., H. Meyer, and J. E. Rives, 1966, *Phys. Rev.* **146**, 110.

Bolef, D. I., 1966, *Physical Acoustics* **4a** (ed.W. P. Mason) (Academic Press, New York).

Bradley, C. C., 1969, *High Pressure Methods in Solid State Research* (Butterworth, London).

Braginsky, V. B., V. P. Mitrofanov, and V. I. Panov, 1985, *Systems with Small Dissipation* (University of Chicago Press, Chicago).

Buchal, C., J. Hanssen, R. M. Mueller, and F. Pobell, 'Platinum Wire NMR Thermometer for Ultralow Temperatures, 1978,' *Rev. Sci. Instr.* **49**, 1360.

Busch, P. A., S. P. Cheston, and D. S. Greywall, 'Properties of Sintered-Silver Heat Exchangers, 1984,' *Cryogenics* **24**, 445.

Byer, N. E., 1967, Ph.D. Thesis, Cornell University.

Carslaw, H. S. and J. C. Jeager, 1959, *Conduction of Heat in Solids* (Oxford Univ. Press, New York).

Cerutti, G., R. Maghenzani, and G. F. Molinar, 1983, *Cryogenics* **23**, 539.

Cesnak, L., and C. Schmidt, 1983, *Cryogenics* **23**, 317.

Cheeke, D., J. Ettinger, and B. Hebral, 'Analysis of Heat Transfer between Solids at Low Temperatures, 1976,' *Can. Jour. Phys.* **54**, 1749.

Clarke, J., C. Tesche, and R. P. Giffard, 1979, *Jour. of Low Temp. Phys.* **37**, 405.

Clarke, J., 1980, *IEEE Trans. Elec. Dev.* **ED-27**, 1896.

Corrucinni, R. J., 1959, *Vacuum* **VII & VIII**, 19.

Corruccini, R. J., and J. J. Gniewek 1960, 'Specific Heat and Enthalpy of Technical Solids at Low Temperatures,' *NBS Monograph 21*.

Corruccini, R. J., and J. J. Gniewek 1961, *Thermal Expansion of Technical Solids at Low Temperatures, NBS Monograph* **29**.

Crooker, B., 1981, *Ben's Compendium*, unpublished Cornell report [the reader is referred to the text by White below for a similar compilation of data].

Dash, J. G., and R. D. Taylor, 1957, *Phys. Rev.* **105**, 7.

de Klerk, J., and E. F. Kelly, 1965, *Rev. Sci. Instr.* **36**, 506.

Denner, H., 1969, *Cryogenics* **9**, 282.

de Vegvar, P. G. N., 1985, Ph.D. Thesis, Cornell University.

Dewar, J., 1907, *Proc. Roy. Soc. Gt. Brit.* **18** , 751.

Dobbs, E. R., 1973, in *Physical Acoustics* **10**, eds. W. P. Mason and R. N. Thurston (Academic Press, New York).

Dushman, Saul, 1962, *Scientific Foundations of Vacuum Technique* (John Wiley & Sons, New York).

Efferson, K., 1967, *Rev. Sci. Instr.* **38**, 1776.

Engel, B. N., G. G. Ihas, E. D. Adams, and C. Fombarlet, 1984, *Rev. Sci. Instr.* **55**, 1489.

Feder, J. D., 1979, Ph.D. Thesis, Ohio State University.

Foster, J., 1984, *Physica B+C* **126**, 199.

Fox, J., 1971, *Proceedings of the Symposium on Submillimeter Waves* **XX** (Polytechnic Press of the Polytechnic Institute of Brooklyn, New York).

Freeman, M. R., R. S. Germain, R. C. Richardson, M. L. Roukes, W. J. Gallagher, and M. B. Ketchen, 1986, *Appl. Phys. Lett.* **48**, 300.

Frossati, G., R. Maynard, R. Rammal, and D. Thoulouze, 1977, *Le Journal de Physique* **38**, L-153.

Frost, H. M., 1979, in *Physical Acoustics* **14**, eds. W. P. Mason and R. N. Thurston (Academic Press, New York).

Fukushima, E., and S. B. W. Roeder, 1981, *Experimental Pulse NMR, a Nuts and Bolts Approach* (Addison-Wesley, Reading).

Gammel, P. L., H. E. Hall, and J. D. Reppy, 1984, *Phys. Rev. Lett.* **52**, 121.

Garrett, M. W., 1951, *Jour. of Appl. Phys.* **22**, 1091.

Giannetta, R. W., A. Ahonen, E. Polturak, J. Saunders, E. K. Ziese, R. C. Richardson, and D. M. Lee, 1980, *Phys. Rev. Lett.* **45**, 262.

Gibson, A. A. V., J. R. Owers-Bradley, I. D. Calder, J. B. Ketterson, and W. P.

324   References

Halperin, 1981, *Rev. Sci. Instr.* **52**, 1509.

Gonano, R. and E. D. Adams, 1971, *Rev. Sci. Instr.* **41**, 71.

Greenberg, A. S., G. Guerrier, M. Bernier, and G. Frossati, 1982, *Cryogenics* **22**, 144.

Greywall D. S., 'Thermal Conductivity of Normal Liquid ^{3}He, 1984,' *Phys. Rev.* **B29**, 4933.

Greywall, D. S., 1986, *Phys. Rev.* **B33**, 7520.

Greywall, D. S., and P. A. Busch, 1982, *Jour. of Low Temp. Phys.* **46**, 451.

Grilly, E. R., 1971, *Jour. of Low Temp. Phys.* **4**, 615.

Hardy, W. N., M. Morrow, R. Jochemsen, and A. J. Berlinsky, 1982, *Physica* **109 & 110B**, 1964.

Harrison, J. P., 'Kapitza Conductance-A Universal Curve, 1974,' *Jour. of Low Temp. Phys.* **17**, 43.

Hartmann, J. B., and T. F. McNelly, 1977, *Rev. Sci. Instr.* **48**, 1072.

Heald S. M. and R. O. Simmons, 1977, *Rev. Sci. Instr.* **48**, 316.

Halperin, W. P., F. B. Rasmussen, C. N. Archie, and R. C. Richardson, 1978, *Jour. Low Temp Phys* **31**, 617.

Hoare, F. E., L. C. Jackson, and N. Kurti, 1961, *Experimental Cryophysics* (Butterworth, London).

Hudson, R. P., 1972, *Principles and Application of Magnetic Cooling* (North Holland, Amsterdam).

Hudson, R., H. Marshak, R. Soulen, D. Utton, 1971, *Journ. Low Temp Phys* **20**, 1.

Hunklinger, S., 1976, *Jour. of Phys.* **C9**, L439.

Jayaraman, A., 1983, *Rev. Mod. Phys.* **55**, 1.

Johnson V. J. and R. B. Stewart, 1960-61, *A Compendium of the Properties of Materials at Low Temperature* (Wadd Technical Report 60-56, Wright Air Development Division).

Johnson, W. L. and A. C. Anderson, 1971, *Rev. Sci. Instr.* **42**, 1296.

Kamper, R. A. and J. E. Zimmerman, 1971, *Journ. Appl. Phys.* **42**, 132.

Kaplan, S. B. 'Accoustic Mismatch Of Superconducting Films to Substrates, 1979,' *Jour. of Low Temp. Phys.* **37**, 343.

Katerberg J. A., C. L. Reynolds, and A. C. Anderson, 'Calculations of the Thermal Boundary Resistance, 1977,' *Phys. Rev.* **B16**, 673.

Kennard, E. H., 1938, *Kinetic Theory of Gases* (McGraw-Hill, New York).

Keith V. and M. G. Ward, 'A Recipe for Sintering Submicron Silver Powders, 1984,' *Cryogenics* **24**, 249.

Ketchen, M. B. and J. M. Jaycox, 1982, *Appl. Phys.Lett.* **40**, 736.

Ketchen, M. B., 1981, *IEEE Trans. Magn* **MAG-17**, 387.

Kinsler, L. E. and A. R. Frey, 1962, *Fundamentals of Acoustics* (John Wiley & Sons, New York).

Kirk, W. P. and M. Twerdochlib, 'Improved method for minimizing vibrational motion transmitted by pumping lines, 1978,' *Rev. Sci. Instr.* **49**, 765.

Kirschman, R., 1985, *Cryogenics* **25**, 115.

Kittel, C., 1971, *Introduction to Solid State Physics (4th ed.)* (John Wiley & Sons, New York).

Kittinger, E. and W. Rehwald, 1977, *Ultrasonics* **15**, 211.

Kubota, Y., H. R. Folle, Ch. Buchal, R. M. Mueller, and F. Pobell, 1980, *Phys. Rev. Lett.* **45**, 1812.

Landau, L. D. and E. M. Lifshitz, 1959, *Fluid Mechanics: Vol 6 of Course of Theoretical Physics*(Pergamon Press, Oxford).

Lawless, W. N., 1972, *Temperature* **4**, 1143.

Lawless, W. N., 1975, *Rev. Sci. Instr.* **46**, 625.

Leck, J. H., 1957, *Pressure Measurements in Vacuum Systems* (Unwin Brothers, London).

Lengeler, B., 1974, *Cryogenics* **14**, 439.

Little, W. A. 'The Transport Of Heat Between Dissimilar Solids at Low Temperatures, 1959,' *Can. Jour. of Phys.* **37**, 334.

Lock, J., 1969, *Cryogenics* **9**, 438.

Lounasmaa, O. V., 1974, *Experimental Principles and Methods Below 1K* (Academic Press, NY).

Mann A. G. and D. G. Blair, 1983, *Jour. of Phys.* **D16**, 105.

Mao, H. K., 1979, *Rev. Sci. Instr.* **50**, 8.

Marshak, H., 1983, *Journal of Research of the National Bureau of Standards* **88**, 175.

Martin, L. H., 1949, *A Manual of Vacuum Practice* (Melbourne University Press, Melbourne).

Mast, D. B., 1982, Ph.D. Thesis, Northwestern University.

Matsumoto, D. S., C. L. Reynolds, and A. C. Anderson, 'Thermal Boundary Resistance at Metal-Epoxy Interfaces, 1977,' *Phys. Rev.* **B16**, 3303.

Mast, D. B., J. R. Owers-Bradley, W. P. Halperin, I. D. Calder, B. K. Sarma, and J. B. Ketterson, 1981, *Physica B+C* **107**, 685.

Matthey, A. P. M., J. T. M. Walraven, and I. F. Silvera, 1981, *Phys. Rev.Lett.* **46**, 668.

Meijer, H. C., G. J. C. Bots, and H. Postma, 1981, *Physica* **107b**, 607.

Mezhov-Deglin, L. P., and A. I. Shalnikov, 1969, *Cryogenics* **9**, 60.

Mikheev, V. A., V. A. Maidanov, and A. I. Shalnikov, 1984, *Cryogenics* **24**, 190.

Montgomery, C. G., R. H. Dicke and E. M. Purcell, 'Principles of Microwave Circuits,' 1948, *MIT Radiation Lab Series* **8** (McGraw-Hill, New York).

Montgomery, C. G., 'Technique of Microwave Measurements,' *MIT Radiation Laboratory Series* **11** (McGraw-Hill, New York).

Moreno, T., 1948, *Microwave Transmission Design Data* (McGraw-Hill, New York).

Morrison, Ralph, 1977, *Grounding and Shielding Techniques in Instrumentation (2nd e.d)*, (John Wiley & Sons, New York).

Moster, P. C., D. F. McQueeney, T. J. Gramila, and R. C. Richardson, 1987, to be published in the *Proceedings of the Electrochemical Society*.

Motchenbacher, C., and F. Fitchen, 1973, *Low-Noise Electronic Design* (John Wiley & Sons, New York).

Mueller, R. M., C. Buchal, T. Oversluizen, and F. Pobell, 1978, *Rev. Sci. Instr.* **49**, 515.

Mueller, R. M., Chr. Buchal, H. R. Folle, M. Kubota, and F. Pobell, 1980, *Cryogenics* **20**, 395.

Naughton, M. J., S. Dickinson, R. C. Samaratunga, J. S. Brooks, and K. P. Martin, 1983, *Rev. Sci. Instr.* **54**, 1529.

Nenuringer, L. J., and L. G. Rubin, 1972, *Temperature* **4**, 1085.

Niinikoski, T. O., 1982, *Nuc. Inst. Meth* **192**, 151.

O'Hara, S. G., and A. C. Anderson, 'Thermal Impedance Across Metallic and Superconducting Foils below 1 K, 1974,' *J. Phys. Chem. Sol.* **35**, 1677.

Ohlmann R. C., 1958, *J. Opt. Soc. America* **48**, 531.

Osborne, D. W., H. E. Hotow, and F. Schreiner, 1967, *Rev. Sci. Instr.* **38**, 159.

Osheroff, D. D., and R. C. Richardson, 'Novel Magnetic Field Dependence of the Coupling of Excitations between Two Fermion Fluids, 1985,' *Phys. Rev. Lett.* **54**, 1178.

Ott, H. W., 1976, *Noise Reduction Techniques in Electronic Systems* (John Wiley & Sons, New York).

Papadikis, E. P., 1976, in *Physical Acoustics* **12** (eds. W. P. Mason and R. N. Thurston) (Academic Press, New York).

Parpia, J. M. and J. D. Reppy, 1979, *Phys. Rev. Lett.* **43**, 1332.

Parpia, J. M., W. P. Kirk, P. S. Kobiela, T. L. Rhodes, Z. Olejniczak, and G. N. Parker, 'Optimization Procedure for the Cooling of Liquid ^{3}He by Adiabatic Demagnetization of Praseodymium Nickel,' 1984, *Texas A&M Technical Report.*

Peshkov, V. P., and A. Ya. Parshin, 1965, *JETP* **21**, 258.

Peshkov, V. P., 1970, *Cryogenics* **10**, 250.

Peterson, R. E., and A. C. Anderson, 1973, *Jour. of Low Temp Phys.* **11**, 639.

Peterson, R. E., and A. C. Anderson, 'The Transport of Heat Between Solids at Low Temperatures, 1972,' *Solid State Comm.* **10**, 891.

Pickens, K. S., 1984, Ph.D. Thesis, Washington University.

Pickens, K. S., G. Mozurkewich, D. I. Bolef, and R. K. Sundfors, 1984, *Phys. Rev. Lett.* **52**, 156.

Pickens, K. S., D. I. Bolef, M. R. Holland, and R. K. Sundfors, 1984, *Phys. Rev.* **B30**, 3644.

Pohl, R. O., V. L. Taylor, and W. M. Goubau, 1969, *Phys. Rev.* **178**, 431.

Pollack, G. 'Kapitza Resistance, 1969,' *Rev. Mod. Phys.* **41**, 48.

Polturak, E., P. G. N. de Vegvar, E. K. Ziese, and D. M. Lee, 1981, *Phys. Rev. Lett.* **46**, 1588.

Radebaugh, R., 1978, *NBS Special Publication* **508**, 93.

Radebaugh, R., J. D. Siegwarth, and J. C. Holste, 'Heat Transfer Between Sub-Micron Silver Powder and Dilute ^{3}He-^{4}He Solutions,' 1974, (IPC Sci. and Tech. Press, Guilford, Surrey, England) 242.

Rawlings, K. C., and J. C. A. van der Sluijs, 'Influence of Dislocations on the Kapitza Conductance of Copper at Temperatures Between 1 and 2 K, 1979,' *Jour. of Low Temp. Phys* **34**, 215.

Raychaudhuri, A. K., 1975, Ph.D. Thesis, Cornell University.

Redhead, P. A., J. P. Hobson, and E. V. Kornelsen, 1968, *Physical Basis of Ultra-high Vacuum* (Chapman and Hall, London).

Reynolds, C. L., and A. C. Anderson, 'Thermal Boundary Resistance to Solid He, H, D, and N, 1976,' *Phys. Rev.* **B14**, 4114.

Richardson, R. C., 1977, *Physica* **90B+C**, 47.

Ried, W. J., 'Vibration Isolation, a Practical Understanding, 1977,' *The Ealing Review* **1**, 4.

Robichaux, J. E., and Anderson, A. C., 1969, *Rev. Sci. Instr.* **40**, 1512.

Robinson, N. W., 1968, *The Physical Principles of Ultra-high Vacuum Systems and Equipment* (Chapman and Hall, London).

Rose-Innes, A. C., 1964, *Low Temperature Techniques* (The English Universities Press, London).

Rose-Innes, A. C., 1973, *Low Temperature Experimental Techniques* (Elliot Bros. and Yoeman, Liverpool).

Rozen, J. R., and D. D. Awschalom, 1986, *Appl. Phys. Lett* **49**, 1649.

Roukes, M. L., R. S. Germain, M. R. Freeman, R. C. Richardson, 1984, *Physica* **126B+C**, 1177.

Rubsy, R. L., 'Temperature Measurement, 1975,' *Inst. Phys. Lond. Conf. Series* **26**, 125.

Rutherford, A. R., J. P. Harrison, and M. J. Stott, 'Heat Transfer between Liquid ^{3}He and Sintered Metal Heat Exchangers, 1984,' *Jour. of Low Temp. Phys.* **55**, 157.

Sagan, D., 1985, Ph.D. Thesis, Cornell University.

Sample, H. H., and L. G. Rubin, 1977, *Cryogenics* **17**, 597.

Saulson, P. R. 'Vibration isolation for broadband gravitational antennas, 1984,' *Rev. Sci. Instr* **55**, 8.

Sauzade, M. D., and S. K. Kan, 1973, *Adv. in Electronics and Electron Physics* **34** (Academic Press).

Schwark, M., F. Pobell, W. P. Halperin, Ch. Buchal, J. Hanssen, M. Kubota, and R. M. Mueller, 1983, *Jour. Low Temp. Phys.* **53**, 685.

Schmidt, C. 'Thermal Boundary Resistance at Niobium-epoxy Interfaces in the Superconducting and Normal States, 1977,' *Phys. Rev.* **B15**, 4187.

Schmidt, C., 1982, *Cryogenics* **22**, 143.

Schmidt, C., and E. Umlauf, 'Thermal Boundary Resistance at Interfaces between Sapphire and Indium, 1976,' *Jour. of Low Temp. Phys.* **22**, 597.

Schubert, H., P. Leiderer, and H. Kinder, 'Saturation of the Anomalous Kapitza Conductance, 1982,' *Phys. Rev.* **B26**, 2312.

Schuberth, E., 1984, *Rev. Sci. Instr.* **55**, 1486.

Schumann, B., F. Nitsche, and G. Paasch, 'Thermal Conductivity of Metal Interfaces at Low Temperatures, 1980,' *Jour. of Low Temp. Phys.* **38**, 167.

Silvera, I. F., 1970, *Rev. Sci. Instr.* **41**, 1513.

Simons, S., 'On the Thermal Boundary Resistance between Insulators, 1974,' *Jour. of Phys.* **C7**, 4048.

Slack, G. A., 1962, *Phys. Rev.* **126**, 427.

Slichter, C. P., 1978, *Principles of Magnetic Resonance, Second Revised and Expanded Edition* (Springer-Verlag, Berlin).

Smith, T.I., 1973, *Jour. Appl. Phys* **4**, 852.

Smithells, Colin J. (ed.), 1976, *Metals Reference Book (5th ed.)* (Butterworths, Boston).

Smythe, W. R., 1939, *Static and Dynamic Electricity* (McGraw-Hill, New York), Chap XI.

Snowdon, J. C., 'Vibration isolation: Use and Characterization, 1979,' *Jour. Acoust. Soc. Am.* **66**, 979.

Soulen, R. J., Noise in Physical Systems, 1978, *Proc. 5th Intl. Conf. on Noise*, 249.

Soulen, R. J. and H. Marshak, 1980, *Cryogenics* **20**, 408.

Soulen, R. J., 1982, *Physica* **B110**, 2020.

Soulen, R. J., D. Van Vechten, and H. Seppa, 1982, *Rev. Sci. Inst.* **53**, 1355.

Star, W. M., J. E. van Dam, and C. van Baarle, 1969, *Jour. Sci. Instr.* **2**, 257.

Steward, G. R., 'Measurement of Low-Temperature specific heat, 1983,' *Rev. Sci. Instr.* **54**, 1.

Straty, G. C., and E. D. Adams, 1969, *Rev. Sci. Instr.* **40**, 1393.

Sundfors, R. K., D. I. Bolef, and P. A. Fedders, 1983, *Hyperfine Interactions* **14**, 271.

Suzuki, T., K. Tsubono, and H. Hirakawa, 1978, *Phys. Lett.* **67A**, 2.

Sverev, A. I., 1967, *Handbook of Filter Synthesis* (John Wiley & Sons, New York).

Swartz, D. L., and J. M. Swartz, 1974, *Cryogenics* **14**, 67.

Swenson, C. A, 1977, *Rev. Sci. Instr.* **48**, 489.

Tait, R. H., 1975, Ph.D. Thesis, Cornell. [On the thermal properties of granular materials].

Taylor, B. N., W. H. Parker, and D. N. Langenberg, 1972, *AIP Handbook of Physics* (McGraw-Hill, New York).

Taylor, V. L., 1968, 'A Study of the Electrocaloric Effect in KCl:Li,' thesis, Cornell University.

Tiggelman, J. L., and M. Durieux, 1972, *Temperature* **4**, 849.

Torre, J. P., and G. Chanin, 1984, *Rev. Sci. Instr.* **55**, 213.

Touloukhian, Y. S., R. W. Powell, C. Y. Ho, and P. G. Klemens, 1970, *Thermophysical Properties of Matter*,Vol **1-2** *(Thermal Conductivity)* (IFI/Plenum, New York).

Touloukhian, Y. S., S. C. Saxena, and P. Hestermans, 1975, *Thermophysical Properties of Matter*,Vol **11** *(Viscosity)* (IFI/Plenum, New York).

Touloukhian, Y. S., R. K. Kirby, R. E. Taylor, and P. D. Desai, 1976, *Thermophysical Properties of Matter*,Vol **12** *(Thermal Expansivity)* (IFI/Plenum, New York).

Touloukhian, Y. S., R. K. Kirby, R. E. Taylor, and T. Y. R. Lee, 1977, *Thermophysical Properties of Matter*, Vol **13** *(Thermal Expansivity)* (IFI/Plenum, New York).

Tward, E., 1981, *NBS Special Publication* **607**, 178.

van Duzer, T., and T. Turner, 1981, *Principles of Superconductive Devices and Circuits* (Elsevier, New York).

Van Sciver, S. W., and J. C. Lottin, 1983, *Rev. Sci. Instr.* **54**, 762.

Viertl, J. R. M., 1973, Ph.D. Thesis, Cornell University.

von Schickfus, M. and Hunklinger, S., 1976, *Jour. of Phys.* **C9**, L439.

Walton, D., 1966, *Rev. Sci. Instr.* **37**, 734.

Webb, R. A., R. P. Giffard, J. C. Wheatley, 1973, *Jour. of Low Temp. Phys.* **13**, 383.

White, G. K., 1979, *Experimental Techniques in Low-Temperature Physics* (Oxford University Press, Oxford).

Wildes, D. G., 1985, Ph.D. Thesis, Cornell University.

Wilkes, J., 1959, *Properties of Liquid and Solid Helium* (Oxford Univ. Press, Oxford).

Williams G. A. and R. E. Packard, 1980, *Jour. of Low Temp. Phys.* **39**, 553.

Wilson, M. N., 1983, *Superconducting Magnets* (Clarendon Press, Oxford).

Wolfmeyer, M. W., G. T. Fox, and J. R. Dillinger, 'An Electron Contribution to the Thermal Conductance across a Metal-Solid Dielectric Interface, 1970,' *Phys. Lett.* **31a**, 401.

Yap, B. C., 1973, Ph.D. Thesis, Cornell University.

Yu, I., A. A. V. Gibson, E. R. Hunt, and W. P. Halperin, 'Observation of Conduction-electron Density Oscillations at the Surface of Platinum Particles, 1980,' *Phys. Rev. Lett.* **44**, 348.

Zemansky, M. W., and R. H. Dittmann, 1981, *Heat and Thermodynamics (6th ed.)* (McGraw-Hill, New York).

Ziman, J. M., 1972, *Principles of the Theory of Solids (2nd ed.)* (Cambridge University Press, Cambridge).

Zimmerman, J., and Weber, G., 1981, *Phys. Rev. Lett.* **46**, 661.

Index